Practical Queueing Analysis

Recent Titles in the IBM McGraw-Hill Series

Details of these titles in the series are available from:

The Product Manager, Professional Books
McGraw-Hill Book Company Europe
Shoppenhangers Road, Maidenhead, Berkshire, SL6 2QL
Telephone: 0628 23432 Fax: 0628 770224

Mike Tanner

Practical Queueing Analysis

McGRAW-HILL BOOK COMPANY

London · New York · St Louis · San Francisco · Auckland
Bogatá · Caracas · Lisbon · Madrid · Mexico
Milan · Montreal · New Delhi · Panama · Paris · San Juan
São Paulo · Singapore · Sydney · Tokyo · Toronto

Published by
McGRAW-HILL Book Company Europe
Shoppenhangers Road, Maidenhead, Berkshire SL6 2QL, England
Tel 0628 23432; Fax 0628 770224

British Library Cataloguing in Publication Data

Tanner, Mike
  Practical Queueing Analysis. — (IBM
  McGraw-Hill Series)
  I. Title   II. Series
  519.82
  ISBN 0-07-709078-0

Library of Congress Cataloging-in-Publication Data

Tanner, Mike
    Practical queueing analysis / Mike Tanner.
        p.   cm. — (IBM McGraw-Hill series)
    Includes bibliographical references and index.
    ISBN 0-07-709078-0
    1. Queueing theory.   I. Title.   II. Series
QA274.8.T36   1995
519.8'2—dc20   94-27972
                                                                CIP

12345 CUP 98765

Typeset by Paston Press Ltd, Loddon, Norfolk
and printed and bound in Great Britain at the University Press, Cambridge

To my wife Rosalind,
and our children Clare and Georgette

Contents

Acknowledgements

I would like to acknowledge the encouragement given to me by IBM colleagues, in particular Bob Steggle and Roger Parr, who have each reviewed much of the material. I was also encouraged by conversations with a number of people including Dr Jeffrey Buzen of BGS, Pat Artis of Performance Associates, Jonathan Flowers of National Westminster Bank, and Steve Barratt of Abbey National. Naturally the errors, omissions, and confusions are mine alone. I am also grateful to Phil Taylor, my manager at IBM when I started the book, for getting permission for me to use IBM facilities for producing the manuscript.

The biggest acknowledgement goes to my wife Rosalind, for her support and encouragement, and for making it possible for me to find the time in the evenings and at weekends to write this book.

Part One
Introductory topics

1
How to use this book

Who this book is aimed at

The author's objective is to help a wide range of professionals to get maximum value from the body of knowledge referred to as queueing theory. A willingness to use mathematical and statistical concepts is essential for using queueing theory, but it is not necessary to be a fluent mathematician. It is hoped this book will be of benefit to the following:

- management sciences groups
- companies and organizations focusing on improving customer service
- students of mathematics, engineering and computing
- application and system designers in the information technology industry.

Intuitive understanding

I hope this book will improve readers' intuitive understanding of how queueing systems behave, and that this will help them to spot the significant characteristics of systems from a performance point of view. Detailed discussion of the intuitively important points is, of necessity, mingled with presentation of the useful analytical results for a particular type of queueing system.

Analytical results

For each type of queueing system, all the formulae that the author considers might be useful in practice are given. This is not a theoretical textbook, and so the derivation of the results is not given. Derivation of the results generally requires mathematical fluency not just in simple algebra and calculus, but in Laplace and other transforms, and probability and statistical theory. Readers of this book are likely to be quite rusty in such things, even if they once studied engineering or mathematical subjects as undergraduates. While it is not realistic to expect to make use of queueing theory without a little mathematical manipulation, the author's aim has been to minimize the demands made on the reader's mathematical fluency. In this book we concentrate on the results. For readers wanting more theoretical background there are a number of excellent books to refer to, and these are listed in a short bibliography.

Program subroutines and Q-Calc

Many queueing theory results that could be useful are not used because the calculations required are non-trivial. Even straightforward formulae can require significant effort to program and debug. This handbook contains subroutines for nearly all the results given. A small amount of programming 'infrastructure' is needed, i.e. some Pascal-type definitions, a very simple error-handling routine, and some simple but convenient file open/close routines. Details of these and some background technical details are in Appendix 3. This is a book about queueing theory, so only straightforward programming will be found here.

A diskette containing the source code of the subroutines is available. Instructions on how to obtain this are in Appendix 3. In addition, readers can obtain copies of a program called Q-Calc. Q-Calc is a menu-driven program that performs most of the queueing calculations described in this book, and operates in a manner analogous to a calculator. So readers who wish to avoid programming effort can still take advantage of the results in this book. Q-Calc is described briefly in Appendix 4.

Overview of book

Part One—Introductory topics

An understanding of basic probability and statistics is essential to understanding and using queueing theory. Chapter 2 provides the necessary groundwork, covering the definition of probability; basic laws of probability; the idea of a probability distribution; random variables; expected values and averages; measures of variation or dispersion, including coefficient of variation; and the moments of a distribution.

Chapter 3 talks about how we describe queueing systems, the significant characteristics of a queueing system, and what we might be interested in about their performance. The Kendall notation for classifying a queueing system, and the mathematical notation used in this book, are described here.

Chapter 4 discusses fluid approximations, which means simple first-pass ways of analysing queueing systems. These methods are simple to apply, and can give valuable insights into rush-hour performance, or systems that start with a backlog of customers, or where the customer arrival rate varies significantly.

Chapter 5 covers fundamental relationships between some of the characteristics of a queueing system, including Little's law. Little's law is one of the few results that applies to almost any queueing system. It is of great practical value for simple queueing systems, and forms the basis for some advanced techniques dealing with networks of queues.

Part Two—The simple queue

Part Two is almost a self-contained description of the best-known queueing model, M/M/1. Together with some general ideas about mathematical modelling, many readers may find Part Two, with the groundwork in Part One, a sufficient basis for applying queueing theory to simple situations.

Chapter 6 deals with random arrivals and exponential service-times, assumptions frequently made in basic queueing theory. We investigate an important property of the exponential distribution that makes it so useful in queueing theory. The meaning of 'random arrivals' is examined, and the relationship between random arrivals, the exponential distribution and the Poisson distribution is discussed. Chapter 7 covers the M/M/1 queueing model. This model is often used as a yardstick against which to compare other queueing systems, and in this chapter is also used to establish some important intuitive ideas about queues. First we see how the time taken to get served increases dramatically as the server gets very busy. Then we look at the conflict between giving good service and utilizing server capacity efficiently. Next we examine the scaling effect, i.e. how, for the same level of service, more powerful servers can be loaded relatively more heavily than less powerful servers.

Chapter 8 talks about mathematical modelling in general. This is done at an early stage in the book, when we have covered only one queueing model, M/M/1. Readers may find it valuable to return to Chapter 8 later, when they are wrestling with questions about which queueing model to apply to a particular problem.

Part Three—Single server with general service-times

Part Three is about single-server queues with random arrivals, but with a variety of service-time distributions. Chapter 9 describes the M/G/1 model, a simple and useful queueing model when the mean and variance of service time are known. This chapter shows how waiting time is affected by variability of service-time, as well as the server utilization. Chapter 10 covers more detailed mathematical results for the M/G/1 model. These are needed to derive results for specific service-time distributions, but are of practical use only with a detailed empirical service-time distribution.

Chapters 11, 12 and 13 are about queueing systems with specific service-time distributions. Chapter 11 deals with the M/Ek/1 model, a useful special case of M/G/1 for service-times that are less variable than exponential service-times. The Erlang-k distribution provides a good fit to many types of service. Chapter 12 covers the M/Ga/1 model, which can be used for service-times with a coefficient of variation squared having any value except zero. M/Ga/1 should be used when the service-time characteristics can be modified. Other special cases of M/G/1 handle only a limited range of coefficient of variation. The M/Ga/1 model is used to show how the distribution of queueing time is affected by the service-time distribution and server utilization. Chapter 13 is about the M/Hk/1 queueing model, which is useful when the coefficient of variation of service-time is greater than 1.

Part Four—Single server with general arrival pattern

Whereas Part Three looked at the effect of different service-time distributions with a single server, in Part Four we look at the effect of different arrival patterns with a single server. Chapter 14 covers G/M/1, where we go back to exponential service-times, but allow the arrival pattern to be general. G/M/1 is one of the few models that does not require random arrivals, while still being susceptible to detailed mathematical analysis. This makes G/M/1 valuable for understanding the effect of arrival patterns on waiting times and other aspects of performance. We also compare the effects of variability in arrivals with the effect of variability in service-times, and examine the difference between the customer's view of the queue and an independent observer's view. In Chapter 15, which deals with the G/G/1 queueing system, we find analytical results in very short supply. Various approximations and bounds for waiting time are given.

Part Five—Multiple servers

In Part Five we move from a single server to multiple servers. Some of the most practically useful queueing models are covered in this part of the book. For the most part we have to revert to assuming random arrivals and exponential service-times in order to get formulae for the queueing characteristics of interest to us, although we do cover briefly general arrival patterns and general service-times.

In Chapter 16 we look at what happens with an unlimited number of servers, i.e. the M/M/infinity model. Although seemingly impractical, this queueing model does have some practical use, and also provides a simple introduction to the 'scaling effect' for multiple-server systems. Chapter 17 covers one of the most widely used models, M/M/m, where there is a single queue for a group of servers. One of the classic comparisons of queueing theory is between a group of servers with a shared queue and servers with separate queues. This comparison is presented here, with an explanation of why it can be misleading. We examine why multiple servers can be more heavily loaded than a single server for the same performance. We also look at why, for fixed total capacity, a single server gives shorter time in system than multiple servers, but a longer wait. In Chapter 18, we investigate the best way of arranging queues for multiple servers, and the fairness and efficiency of different methods. We demonstrate why a single queue is often the best, especially when fairness between customers is important. On the other hand, we show that separate queues are just as efficient as a single queue if customers hop between queues, and that in some cases separate queues can be more efficient than a single queue.

In Chapter 19 we look at $M/M/m/m$, i.e. a group of servers for which no waiting is allowed. This is a widely used queueing model in telephone systems. Chapter 20 looks at a more general case of this where waiting is allowed but there is a limit on how many customers may be waiting. This is the $M/M/m/K$ model.

Chapter 21 covers $G/G/m$, where a general arrival pattern and general distribution of service-times are assumed. Some bounds and approximations are available but, even more than for $G/G/1$, analytic results are scarce. Reference is made to an extensive set of tables of numerical results for inter-arrival and service-time distributions having Erlang or hyperexponential form.

Part Six—Priority queues

Chapter 22 is about priority queues. Priority mechanisms are very common in computer and communication systems. Exact analytic results are available only for single-server systems. For multiple servers the effect of priorities can be approximated, but this needs some care, so simulation is probably a safer approach.

Part Seven—Limited number of customers

In Part Seven we look at the effect of the customer population size being limited, i.e. where there is a fixed number of customers in the system. Chapter 23 covers the single-server case, known as the 'single repairman' model, which gives valuable insights into the effects of a finite customer population. This chapter gives a definition of offered traffic and efficiency for a queueing system, and introduces the idea of latent demand. The multiple-server case is dealt with in Chapter 24, which also covers the scaling effect and the optimization of the number of servers. These queueing models are, in the author's experience, overlooked by performance analysts and system designers. This is a pity, since these models have some important intuitive ideas to offer, particularly about efficiency and latent demand.

Part Eight—Some important topics

In Chapter 25 we look at server-sharing. This is where several customers are to be served in parallel by a single server, and a way is needed to share the capacity of the server between the active customers. This is called server-sharing, usually encountered as scheduling algorithms for computer systems. This chapter describes the round-robin processor sharing algorithm.

Chapter 26 discusses networks of queues. While networks of queues occur in many fields, this branch of queueing theory has been greatly developed specifically for the analysis of computer system performance. This chapter serves as an introduction to the different types of queueing network, i.e. open, closed and mixed. The idea of a product-form solution to a queueing network is introduced. Jackson networks, an important easy-to-solve type of open queue network, are described. The central-server model, a simple closed queueing network, is explained. Serious application of advanced queueing-network theory is probably beyond the non-specialists at whom this book is aimed. Non-specialists need to be aware of the existence and scope of this branch of queueing theory, since it forms the basis of some important commercial performance analysis packages.

Chapter 27 discusses simulation, as an alternative or complementary technique to analytical queueing theory. The objective is to introduce the basic concepts of simulation, not to provide a detailed practical guide to actually using simulation.

Part Nine—Applications to computer systems and telecommunications

Part Nine describes how queueing theory can be applied to computer systems and telecommunications.

Chapter 28 is about the performance of computer systems. Some of the queueing models described previously are used to analyse aspects of computer system performance such as disk and channel performance (with rotational position sensing), the effect of paging on cpu throughout, the interaction of cpu and disk performance, and a model of transaction processing. The objective is to show how fairly simple queueing and queueing-network models can be combined to explain the behaviour of computer systems.

Chapter 29 is about voice, i.e. telephone, networks. Topics covered are the use of Erlang-B and Erlang-C multiserver models for voice networks, Engset limited sources model, and extended Erlang-B for retries of failed calls. The emphasis is on results that can be applied to moderate-sized voice networks, so that considerations that apply to countrywide public networks are excluded. A brief introduction to the analysis of call centres is included.

Chapter 30 is about data communication networks. Simple queueing theory models are suggested for the main components of data networks. These are aimed at estimating each component's contribution to overall network delay. Topics covered are simple models of links and data switches, outline of data flow in a network, packet delay for CSMA/CD and token-ring LANs, and the throughput of virtual circuits or connections with flow-control mechanisms.

Part Ten—Appendices

Appendix 1 contains reference information for the probability distributions commonly encountered in queueing theory, including subroutines for calculating distribution functions, moments, percentiles, and so on. Appendix 2 covers some mathematical functions that are needed for calculating some of the results in this book. Relevant subroutines are provided.

Appendix 3 contains technical background information about the computer programs in this book. Instructions and order forms for obtaining diskette copies of the programs are also in this appendix. Appendix 4 is a description of the Q-Calc package. Q-Calc is an easy-to-use program for performing the calculations for nearly all the queueing models in this book.

Appendix 5 contains a bibliography and references. As this book is intended as a practical manual rather than an academic treatise, the references are not extensive.

Subjects not covered

There are some subjects that might have been included but have been omitted. In practice, simulation would often be used to study situations involving these subjects. The omissions include the following.

Transient analysis

This book is almost entirely about steady-state behaviour of queueing systems. Chapter 4, on fluid approximations, gives an elementary technique for handling very simple transient behaviour, but readers are referred to Kleinrock (1976) and Newell (1982) for more incisive material on transient behaviour.

Diffusion approximations

Diffusion approximations will seem curiously-named to non-engineers. They are really a more sophisticated form of fluid approximations, which are described in this book. The term 'diffusion approximation' comes from the fact that some of the mathematical results used were developed in the study of diffusion of gases. Again, readers are referred to Kleinrock (1976) and Newell (1982) for details.

Bulk queues

In some queueing systems customers arrive in bulk, and/or are served in bulk. This occurs particularly in transportation, where customers will literally arrive or depart 'by the bus-load'. Readers are referred to Medhi (1984) for a survey of results in bulk queues, and to Chaudry and Templeton (1983) for more detailed material.

Discrete approximations

This is a branch of queueing theory that uses discrete distributions to approximate service-time and inter-arrival time distributions. This allows numerical methods to be used to study queueing systems that otherwise are very difficult to analyse. The essential idea is to provide an exact solution to an approximation to the original problem. Again readers should consult Kleinrock (1976) for an introduction to this technique and references to other authors.

2
Probability and statistics

Introduction

We use the ideas of probability in a number of different ways. What is the probability that our team will win the game? How likely is this horse to win the race? What are my chances of being promoted? What is the probability that my plane will be delayed? How likely is it to rain today? To some of these questions we expect only an imprecise answer, e.g. I do not think there is much chance that it will rain today. In probability and statistics we use a precise definition of probability. We seek to make precise statements about uncertainty.

Probability

There are several approaches to defining probability. The idea of probability is a rather deep philosophical topic. Luckily, for practical applications we need only a simple definition of probability. We have to think of an 'experiment' or procedure that has a number of possible outcomes, e.g. drawing a card from a well-shuffled pack of cards. Then we have to define 'success' as a subset of the possible outcomes, e.g. drawing an ace. The probability of 'success' is then defined to be

$$\text{Probability(success)} = \frac{\text{number of successful outcomes}}{\text{total number of outcomes}} \tag{2.1}$$

Two simple examples of this are

$$\text{Prob(drawing an ace from full pack of cards)} = \frac{4 \text{ aces}}{52 \text{ cards}} \approx 0.077 \tag{2.2}$$

$$\text{Prob(getting a six from throwing a die)} = \frac{1 \text{ six per die}}{6 \text{ faces per die}} \approx 0.167 \tag{2.3}$$

Probability values range from 0, for impossible outcomes, to 1 for definite outcomes. Quite often probabilities are given as percentages. As a general rule it is best to use percentages only for the presentation of final results.

Readers will notice that many illustrations and exercises on probability are based on packs of cards, dice, and similar gambling games. This is because gambling games provide well-defined examples. Another reason is that an event such as throwing a die can easily be repeated a large number of times. We can actually demonstrate that the probability of throwing a six is 1/6, by throwing a die many times and observing the ratio of the number of sixes obtained to the number of throws. In contrast, if we make the statement 'there is a 60 per cent chance it will rain tomorrow' we mean that past experience shows that 60 per cent of the time similar weather conditions have been followed by rain. We cannot 'repeat the experiment' by arranging for exactly the same weather conditions and then seeing if it rains.

Independent events

A vital concept in probability is the independence of events. Suppose event A is drawing an ace from a full pack of cards, and event B is tossing a coin and getting 'heads'. Common sense tells us that these two events are not in any way dependent on each other. Let us work out, from our basic definition, the probability of the composite event of getting an ace and a head. First of all, how many possible outcomes are there? These are illustrated in Table 2.1, and there are 104 possible outcomes.

Of these, four would be classed as a 'success', i.e. an ace with heads. So the probability of an ace with heads is 4/104 = 0.0385. Instead of working out the probability from first principles by enumerating all the possible composite outcomes, we can use one of the most important laws of probability. This is that if two events A and B are independent, then the probability of both A and B happening is given by

$$\text{Prob(A and B)} = \text{Prob(A)} \times \text{Prob(B)} \tag{2.4}$$

In the example we are using, this means

$$\text{Prob(ace drawn AND heads)} = \text{Prob(ace drawn)} \times \text{Prob(heads)} \tag{2.5}$$

$$= \left(\frac{4}{52}\right) \times \left(\frac{1}{2}\right) = 0.0385 \tag{2.6}$$

Table 2.1. Possible outcomes from drawing card and tossing coin

	Card drawn	Heads or tails
1	Ace of spades	Heads
2	Ace of spades	Tails
3	King of spades	Heads
4	King of spades	Tails
...	...	...
103	2 of hearts	Heads
104	2 of hearts	Tails

So the probability of the joint occurrence of independent events can be calculated by multiplying the probabilities of the separate events.

Suppose now we want to know the probability of either event A or event B, or both, happening. Still using the playing cards and coin example, what is the probability that either we get an ace or we toss the coin and get heads? How many of the possible outcomes would we count as a success? There are four outcomes that include both events, i.e. drawing an ace and getting heads. There are a further four outcomes in which we draw an ace but get tails. Finally, there are 48 outcomes in which we fail to draw an ace but get heads. So we have

$$\text{Prob(ace drawn OR heads)} = \frac{4 + 4 + 48}{104} = 0.5385 \qquad (2.7)$$

How do we relate this to Prob(ace drawn) and Prob(heads)? In counting the number of outcomes that constitute a success, we need to add the number of outcomes that include drawing an ace and the number of outcomes that include getting heads. However, we must avoid double-counting the outcomes that include both events. So for independent events A and B, we have

$$\text{Prob(A OR B)} = \text{Prob(A)} + \text{Prob(B)} - \text{Prob(A AND B)} \qquad (2.8)$$
$$= \text{Prob(A)} + \text{Prob(B)} - \text{Prob(A)} \times \text{Prob(B)} \qquad (2.9)$$

In our example this means

$$\text{Prob(ace drawn OR heads)}$$
$$= \text{Prob(ace drawn)} + \text{Prob(heads)} - \text{Prob(ace drawn)}$$
$$\times \text{Prob(heads)} \qquad (2.10)$$
$$= \left(\frac{4}{52}\right) + \left(\frac{1}{2}\right) - \left(\frac{4}{52}\right) \times \left(\frac{1}{2}\right) = \frac{56}{104} = 0.5385 \qquad (2.11)$$

These probability calculations can be illustrated with a diagram called a Venn diagram, shown in Fig. 2.1. The rectangle represents all the possible outcomes of an 'experiment'. The left-hand circle represents the collection of outcomes that make up an event A, and the right-hand circle represents event B. The area of overlap contains the outcomes that mean both events A and B have occurred. If we want to calculate the area in which either A or B has occurred, then we start by adding the areas of the two circles. In doing so we have included the 'A and B' area twice, so we have to subtract it to get the correct area.

Probability distribution functions

Suppose now we throw two dice. Figure 2.2 shows all the possible outcomes of this 'experiment': there are 36 ways in which a pair of dice can fall.

If we are interested in the sum of the two dice, we can represent the possible outcomes with Fig. 2.3. We can then count the number of ways in which each possible total can occur. A total of 7 is the most likely, since there are six possible ways of

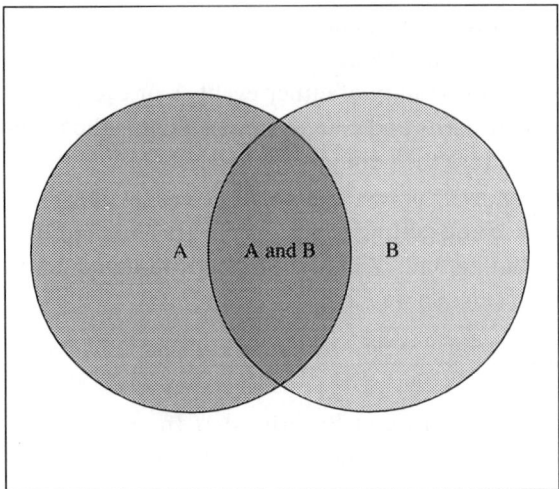

Figure 2.1. Venn diagram.

Figure 2.2. Possible outcomes of throwing two dice.

Figure 2.3. Possible outcomes for sum of two dice.

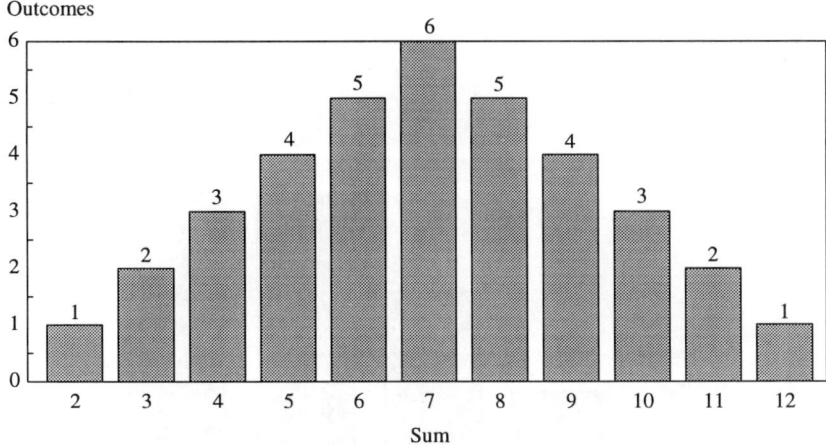

Figure 2.4. Graph of possible results for sum of two dice.

getting 7. A total of 4 can occur in three ways, and so on. If we draw a histogram of the number of ways in which each possible total can occur, we get Fig. 2.4.

This graph shows how the total number of outcomes, in this case 36, is distributed across each of the possible sums. If we converted number of outcomes to probability, by calculating

$$\text{Prob}(\text{sum} = k) = \frac{\text{number of outcomes for which sum} = k}{\text{total number of outcomes}} \qquad (2.12)$$

we would have a graph that showed how the total probability, 1 by definition, is distributed across each of the possible sums. This would be called the 'probability distribution function', often abbreviated to pdf.

Now suppose that instead of the sum of the two dice, we are interested in the maximum of the two dice. Figure 2.5 shows how each possible maximum value

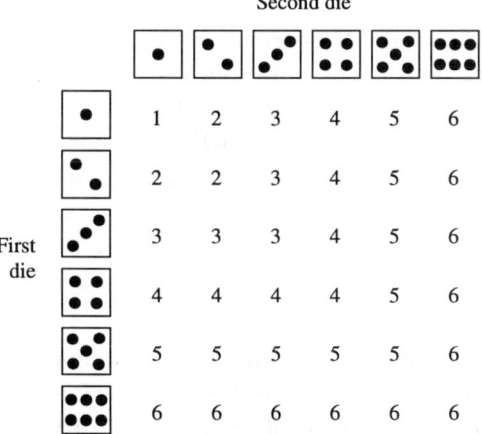

Figure 2.5. Possible outcomes for the maximum of two dice.

Outcomes

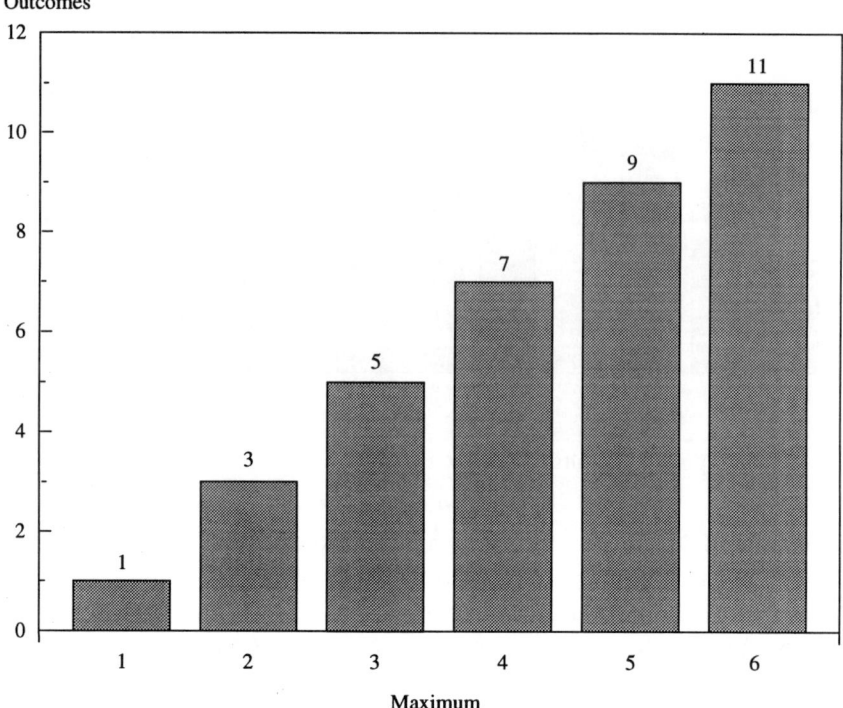

Figure 2.6. Graph of possible outcomes for the maximum of two dice.

can occur, and Fig. 2.6 is a histogram showing how the total number of possible outcomes is distributed across the possible maximum values. As before, if we converted number of outcomes into probabilities we would have probability distribution function, this time for the maximum of two dice.

Cumulative distribution functions

So far we have established the idea of a probability distribution function. Figure 2.7 shows the pdf for the sum of two dice. This is a plot of $\mathrm{Prob}(N = k)$ where N is the sum obtained when throwing two dice, and k is a particular value.

We can also draw a histogram that shows $\mathrm{Prob}(N \leqslant k)$, which is the probability that we will get a sum that is no greater than k. This is shown in Fig. 2.8, and the calculation required is

$$\mathrm{Prob}(N \leq k) = \sum_{i=2}^{i=k} \mathrm{Prob}(N = i) \tag{2.13}$$

(Where the Greek letter Σ (sigma) means 'sum the following expression over the range of values defined by the expressions above and below the Σ. Sometimes the range is not given, in which case we assume the range of all possible values.) This is

PDF

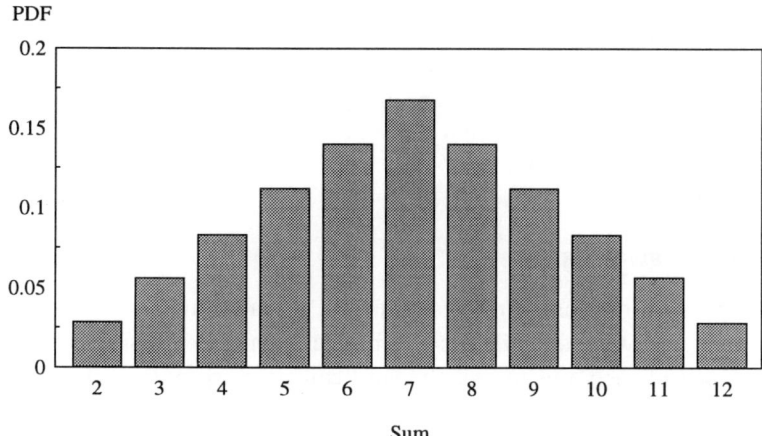

Figure 2.7. PDF for sum of two dice.

CDF

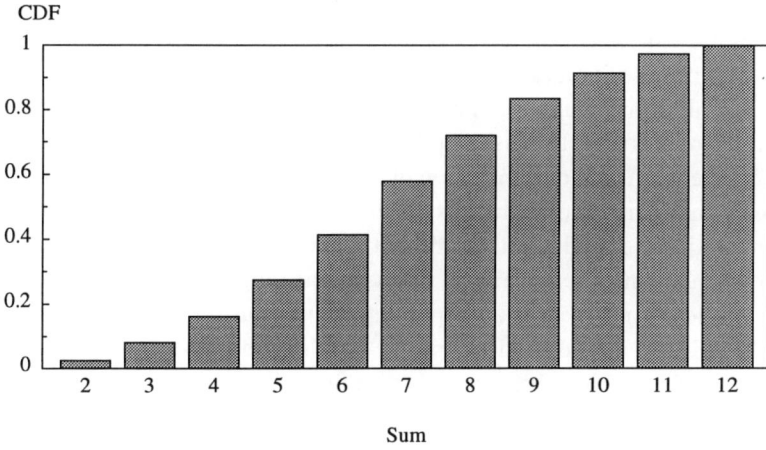

Figure 2.8. CDF for sum of two dice.

called the 'cumulative distribution function', abbreviated to cdf. The term 'cumulative' is used because the probability associated with each particular value is the cumulative probability of all the values up to and including that particular value. For example, the probability that on a particular throw we get a sum that does not exceed 8 is 0.72

Random variables

Let us use the symbol X to stand for, say, the result of throwing a die. Since X takes on values at random, according to the relevant probability distribution, we say that X is a 'random variable'. This is a piece of statistician's jargon, but it is

very useful jargon so we might as well use it. In statistical theory it is very important to distinguish between a random variable and particular instances or values of
that random variable. The random variable is usually represented by an upper-
case letter, such as 'X', while particular values are represented by the corresponding lower-case letter, such as 'x'. We shall not need to stress this distinction a great
deal, but readers should be aware of it.

Continuous random variables

So far we have dealt with random variables that have a finite number of possible
values. For example, throwing a single die can produce only six possible values,
while throwing a pair of dice and taking the sum can produce only 11 possible
values. Random variables like these are called 'discrete' or 'discrete-valued'
random variables, since their value can change only in discrete steps. Often the
possible values are integers, but this is not always the case. Other random variables can take on an infinite number of possible values. Examples might be the
height of a randomly chosen person, or the duration of a telephone call. Such
random variables are called 'continuous' random variables because they can vary
continuously over a range of values.

For continuous random variables, the probability distribution function does
not have as simple an interpretation as the pdf of a discrete random variable. Consider the pdf for the duration of a phone call. A typical distribution for this is
called the exponential distribution, which is illustrated in Fig. 2.9(a)—the meaning
of the shaded area will be explained in a moment. (The exponential distribution is
very important in queueing theory and is discussed at length in later chapters.)
When we say a random variable has an exponential distribution, we have defined
the shape of the pdf, and all we need to specify in addition is the average value.
Note that the range of possible values extends to infinity, so only part of the pdf is
actually shown. As X gets larger, the pdf gradually approaches zero, but in theory
never reaches zero.

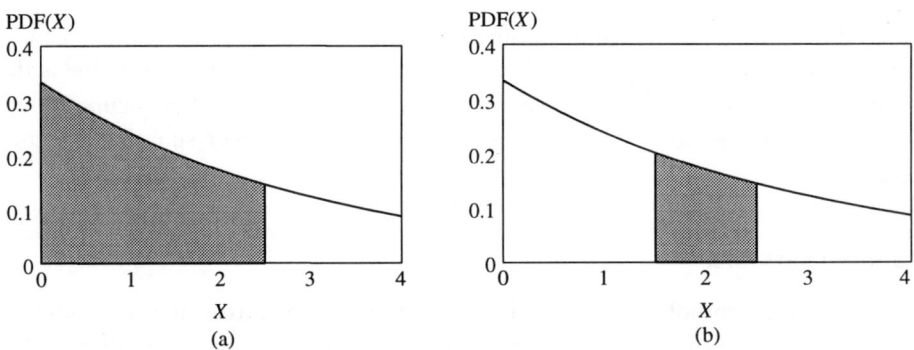

Figure 2.9. PDF of exponential distribution with average of 1.

For a discrete distribution, the pdf tells us the probability of a particular value occurring. For a continuous distribution, the probability of a particular value is zero (or at least infinitesimally small). So how do we interpret a continuous pdf? The total probability that is to be distributed over the possible values of X is, of course, 1. If we calculated the area enclosed by the pdf and the vertical and horizontal axes, we would find it came to 1. Probabilities are represented by areas, and we can interpret the pdf by saying that the probability that X is less than or equal to, say, 2.5 is the area enclosed by the pdf, the axes, and the line $X = 2.5$. This idea is illustrated in Fig. 2.9(a). In a similar way, the probability that X lies between 1.5 and 2.5 is the area enclosed by the pdf, the x-axis, and the lines $X = 1.5$ and $X = 2.5$. This is shown in Fig. 2.9(b).

The cumulative distribution function, or cdf(x), for a continuous random variable is in fact the area enclosed by the pdf up to the value $X = x$. So for making probability statements about a continuous random variable we need the cdf rather than the pdf. The pdf and cdf are related mathematically by

$$F(x) = \int_0^x f(w) \, dw \tag{2.14}$$

where $F(x)$ is the cdf and $f(x)$ is the pdf. (The lower limit of integration should in general be minus infinity, but in queueing theory we do not encounter negative variables.) Since many readers may not be comfortable with integration, we shall avoid it except for occasional mention. The pdf is essential in the mathematical derivation of many statistical and queueing theory results, but for practical purposes its main value is to show the shape of how probability is spread over the range of values. A graph of the pdf allows us to see clearly whether values a long way from the average are at all likely, or whether most of the probability is concentrated around the average, or whether the probability distribution is 'skewed' towards high or low values. Appendix 1 contains details of most of the probability distributions likely to be encountered in queueing theory, including illustrations of the shape of the pdf.

Expected value of a random variable

Another piece of statistician's jargon is 'expected value'. The expected value of a random variable is the average value for the random variable. Let us illustrate what this means. Suppose X is the result of throwing a single die. What is the expected value of X? We know that X can take on the values 1 through 6. The expected value, which is written $E(X)$, is

$$E(X) = 1 \times \text{Prob}(X = 1) + 2 \times \text{Prob}(X = 2) + \ldots + 6 \times \text{Prob}(X = 6) \tag{2.15}$$

$$E(X) = (1 + 2 + 3 + 4 + 5 + 6) \times \frac{1}{6} = \frac{21}{6} = 3.5 \tag{2.16}$$

There is nothing complicated about this definition, it is just a useful way of expressing how to calculate the average value. In general, if a random variable X can take on values $x_1, x_2, \ldots, x_n$ then the expected value of X is given by

$$E(X) = \sum_{k=1}^{k=n} x_k \, \text{Prob}(X = x_k) \tag{2.17}$$

The idea of expected value applies not just to X itself, but to any mathematical function whose value depends on X. So if $h(X)$ is a function of X, then

$$E[h(X)] = \sum_{k=1}^{k=n} h(x_k)\text{Prob}(X = x_k) \tag{2.18}$$

The random variable X may be continuous, so that it can take on an infinite possible number of values, e.g. any value in the range 0 to 1. We must then define and calculate expected value in a slightly different way. Mathematical integration for continuous variables corresponds to summation for discrete variables. For a continuous random variable X, expected value is defined as

$$E(X) = \int_{x=a}^{x=b} xf(x) \, \mathrm{d}x \tag{2.19}$$

where x can take on values from $x = a$ to $x = b$, and $f(x)$ is the probability distribution function. Readers will not have to deal with integration to use the queueing theory results in this book.

Averages (mean, mode, median)

Most people, when they say the average of a set of numbers, are using the word average in the sense of the arithmetic mean, or 'mean'. The mean of a set of numbers $x_1, x_2, \ldots, x_n$, which is denoted by μ, is simply

$$\text{mean} = \mu = \frac{x_1 + x_2 + \ldots + x_n}{n} \tag{2.20}$$

The mean value of a random variable is the same as its expected value, as defined in Eq. (2.17). The mean is usually the most representative value of a set of numbers, but sometimes other values are typical for some purposes.

The mode of a set of numbers is the most common value. For a random variable, the mode is the value for which the pdf has a maximum value. For some probability distributions the pdf may have more than one maximum value, or no specific maximum at all. So the mode may not be unique, or may not exist (although such cases are unusual).

The median of a set of numbers is the middle value, so that half the numbers are less than the median and half the numbers are greater than the median. For a set of discrete numbers, this definition requires an odd number of values. With an even number of values it is usual to take the average of the two middle values and

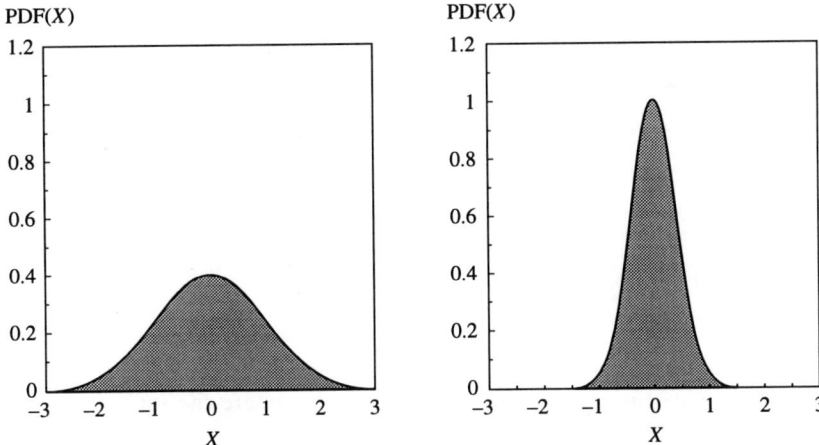

Figure 2.10. Normal distributions with high and low variance.

call that the median. We can also define the median of a probability distribution as the value for which the cumulative distribution function is 0.5.

Measures of variation or dispersion

As well as the average or mean value of a set of numbers, we want to know how spread out the numbers are. This idea is illustrated in Fig. 2.10, which shows the pdfs of two Normal distributions. Both distributions have the same mean value, but the left-hand one has a lot of values that differ from the mean by a large amount. The right-hand distribution has its values more tightly clustered around the mean. The most useful measure of spread or dispersion is 'variance', which is defined as

$$\text{variance of } X = \sigma^2 = E[(x - \mu)^2] \tag{2.21}$$

$$\text{variance} = \sigma^2 = \frac{1}{n}\sum_{i=1}^{i=n}(x_i - \mu)^2\, \text{Prob}(X = x_i) \tag{2.22}$$

One difficulty with variance is that it is often a very large value, being based on squared values. For presentation of results we often prefer to use the 'standard deviation', which is the square root of the variance:

$$\text{standard deviation} = \sqrt{\text{variance}} = \sigma \tag{2.23}$$

Coefficient of variation

Yet another way of expressing the amount of variation in a set of numbers is the coefficient (coeff) of variation. This is defined by

$$\text{coeff of variation squared} = C^2 = \frac{\text{variance}}{\text{mean}^2} \tag{2.24}$$

Readers are cautioned to be careful about whether C or C^2 is being quoted. Generally, and in this book, we shall always quote C^2. The squared coeff of variation is useful because it appears in many queueing theory results. Also, it is easy to relate to, since $C^2 = 0$ means zero variation, i.e. a constant value, and $C^2 = 1$ corresponds to the exponential distribution, which plays a large part in queueing theory.

Sums of random variables

Sometimes we are interested in the sum of two or more random variables. For example, the total time a customer spends in a queueing system is the sum of waiting time and service-time. Other examples might be where the service-time is made up from several stages or phases of service. If

$$X = Y + Z \tag{2.25}$$

then the variances can simply be added together

$$\sigma_X^2 = \sigma_Y^2 + \sigma_Z^2 \tag{2.26}$$

Note that we add variances, not standard deviations.

Moments of a probability distribution

When we describe probability distributions, we usually talk about the mean and variance. In queueing theory we tend to use the coefficient of variation rather than variance itself. These parameters, together with the particular 'shape' of probability distribution, e.g. exponential or Normal, etc., are sufficient to define the distribution. Mean and variance are the first two 'moments of a distribution'. For some purposes in queueing theory we need to make use of other, higher, moments. There are two ways of giving the moments of a distribution, called 'moments about the mean' and 'moments about zero'. Moments about the mean can be related to the 'shape' of the probability distribution, and so we describe them first:

$$a_r = r\text{th moment about the mean} = E[(X - \mu)^r] \tag{2.27}$$

$$a_r = \frac{1}{n} \sum_{i=1}^{i=n} (x_i - \mu)^r \, \text{Prob}(X = x_i) \tag{2.28}$$

We can see immediately that $a_1 = 0$, and that $a_2 = E[(X - \mu)^2]$ is identical with the variance of the distribution. The term 'moment' might suggest, to those with some mechanical engineering in their past, an analogy with 'moments of inertia'. The variance of a distribution can indeed be regarded as analogous to the moment of inertia of a solid object about its centre of gravity, but higher moments are hard to relate to mechanical phenomena. The third moment about the mean is called the 'skewness' of the distribution, and is defined as $a_3 = E[(X - \mu)^3]$. As the name suggests, skewness is related to whether the shape of the probability distribution is skewed to one side rather than being symmetrical. The fourth moment, $a_4 = E[(X - \mu)^4]$, is called 'kurtosis'. Kurtosis is to do with how sharply peaked a probability distribution is, but is not a very useful concept in applying statistics or queueing theory. Higher moments than these are not named, and do not lend themselves to interpretation in terms of the shape of a distribution.

Moments about zero, or the origin, are defined in the same way as moments about the mean, except that we use 0 as the reference point instead of the mean. (In mechanical terms we might consider we are holding, say, a wooden shape at one end instead of balancing it on our hand underneath its centre of gravity.) The rth moment about the origin is defined as

$$z_r = r\text{th moment about zero} = E[X^r] \tag{2.29}$$

In subsequent chapters, and also in Appendix 1, formulae for the moments of individual distributions are given.

Defining a distribution of moments

We usually expect to define a probability distribution by specifying the probability distribution function, i.e. a mathematical formula that describes the shape of the probability curve. An alternative way is to specify all the moments of the distribution. It does not matter whether we give moments about zero or about the mean (or indeed about any other arbitrary value). If all the corresponding moments for two distributions are equal, then the two distributions must be identical. In practice we may specify a distribution in terms of its moments if we have empirical data. Also, in queueing theory it is sometimes possible to work out the moments of, say, waiting time, when it is mathematically impractical to derive the pdf or cdf.

Normalizing moments about zero

Sometimes we may wish to compare or report the moments about zero of two distributions, in order to make a judgement about how similar or different they are. Moments tend to be very high values, because of the high power to which all the values are raised. To combat this we can 'normalize' the moments by calculating

$$z'_r = (z_r)^{1/r} \tag{2.30}$$

so that we take the rth root of the rth moment. This has the effect of reducing all the moments to the same 'dimensions', and eliminates very high values. As an illustration of this, Table 2.2 gives the moments and normalized moments for a

Table 2.2. Normalized moments of the gamma distribution

r	rth moment about zero	Normalized rth moment about zero
1	1.0	1.000
2	3.0	1.732
3	15.0	2.466
4	105.0	3.201
5	945.0	3.936
6	10395.0	4.672
7	135135.0	5.407
8	2027025.0	6.143
9	34459425.0	6.878
10	654729075.0	7.614

gamma distribution with a mean of 1.0 and a squared coefficient of variation of 2.0.

Converting moments about zero to moments about mean

In the mathematics of queueing theory we usually work with moments about zero of the service-time distribution, and use these to calculate moments of waiting time and time-in-system. For reporting purposes we shall need to convert these into moments about the mean, so that we can quote variance or standard deviation, and perhaps skewness. Moments about the mean are defined by Eq. (2.27). Expanding this algebraically, we get the conversion formula that we want:

$$a_r = E\left[\sum_{i=0}^{r} \binom{r}{i} x^{r-i}(-\mu)^r\right] \tag{2.31}$$

$$a_r = \sum_{i=0}^{r} \binom{r}{i} (-z_1)^r z_{r-i} \tag{2.32}$$

Coefficients of variation, skewness and kurtosis

We shall probably want to normalize the moments about the mean. In this context 'to normalize' means to convert the moments into coefficients of variation and skewness. As we have seen with the coefficient of variation, coefficients of this kind are easier to relate to than actual moments. An important reason why coefficients are useful is that they do not alter their value if all the values of a distribution are multiplied by a constant. You may, for example, have service-time data that is measured in minutes, and you wish instead to work in seconds. This means all the values will be multiplied by 60. The moments themselves will now have different values, but the coefficients will stay the same. The definitions of coefficients that achieve this aim are as follows:

$$\text{coeff of variation squared} = C^2 = \frac{E[(x - \mu)^2]}{[E(x)]^2} = \frac{\sigma^2}{\mu^2} = \frac{a_2}{z_1^2} \tag{2.33}$$

$$\text{coeff of skewness} = \frac{E[(x - \mu)^3]}{E[(x - \mu)^2]^{3/2}} = \frac{a_3}{a_2^{3/2}} = \frac{a_3}{\sigma^3} \tag{2.34}$$

$$\text{coeff of kurtosis} = \frac{E[(x - \mu)^4]}{E[(x - \mu)^2]^2} - 3 = \frac{a_4}{a_2^2} - 3 = \frac{a_4}{\sigma^4} - 3 \tag{2.35}$$

The coefficient of kurtosis is conventionally defined with the term -3 included so that for the Normal distribution kurtosis is zero. A negative coeff of kurtosis means the pdf is flatter than the normal; a positive kurtosis means the pdf is more peaked than the normal.

As an illustration of typical values of the coefficients, Table 2.3 gives the coefficients of variation, skewness and kurtosis for the gamma distribution. In this case a mean of 1.0 has been assumed, and the coeff of variation squared has been given a range of values. The mean and variance (equivalently the coeff of variation) completely define a particular gamma distribution. The probability distribution functions are illustrated in Chapter 12, except for $C^2 = 1$, which is the exponential distribution.

Table 2.3. Coeffs of variation, skewness, kurtosis of gamma distribution

Coeff of variation squared	Coeff of skewness	Coeff of kurtosis
0.1	0.63	0.60
0.5	1.41	3.00
1.0	2.00	6.00
1.5	2.45	9.00
2.0	2.83	12.00
5.0	4.47	30.00

Percentiles of a distribution

We have already come across the median, which is the middle value of a set of numbers. The median is just a particular case of a percentile. The rth percentile of a distribution is defined by

$$\text{Prob}(X \leq r\text{th percentile}) = \frac{r}{100} \tag{2.36}$$

so that the median is the 50th percentile. Percentiles are very important in queueing theory. Many of the useful queueing-theory models require us to assume that very long service-times may occur even if in the real situation being modelled there is an upper limit on the service-time. The probability associated with these unrealistically long service-times may be extremely small, so to that extent the assumptions we have to make are not unreasonable. However, the theory will still say that extremely long waiting times may occur, albeit rarely. If we tried to answer the perfectly reasonable practical question of 'what is the maximum time a customer may have to wait?', then the answer we would have to give would be 'an infinite length of time!'. Instead, we might say that 95 per cent of customers would not have to wait longer than Y seconds. So percentiles perform in practice the role of a maximum. Usually the 90th or 95th percentile is used. Figure 6.1 shows the pdf of the exponential distribution with the 90th and 95th percentiles marked.

Programs for probability and statistics

The routines for manipulating moments are in the file NMLSEMOM.PAS, and a routine for calculating statistics from data is in DATASTAT.PAS, so the following statements are needed at the start of a program.

```
{$N+,E+ }
{$I \qthprog\qthtypes.pas    }
{$I \qthprog\qtherror.pas    }
{$I \qthprog\qthfile.pas     }
{$I \qthprog\bicoeff.pas     }
{$I \qthprog\nmlsemom.pas    }
{$I \qthprog\datastat.pas    }
```

In addition, routines are provided for calculating the moments for a number of distributions. These are described in Appendix 1.

Data structures

A simple Pascal record type is defined for holding a set of moments. The definitions are in the file QTHTYPES.PAS. First there is a constant that specifies the maximum number of moments to be used.

```
Const MaxMoments=10;
```

Next there is the record definition itself. This consists of an array to hold the moments, and a variable that says how many entries in the array are meaningful. This may be less than MaxMoments when, for example, moments of waiting time are calculated from moments of service-time. The zeroth moment is included and set to 1 for programming convenience.

```
Type Moments=record
        MOM:array[0..MaxMoments] of double;
        MVM:integer;
     end;
```

In addition, a simple array type is defined for holding data. Readers should adjust the value of MaxDataPts to an appropriate value for their application.

```
Const MaxDataPts=1000;
Type QTHDataVec=array[1..MaxDataPts] of QTHreal;
```

DataStats—calculate basic statistics for a set of data

This routine takes a set of values, together with an associated probability for each value, and calculates moments about zero and about the mean, and coefficients of variance, skewness and kurtosis. Moments about zero and the mean are calculated independently, rather than converting moments about zero to moments about the origin. This is to reduce difficulties that might arise from rounding errors with data that has a low variance and high mean.

```
{-------------------------------------------------------}
{> DataStat -- Calculate descriptive statistics         }
{>              for a set of numbers.                    }
{ Inputs:                                                }
{      V[.]    values of the random variable             }
{      P[.]    probability of each value                 }
{      N       number of values                          }
{ Outputs:                                               }
{      MZRO    first 4 moments about zero                }
{      MAVG    first 4 moments about mean                }
{      MEAN    mean                                      }
{      STDV    standard deviation                        }
{      CVAR    coeff of variation squared                }
{      CSKW    coeff of skewness                         }
{      CKUR    coeff of kurtosis                         }
{   Copyright Mike Tanner 1993                           }
{-------------------------------------------------------}
Procedure DataStat(Var V,P:QTHdatavec;N:integer;
                   Var MZRO,MAVG:Moments;
                 Var MEAN,STDV,CVAR,CSKW,CKUR:QTHreal);
Var I,J:integer;X,D:QTHreal;
begin
    MEAN:=0;
    For I:=1 to N do MEAN:=MEAN+V[I]*P[I];
    MAVG.MVM:=4;  MAVG.MOM[0]:=1; MAVG.MOM[1]:=MEAN;
    For J:=2 to 4 do MAVG.MOM[J]:=0;
    For I:=1 to N do begin
       D:=V[I]-MEAN;  X:=D;
       For J:=2 to 4 do begin
          X:=X*D;
          MAVG.MOM[J]:=MAVG.MOM[J]+X*P[I];
       end;
    end;
    MZRO.MVM:=4;  MZRO.MOM[0]:=1;
    For J:=1 to 4 do MZRO.MOM[J]:=0;
    For I:=1 to N do begin
       X:=1;
```

```
      For J:=1 to 4 do begin
         X:=X*V[I];
         MZRO.MOM[J]:=MZRO.MOM[J]+X*P[I];
      end;
   end;
   STDV:=Sqrt(MAVG.MOM[2]);
   CoeffMoments(MAVG,CVAR,CSKW,CKUR);
end;
```

NmlseMoments—normalize moments about zero

This routine normalizes moments about zero by taking the rth root of the rth moment except for the zeroth and 1st moment, which are not modified.

```
{-----------------------------------------------------}
{> NmlseMoments -- Normalise a set of moments         }
{>                 about the origin.                  }
{ Inputs:  BIN  moments about the origin              }
{ Outputs: BOUT normalised moments about origin       }
{  Copyright Mike Tanner 1993                         }
{-----------------------------------------------------}
Procedure NmlseMoments(Var BIN,BOUT:Moments);
Var I:integer; RI:real;
begin
   For I:=0 to 1 do BOUT.MOM[I]:=BIN.MOM[I];
   For I:=2 to BIN.MVM do begin
      RI:=I;
      BOUT.MOM[I]:=Exp(Ln(BIN.MOM[I])/RI);
   end;
   BOUT.MVM:=BIN.MVM;
end;
```

MeanMoments—convert moments about zero to moments about mean

This routine converts moments about zero to moments about the mean. The BiCoeff routine is required for calculating binomial coefficients, and this is included by means of the {$I\qthprog\bicoeff.pas} statement. Caution is required when using this for moments derived from empirical data, in case rounding errors in calculating moments about zero result in relatively high error for the moments about the mean.

```
{-----------------------------------------------------}
{> MeanMoments -- Convert moments about origin to     }
{>                moments about the mean.             }
{ Inputs:  BIN    moments about the origin            }
{ Outputs: BOUT   moments about the mean              }
{  Copyright Mike Tanner 1993                         }
{-----------------------------------------------------}
Procedure MeanMoments(Var BIN,BOUT:Moments);
Var N,K:integer; MEAN,U:qthreal;
```

```
begin
    MEAN:=BIN.MOM[1];   BOUT.MOM[1]:=MEAN;
    For N:=2 to BIN.MVM do begin
        BOUT.MOM[N]:=0; U:=1;
        For K:=0 to N do begin
            BOUT.MOM[N]:=BOUT.MOM[N]
                            +BiCoeff(N,K)*BIN.MOM[N-K]*U;
            U:=U*(-MEAN);
        end;
    end;
    BOUT.MVM:=BIN.MVM;
end;
```

CoeffMoments—Calculate coefficients from moments

This routine calculates the scale-independent coefficients of variation, skewness and kurtosis.

```
{-----------------------------------------------------}
{> CoeffMoments -- Calculate coefficients of          }
{>                 variation, skewness, and           }
{>                 kurtosis, from moments about        }
{>                 the mean.                           }
{ Inputs:                                              }
{     BIN     moments about the mean                   }
{ Outputs:                                             }
{     CVAR    coeff of variation squared               }
{     CSKW    coeff of skewness                        }
{     CKUR    coeff of kurtosis                        }
{  Copyright Mike Tanner 1993                          }
{-----------------------------------------------------}
Procedure CoeffMoments(Var BIN:Moments;
                       Var CVAR,CSKW,CKUR:qthreal);
Var MEAN,SIGMA1,SIGMA2:qthreal;
begin
    MEAN:=BIN.MOM[1];
    SIGMA2:=BIN.MOM[2];
    SIGMA1:=Sqrt(SIGMA2);
    CVAR:=SIGMA2/Sqr(MEAN);
    CSKW:=BIN.MOM[3]/(SIGMA1*SIGMA2);
    CKUR:=BIN.MOM[4]/Sqr(SIGMA2)-3;
end;
```

Exercise 2A—Throw a single die

This exercise is very simple. Get a die and throw it 24 times. Count how many 1s, 2s, etc., you get. Calculate the proportion of throws that are 1s, 2s, etc. Use a table layout like Table 2.4 to record your results. Actually throwing a die is not a childish diversion, it helps to reinforce the distinction between probability statements that relate to the proportion of successes over an 'infinite' number of trials or experiments and the results from a small sample of trials. Readers may wish to

Table 2.4. Worksheet for Exercise 2A

Result of throw	Number of throws	Proportion of throws
1		
2		
3		
4		
5		
6		
Total	24	1

consider whether their particular set of results seem reasonable. How different from the average (same number of 1s, 2s, etc.) would your results have to be before you suspected that the die was biased?

Exercise 2B—Throw a pair of dice

Throw a pair of dice 24 times. Use Tables 2.5 and 2.6 to record the results. In Table 2.5 record how many times the maximum of the two dice for each throw has each possible value. In Table 2.6 record how many times the sum of the two dice has each possible value. For example, if you throw a 5 and a 3, then the maximum is 5 and the sum is 8. Plot graphs of the results and compare them with the theoretical results given earlier in the chapter.

Example 2C—Throwing a die until getting a 6

A rule in some board games is that a player must throw a 6 in order to start. In this example we shall work out the average number of throws required to get a 6, and

Table 2.5. First worksheet for Exercise 2B

Maximum of the two dice	Number of throws	Proportion
1		
2		
3		
4		
5		
6		
Total	24	1

Table 2.6. Second worksheet for Exercise 2B

Sum of the two dice	Number of throws	Proportion
2		
3		
4		
5		
6		
7		
8		
9		
10		
11		
12		
Total	24	1

also the probability of getting a 6 within three throws, and of needing 10 or more throws.

The probability of needing only one throw is the probability of getting a 6 with a single throw, which is 1/6. To get a 6 in exactly two throws means the first throw must not be a 6 while the second throw is a 6. Similarly, to get a 6 in exactly k throws means the first $k - 1$ throws must be non-6s and the kth throw must be a 6. We shall denote the probability of this by

$$p_k = \text{Prob(first 6 occurs on } k\text{th throw)}$$

$$= [\text{Prob(non-6)}]^{k-1} \times \text{Prob(6)} \tag{2.37}$$

$$p_k = \left(\frac{5}{6}\right)^{k-1} \times \frac{1}{6} \tag{2.38}$$

The random variable X is the number of throws required to get a 6. There is no upper limit on the values of X, so X can range from 1 to infinity. The average, or expected value, of X is given by

$$E(X) = \sum_{k=1}^{k=\infty} k p_k = \sum_{k=1}^{k=\infty} k \cdot \left(\frac{5}{6}\right)^{k-1} \cdot \frac{1}{6} \tag{2.39}$$

Readers can use standard mathematical results for the sums of series, from reference books such as Spiegel, giving immediately $E(X) = 6$. In general, if the probability of success on a particular turn or trial is p, then

$$\text{average attempts until success} = \frac{1}{p} \tag{2.40}$$

Table 2.7. Numerical evaluation for Example 2C

k	p(k)	kp(k)	Sum kp(k)
1	0.166667	0.166667	0.166667
2	0.138889	0.277778	0.444444
3	0.115741	0.347222	0.791667
4	0.096451	0.385802	1.177469
5	0.080376	0.401878	1.579347
6	0.066980	0.401878	1.981224
7	0.055816	0.390714	2.371939
8	0.046514	0.372109	2.744047
9	0.038761	0.348852	3.092900
10	0.032301	0.323011	3.415911
...	...	...	...
20	0.005217	0.104336	5.32815
...	...	...	...
40	0.000136	0.005443	5.968703
etc.	etc.	etc.	etc.

Alternatively, direct numerical evaluation is shown in Table 2.7. With computers now commonplace, sometimes it is easier to use direct calculations than even simple mathematical manipulation.

The probability of getting a 6 within three throws is $p_1 + p_2 + p_3$, which from the values in Table 2.7 is 0.42, so there is just over a 40 per cent chance of throwing a 6 within three throws.

The likelihood of needing 10 or more throws is 1 minus the probability of getting a 6 within nine throws, i.e. $1 - (p_1 + p_2 + \ldots + p_9)$. Calculating this directly from Table 2.7, we find the probability of requiring 10 or more throws is 0.19.

Exercise 2D—Throwing a die, throwing again on 6

There are many board games in which players move around the board based on throws of a die. Often the rule is that throwing a six entitles the player to an extra throw. In this example we shall first work out the mean and standard deviation of the number of throws a player gets in each turn, and then the mean of the die throws for a turn.

Let T be the number of throws of the die on a single turn. If the player throws 5 or less on the first throw, then he or she gets just that one throw. If the player throws a 6, then he or she gets that first throw plus a number of further throws. A little thought will show that the number of extra throws obtained, excluding the first throw, is actually the same as the average number of throws for the turn as a whole. So we can write:

$$E(T) = 1 + \text{Prob}(T = 6) \times E(T) \tag{2.41}$$

$$E(T) = 1 + \frac{1}{6} \times E(T) \tag{2.42}$$

so that

$$E(T) = \frac{6}{5} = 1.2 \qquad (2.43)$$

Let us define X as the total die throws from one turn. The player throws the die once. For the first throw the result can be 1 through 6. If the first throw is 5 or less, then that is the end of the turn. If the player gets a 6, then he or she throws again. Reasoning as before, the value of all the extra throws will be the same as the value of the turn as a whole, so we have

$$E(X) = \left(1 \times \frac{1}{6}\right) + \left(2 \times \frac{1}{6}\right) + \dots + \left(5 \times \frac{1}{6}\right) + \left\{[6 + E(X)] \times \frac{1}{6}\right\}$$

$$(2.44)$$

Simple algebraic rearrangement gives

$$6 \times E(X) = (1 + 2 + 3 + 4 + 5 + 6) + E(X) \qquad (2.45)$$

so that

$$E(X) = \frac{21}{5} = 4.2 \qquad (2.46)$$

So the rule giving an extra throw on a 6 raises the average value of a turn from 3.5 to 4.2.

Exercise 2E—Mean, variance of a single die

A single die is thrown. Calculate the mean, variance, standard deviation, and squared coefficient of variation for the value of the throw.

The numerical working required is set out in Table 2.8. Calculating the mean is trivial, of course, simply adding up the possible values and dividing by six, since all values are equally likely. In order to calculate the variance we first need the

Table 2.8. Worksheet for Exercise 2E

	x	$(x - 3.5)^2$
	1	6.25
	2	2.25
	3	0.25
	4	0.25
	5	2.25
	6	6.25
Total	21	17.5
/6	3.5	2.917

squares of the differences from the mean. These are set out in the right-hand column, and we then simply take the average of the squared differences. The results are

$$\text{mean} = \mu = 3.5$$

$$\text{variance} = \sigma^2 = 2.917$$

The standard deviation and squared coefficient of variation can then be obtained:

$$\text{standard deviation} = \sigma = \sqrt{2.917} = 1.708$$

$$\text{squared coeff of variation} = C^2 = \frac{\sigma^2}{\mu^2} = \frac{2.917}{3.5^2} = 0.238$$

Exercise 2F—Mean, variance of sum of two dice

Two dice are thrown. Calculate the mean, variance, standard deviation, and squared coefficient of variation for the sum of the two dice.

   In this example the probabilities of each of the values are different. Table 2.9 gives the possible values and their associated probabilities, and provides columns for the numerical work. The results are

$$\text{mean} = 7.000$$

$$\text{variance} = 5.833$$

$$\text{standard deviation} = 2.415$$

$$\text{squared coeff of variation} = 0.119$$

Table 2.9. Worksheet for Exercise 2F

x	p	xp	$(x - \text{mean})p$
2	1/36		
3	2/36		
4	3/36		
5	4/36		
6	5/36		
7	6/36		
8	5/36		
9	4/36		
10	3/36		
11	2/36		
12	1/36		
	Total		

Exercise 2G—Mean, variance of maximum of two dice

Two dice are thrown. Calculate the mean, variance, standard deviation, and squared coefficient of variation for the maximum of the two dice.

Table 2.10 gives the possible values and their associated probabilities, and provides columns for the numerical work. The results are

mean = 3.917

variance = 1.742

standard deviation = 1.320

squared coeff of variation = 0.114

Table 2.10. Worksheet for Exercise 2G

x	p	xp	$(x - \text{mean})^2 p$
1.00	1/36		
2.00	3/36		
3.00	5/36		
4.00	7/36		
5.00	9/36		
6.00	11/36		
	Total		

3
Describing a queueing system

The components of a queueing system

Before we start analysing the performance of queueing systems, we have to have a systematic way of describing the system and identifying its components. A fairly general structure of a queueing system is shown in Fig. 3.1.

Customer population

The customer population is the pool of people, or other kinds of customers, from which individual customers may arrive at the queue or waiting-line for the server. The important factor is whether the size of this population is infinite (or at least very big), or finite and quite small. If there is an infinite number of potential customers, then the average arrival rate will not be affected if there are already some customers waiting or being served. With a small number of potential customers,

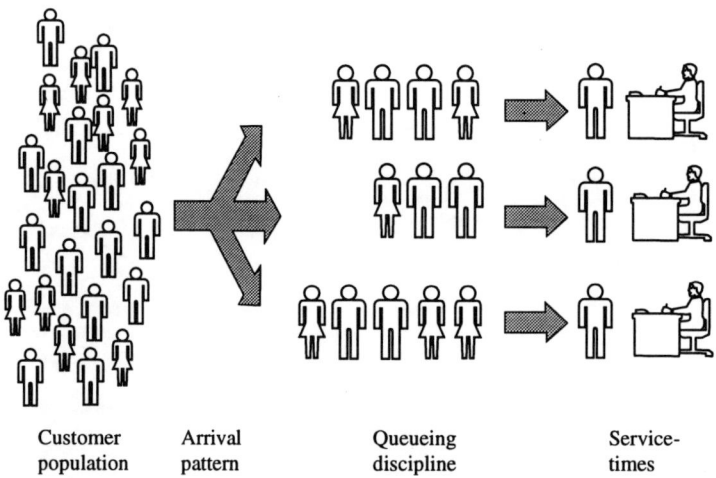

| Customer population | Arrival pattern | Queueing discipline | Service-times |

Figure 3.1. Components of a queueing system.

the more that are already waiting the fewer there are to join the queue, so the arrival rate will reduce as the system gets busy.

Arrival pattern

As well as an average arrival rate of customers, the arrival pattern will have other characteristics. We often assume 'random' arrivals, which is defined more precisely in Chapter 6. But customers may arrive in bunches, or at more regular intervals. Customers may be encouraged or discouraged from requesting service by the number of customers already waiting.

Queueing discipline

The queueing discipline is the method or logic by which the server selects the next customer to serve from among those waiting. The simplest method is first-come-first-served. Alternatively, a priority scheme may be used. Customers might be selected at random. For multiple servers there may be a single common queue or a separate queue for each server. Even a single server could have several queues, and serve one customer from each queue in cyclic order. There are many possible queueing disciplines, and manipulating the queueing discipline is one way of getting the best possible service from a limited server capacity.

Number of servers

The number of servers is important, because a system with a single fast server has different performance characteristics from one with several slower servers, even if the total 'service capacity' is the same in each case.

Service-time

Obviously, a key characteristic of any queueing system is how long it takes a server to deal with a single customer. As well as the average time, we need to know the distribution of service-times around this average. For many queueing models we assume an exponential distribution of service-time, but we need to understand the effect of non-exponential service-times.

Kendall's notation for describing queues

There is a convenient shorthand for describing a queueing system, called Kendall's notation. The description of a queue takes the form

$$A/B/m/K/P/Z$$

or, if there is no limit to the queue size, the customer population size is infinite and service is first-come-first-served, the shorter form A/B/M is used. The 'A' is a letter describing the arrival pattern, by describing the distribution of time intervals between successive customer arrivals. The most common values are as follows.

M means the inter-arrival time has an exponential distribution, and the arrival pattern is random, or Poisson (the M stands for Markov, a famous probability theorist who established important results about series of events with exponential inter-arrival times).

D means deterministic arrivals, i.e. the time between successive arrivals is constant, so that arrivals are regular.

GI means inter-arrival times are 'general independent', i.e. the arrival pattern is quite general but the inter-arrival times are independent, e.g. a long interval between customers k and $k + 1$ arriving does not imply anything about the interval between customers $k + 1$ and $k + 2$.

G stands for a general distribution of inter-arrival times. This is different from GI in that some relationship may exist between successive inter-arrival times, e.g. a long interval between customers k and $k + 1$ may mean the interval between customers $k + 1$ and $k + 2$ is more likely to be short than long.

The 'B' is a letter describing the distribution of service-times, and the most common values are as follows.

M means an exponential distribution.

D means deterministic arrivals, i.e. constant service-time.

G means a general distribution of service-times. Service-times are almost always assumed to be independent, so a distinction between G and GI is seldom made.

E_k means the Erlang-k distribution.

H_k means the hyperexponential distribution of degree k.

The 'm' is the number of servers, which can sometimes be infinite. The symbol 'c' is used in many texts, but this may cause confusion with 'C' which is used for coefficient of variation.

The 'K' is the maximum capacity of the queueing system, i.e. the maximum number that may either be waiting or being served. If this value is not given, it is assumed there is no capacity limitation.

The 'P' is the size of the customer population. If no value is given, it is assumed to be infinite.

The 'Z' is the order of service. Sometimes this is not specified as part of the Kendall notation, but explained alongside the Kendall description, e.g. 'M/M/1 with LIFO service'. If order of service is not given, it is assumed to be FIFO. The abbreviations for order of service commonly encountered are

FIFO or first-in-first-out
FCFS or first-come-first-served, identical to FIFO
LIFO or last-in-first-out
SIRO or service-in-random-order

The use of priority, either pre-emptive or non-preemptive, or cyclic or random service is usually explained without the use of abbreviations.

Steady-state and transient performance

In this book we deal with steady-state behaviour, i.e. the performance of the system once it has been operating for a long time. 'Long' means long enough for a large number of customers to have passed through the system, so that any effects from starting up with zero customers present, or possibly with a backlog of customers who arrived before the server started working, have been reduced to negligible proportions. For computer systems and telecommunication networks this is almost always reasonable, since the 'customers' are typically small pieces of data or processing to be handled, and even a modest system will be processing large numbers of 'customers'. On the other hand, when it comes to queueing situations such as shops and banks, it may take a large proportion of the working day before a steady state is achieved, if indeed it is achieved before closing time. Readers must exercise judgement about whether a steady state can sensibly be assumed.

If a steady state cannot be assumed, then an analysis of transient behaviour is required. Transient states include not just start-up, but temporary periods of over-load such as rush-hour. Analysis of transient behaviour is possible with queueing theory, but requires techniques that are unlikely to be used by the non-specialist. Fluid approximations provide a simple approach for some problems, but even the specialist is likely to resort to simulation to study transient performance.

Performance characteristics of a queueing system

When we set out to use queueing theory, what exactly are we trying to find out? There are a number of obvious things we may want to know, and a few that are perhaps not so obvious. When looking at performance, there are three different points of view that might be taken. First there are the customers who wish to obtain service. For them a good system is one with very small waiting times, and preferably also a short service-time. Secondly, there are the servers themselves. Where the servers are people rather than machines, the server would probably think that being busy, but not too busy, makes a good queueing system. Being not very busy is likely to be boring, while being overworked is not only hard work but will mean that the customers are irritated with long waits and more difficult to deal with. Finally, there is the owner, manager or supervisor of the system. For him or her a good system may be one with a good balance between keeping the servers fairly busy and not causing customers to wait unduly. So queueing theory can be used to tell you how a system will perform, but what constitutes 'good' performance will always be a matter of judgement. The performance characteristics that queueing theory may be able to evaluate are as follows.

Waiting time and waiting-line size

The average waiting time and distribution of waiting time are usually the most important statistics required. Allied to this is the distribution of the number of customers waiting.

Time-in-system and number-in-system

The time a customer spends in the system is waiting time plus service-time. Depending on the situation being analysed, either waiting time or time-in-system may be the more relevant or natural measure to use. Similarly, the number of customers in the system is the number waiting plus the number being served.

Lost customers

In some systems customers may not be allowed to wait if the servers are all busy, or perhaps only a limited number are allowed to wait because of physical limitations such as the space in a shop or office. The number of lost customers then becomes one of the key performance statistics required.

Server utilization

We shall almost always want to know how busy the servers are, either because we are interested in this characteristic for its own sake or just because server utilization is a factor in practically all the formulae for calculating other statistics. Sometimes we also want to know about the 'busy period' distribution for servers. A busy period starts when the server is idle and a customer arrives, and finishes when the server again becomes idle because there are no customers to serve. In one busy period a server may deal with anything from one customer to many customers.

Terminology difficulties

There are some problems with the terminology used in queueing theory. The main one is between the terms 'queueing' and 'waiting'. A 'queue' usually refers to the queueing system, i.e. the waiting-line plus the server and the customer being served. On the other hand, we usually say that we had to queue if we mean we did not get served as soon as we arrived, i.e. if we had to wait. So when we say 'queue size' what do we actually mean? Do we mean the number of customers waiting, or the number waiting plus the number being served? It is no use trying to be pedantic about the precise meaning of words, since different authors have used words to mean different things.

The reader needs to do three things to avoid confusion. First, be aware that confusion is possible, and do not assume that the other person is using words in the same sense as you are. Secondly, whichever book on queueing theory you use, make sure you know how that particular author uses the words queueing, waiting, etc. Thirdly, be aware of the terms used in this book, I have sought to minimize confusion by using the terms as follows. 'Time-in-system' is the phrase used to include both time waiting and time being served. However, since this is a long and unbeautiful phrase, sometimes 'queueing-time' has been used where I hope the context avoids confusion with time spent waiting but excluding time being served. 'Waiting time' is used to mean exactly what it says, i.e. time spent waiting, but not time being served. (Just to emphasize the point, at least one author uses the term 'waiting time' for both time spent waiting and time being served!)

Notational difficulties

Some confusion arises in the notation used in queueing-theory formulae. One source of confusion is the terminology problem discussed in the previous section. Other sources of confusion are the use or non-use of the Greek alphabet, and the limited character set that until recently was typical of many computer systems and languages. Before computers and word processors came along, mathematicians found they needed to represent more variables than there were letters in the English alphabet. Their solution was to use Greek letters. In contrast, computer programming languages, which in many cases do not (or at least did not) recognize upper and lower case letters as being different, represent variables by 'words', i.e. strings of letters, rather than a single letter. The net result of all this is that notation is not standardized, although a reasonable degree of commonality will be found. To help the reader, the full Greek alphabet is given in Table 3.1, and the notation used for the formulae in this book follows. In most books on queueing theory, the notation is listed as an appendix. The present author's experience is that this causes much page-turning and cross-checking, and makes such books more difficult to use than they need be. I have tried to present all results using explanatory phrases alongside the formulae, in the hope of making this book easier to use.

Table 3.1. The Greek alphabet

Greek name	Greek letter		Greek name	Greek letter	
	Lower case	Capital		Lower case	Capital
alpha	α	A	nu	ν	N
beta	β	B	xi	ξ	Ξ
gamma	γ	Γ	omicron	o	O
delta	δ	Δ	pi	π	Π
epsilon	ε	E	rho	ρ	P
zeta	ζ	Z	sigma	σ	Σ
eta	η	H	tau	τ	T
theta	θ	Θ	upsilon	υ	Υ
iota	ι	I	phi	ϕ	Φ
kappa	κ	K	chi	χ	X
lambda	λ	Λ	psi	ψ	Ψ
mu	μ	M	omega	ω	Ω

Notation for formulae

$$T_S = \text{mean service-time} \tag{3.1}$$

$$T_{S(j)} = \text{mean service-time for } j\text{th priority class} \tag{3.2}$$

$$\sigma^2_{T_S} = \text{variance of service-time} \tag{3.3}$$

$$\sigma^2_{T_{S(j)}} = \text{variance of service-time for } j\text{th priority class} \tag{3.4}$$

$$C^2_S = \frac{T^2_S}{\sigma^2_{T_S}} = \text{squared coefficient of variation for service-time} \tag{3.5}$$

$$C^2_{S(j)} = \frac{T^2_{S(j)}}{\sigma^2_{T_{S(j)}}} = \text{squared coeff of variation of service-time } j\text{th priority class} \tag{3.6}$$

$$\lambda = \text{arrival rate to system} \tag{3.7}$$

$$\lambda_A = \text{actual arrival rate to servers, where some customers are lost} \tag{3.8}$$

$$\lambda_O = \text{offered, or potential, arrival rate} \tag{3.9}$$

$$\lambda_{(j)} = \text{arrival rate for } j\text{th priority class} \tag{3.10}$$

$$T_A = \text{mean inter-arrival time, operating time} \tag{3.11}$$

$$z = \frac{T_A}{T_S} = \text{service ratio for finite number of customers} \tag{3.12}$$

$$\sigma^2_{T_A} = \text{variance of inter-arrival time} \tag{3.13}$$

$$C^2_A = \frac{T^2_A}{\sigma^2_{T_A}} = \text{squared coefficient of variation for inter-arrival time} \tag{3.14}$$

m = number of servers (3.15)

ρ = server utilization (3.16)

u = traffic intensity (3.17)

$u_{(j)}$ = traffic intensity for jth priority class (3.18)

T_Q = mean time in system (queueing time) for all customers (3.19)

$T_{Q(j)}$ = mean time in system for jth priority class (3.20)

$\sigma^2_{T_Q}$ = variance of time in system (3.21)

T_{QD} = mean time in system for delayed customers (3.22)

$\pi(r)_{T_Q}$ = rth percentile of time in system (3.23)

$W(t)$ = Prob(waiting time $\leq t$) (3.24)

T_W = mean waiting time for all customers (3.25)

$T_{W(j)}$ = mean waiting time for jth priority class (3.26)

$\sigma^2_{T_W}$ = variance of waiting time (3.27)

T_{WD} = mean waiting time for delayed customers (3.28)

$\pi(r)_{T_W}$ = rth percentile of waiting time (3.29)

p_k = Prob(exactly k customers in the system) (3.30)

p_0 = Prob(system is empty) (3.31)

L_Q = mean number in system including customers being served (3.32)

$\sigma^2_{L_Q}$ = variance of number of customers in the system (3.33)

L_A = mean number in system at customer arrival instants (3.34)

$\sigma^2_{L_A}$ = variance of number in system at arrival instants (3.35)

L_W = mean number of customers waiting (3.36)

$\sigma^2_{L_W}$ = variance of number of customers waiting (3.37)

L_S = mean number of customers being served (3.38)

g_1 = mean duration of busy period (3.39)

σ^2_g = variance of duration of busy period (3.40)

h_1 = mean number served during a busy period (3.41)

σ^2_h = variance of number served during busy period (3.42)

$s_1, s_2, s_3, \ldots$ = moments of service-time about the origin (3.43)

$s_{1(j)}, s_{2(j)}, \ldots$ = moments of service-time about the origin for jth priority class (3.44)

$w_1, w_2, w_3, \ldots$ = moments of waiting time about the origin (3.45)

$q_1, q_2, q_3, \ldots$ = moments of time-in-system about the origin   (3.46)

$R(m, u)$ = Poisson ratio function   (3.47)

$E_B(m, u)$ = Erlang-B function   (3.48)

$E_C(m, u)$ = Erlang-C function   (3.49)

$\binom{n}{k}$ = binomial coefficient   (3.50)

4
Fluid approximations

Introduction

Fluid approximations are so named because we approximate the flow of customers through a queueing system with a flow of water through pipes and reservoirs of restricted size. We take account only of average arrival rates and average service-times. This means we ignore waiting-lines and waiting times that arise from the 'stochastic' or random nature of arrival rates and service-times. On the other hand, we can use a fluid approximation to find out at least something about transient and rush-hour behaviour that would be very difficult to analyse otherwise.

The idea is illustrated in Fig. 4.1, where arrivals are represented by water being poured from a beaker. The server is represented by the tap at the bottom of the funnel. The degree to which this tap is opened corresponds to the speed of the server. The waiting-line is the water accumulating in the top of the funnel. Obviously, the rate at which water is poured in at the top must not exceed the

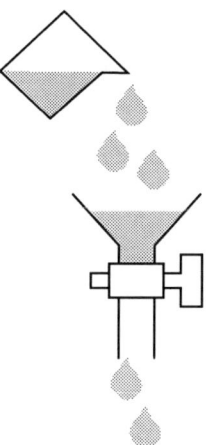

Figure 4.1. Fluid approximation.

speed at which water can drain out through the tap, unless we are prepared for water to be spilt, i.e. for customers to be lost.

Rush-hour

Fluid approximation is not going to tell us much about a single server operating in a steady state, so let us look at a server that is not in a steady state. Most of us suffer from traffic congestion on our route to the office. We shall assume that a particular road or intersection is the main bottleneck. Figure 4.2 illustrates the situation. Suppose this road can handle one car every 4.2 seconds, or about 13.6 cars per minute. The rate at which cars arrive is defined by Table 4.1. Arrival rates are assumed to change linearly between the times given.

Figure 4.3 shows a graph of the arrival and throughput rates during the hours we shall be analysing. The throughput rate is a constant, and is the maximum rate at which cars can pass through the bottleneck. Before and after the rush-hour the rate at which cars arrive is less than the throughput, so no hold-ups will occur. However, there is a period when cars arrive at a greater rate than they can get through the bottleneck. Obviously a tailback will occur, but how big will this tailback be, and how long will it take to clear? The key to answering these questions is to track the cumulative number of cars that have arrived up to some point in time, and the cumulative number that have got through the bottleneck. The difference is the size of the tailback at that time. Figure 4.4 shows a graph of cumulative arrivals and departures for our example. Naturally the number of departures

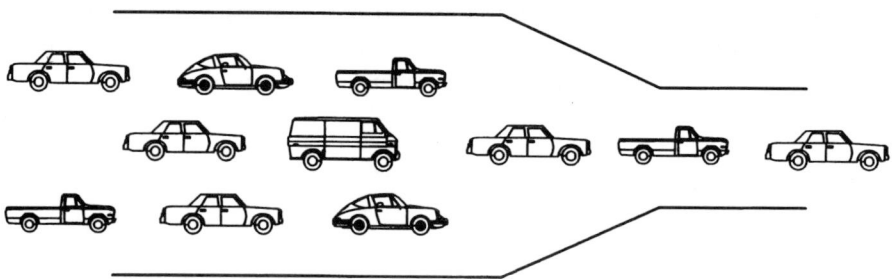

Figure 4.2. The rush-hour bottleneck.

Table 4.1. Car arrival rates during the rush-hour

Time	Cars per minute at this time
Until 07:00	1.00
07:30	3.00
08:00	10.00
08:30	15.00
08:45	25.00
08:55	15.00
After 09:00	10.00

Cars/min

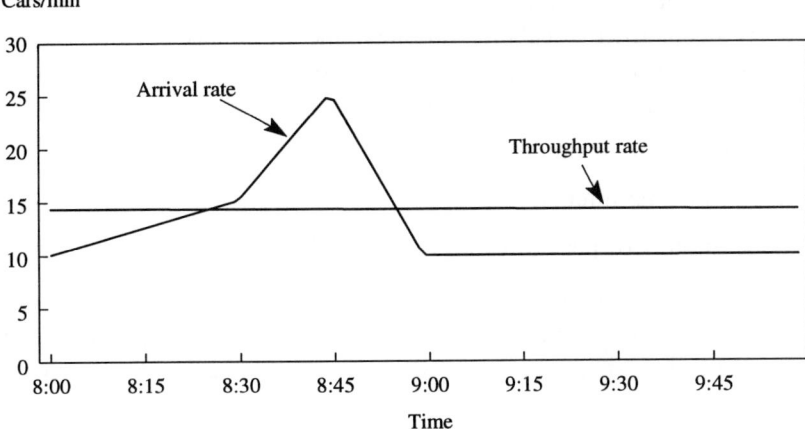

Figure 4.3. Arrival and throughput rates during rush-hour.

cannot exceed the number of arrivals, assuming there is no tailback at the time our analysis starts.

Exactly how we do these calculations can vary. In order to work out the cumulative number of arrivals, we may need just some simple arithmetic if the arrival rate changes in a simple way. For example, if we assume that the arrival rate is constant for a period of time and then changes abruptly to another rate at which it stays constant for a further period, the calculations are easy. If arrival rate is assumed to change smoothly and linearly, then calculations are still fairly

Cars

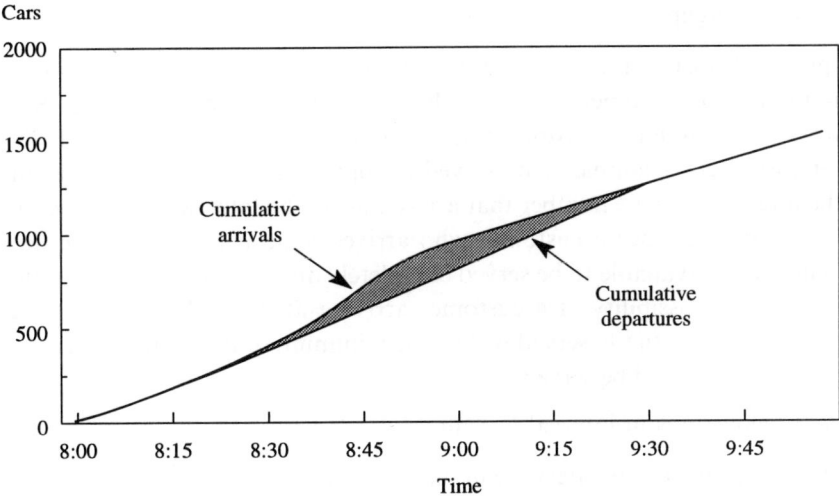

Figure 4.4. Cumulative arrivals and departures during the rush-hour.

straightforward. On the other hand, if arrival rates follow even simple mathematical functions then mathematical integration is called for. Since one of the aims of this book is to avoid the need for mathematical skills readers may no longer possess, we instead take a programming approach.

$$R(t) = \text{arrival rate at time } t \qquad (4.1)$$

$$\delta = \text{length of time interval} \qquad (4.2)$$

$$t_0 = \text{time at start of period to be analysed} \qquad (4.3)$$

We shall assume that the arrival rate during a specific interval is constant, and equal to the value of $R(t)$ at the midpoint of the interval. Interval k starts at time $t_0 + (k-1)\delta$ and finishes at time $t_0 + k\delta$. So the number of customers arriving during interval k is

$$a_k = \text{arrivals during interval } k = \delta R[t_0 + (k - 1/2)\delta] \qquad (4.4)$$

We also need to know the backlog at the end of each interval, and the backlog, if any, that exists at the time the analysis starts:

$$N_k = \text{backlog at end of interval } k \qquad (4.5)$$

$$N_0 = \text{backlog at } t_0 \qquad (4.6)$$

The final quantity we have to deal with is the number of departures in each time interval. Throughput of customers is the maximum possible number of customers that can be processed in a time interval. In many cases this will be constant, but in general we want throughput to be able to take on different values for each interval. We shall see an example later where we represent the effect of a traffic incident by changing the throughput value for some of the intervals. So we shall represent throughput by

$$\mu_k = \text{throughput in interval} \qquad (4.7)$$

Throughput is an instantaneous rate rather than a number per interval, so the maximum number of customers that could be dealt with in an interval is $\delta\mu_k$. Now we can work out how many customers are actually served in each interval. The number of customers available to be served is approximately the backlog at the start of the interval plus the number that arrive during the interval. Of course this is not strictly correct, since a customer who arrives right at the end of the time interval is not really available to be served completely during that interval, but this is balanced by the possibility of a customer arriving at the end of the previous interval. The number actually served will be the minimum of the number available and the number that could be served:

$$d_k = \text{departures in interval } k = \min(N_{k-1} + a_k; \ \delta\mu_k) \qquad (4.8)$$

The backlog at the end of the interval will therefore be

$$N_k = \text{backlog at end of interval } k = N_{k-1} + a_k - d_k \qquad (4.9)$$

The cumulative number of arrivals and departures is then

$$A_k = \text{cumulative arrivals at end of interval } k = \sum_{i=1}^{k} a_k \qquad (4.10)$$

$$D_k = \text{cumulative departures at end of interval } k = \sum_{i=1}^{k} d_k \qquad (4.11)$$

and we have the relationships

$$D_k \leq N_0 + A_k \qquad (4.12)$$
$$N_k = N_0 + A_k - D_k \qquad (4.13)$$

These definitions and relationships are used in the program RushHour, described later in this chapter, to analyse the traffic behaviour at the bottleneck we are interested in. Figure 4.4 shows the cumulative arrivals and departures. The size of the tailback is plotted in Fig. 4.5. This shows a pattern that will be familiar to most of us. As soon as the arrival rate exceeds throughput at just after 8:30, the tailback appears. It continues to grow until the arrival rate drops below throughput at 8:55. However, the tailback takes quite a while to clear, and it is not until 9:30 that the road is again clear, half an hour after the arrival rate dropped to below the capacity of the road.

In practice the situation would be much worse than this 'fluid approximation' suggests. We know that even if the arrival rate is less than the throughput capacity of the road, hold-ups will occur because of the randomness of arrivals. The fluid approximation assumes arrivals are completely regular. Even so, with the above simple analysis we have been able to demonstrate the type of congestion that occurs during a rush-hour.

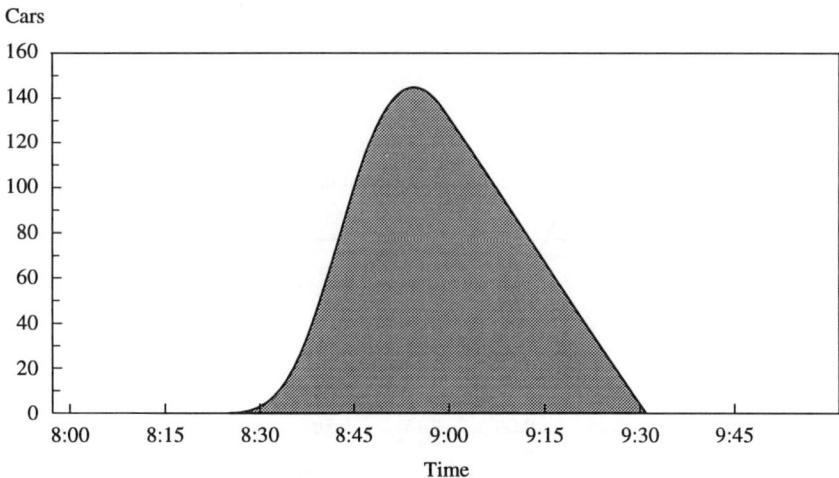

Figure 4.5. Backlog of cars during the rush-hour.

Effect of a traffic incident

Now we shall make the analysis slightly more complicated, by asking what will happen if there is a traffic incident. Suppose a car breaks down or there is a minor collision, and this causes the road to be blocked for five minutes. We can easily represent this by saying that the throughput rate is zero for those five minutes. Assuming the incident occurs at 8:30, the graph of arrival and throughput rates is now shown in Fig. 4.6. The arrival curve is the same as before, but the throughput curve goes to zero for a short period.

The program RushHour deals with this case as well as the original problem. The graph of cumulative arrivals and departures is shown in Fig. 4.7. The cumulative arrivals curve is the same as before, but the cumulative departures curve remains flat for the five minutes during which the road is blocked.

Looking now at Fig. 4.8, we can see the effect of the incident on the tailback. Two differences are readily apparent. First, the tailback reaches well over 200 cars, whereas without the incident it was fewer than 150 cars. Secondly, it is not now until after 9:45 that the tailback clears, compared to just after 9:30 previously. So the five minute incident has caused the tailback to last an additional 15 minutes.

The RushHour program

Calculations for the rush-hour example are performed by the program RushHour, which is listed below. This is a complete program rather than a subroutine, and makes use of the simple routines in QTHFILE.PAS for opening, writing to, and closing a file of results. Arrival rates are defined by the table **ARF**, and this table is

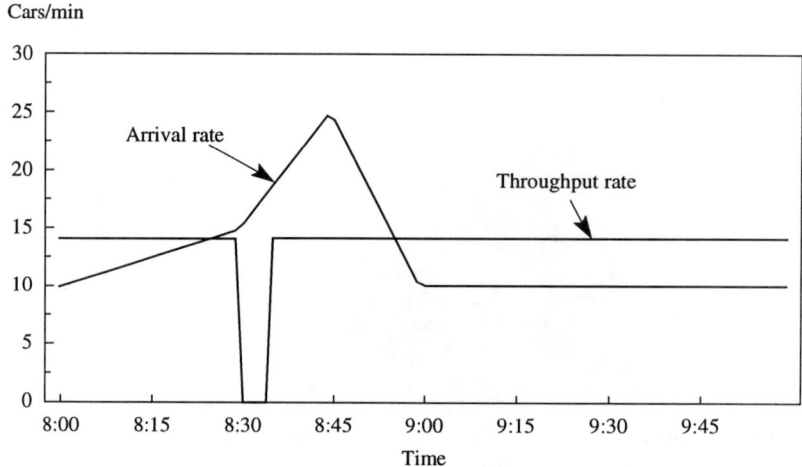

Figure 4.6. Arrival and throughput rates with incident.

Cars

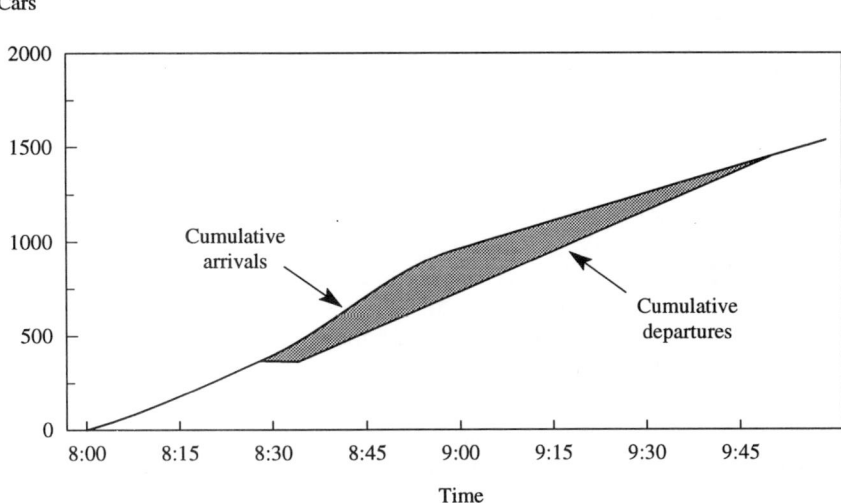

Figure 4.7. Cumulative arrivals and departures with incident.

Cars

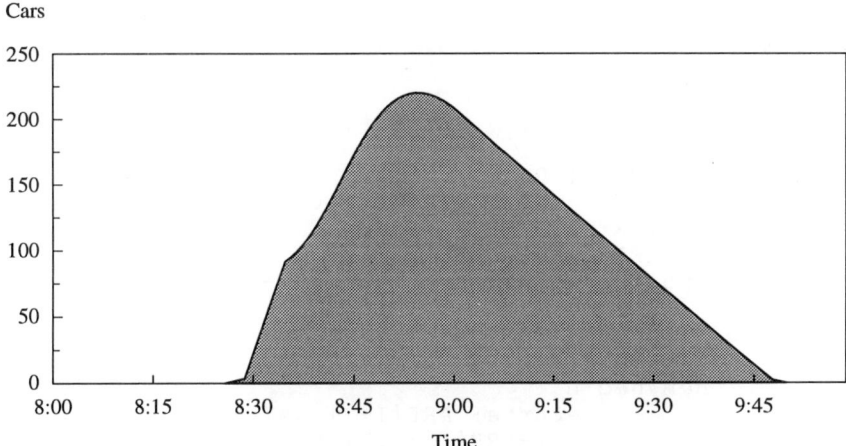

Figure 4.8. Backlog of cars with incident.

used by the subroutine **GetArte**, which performs simple interpolation to get the arrival rate at any particular time. For times before the first entry in **ARF**, GetArte returns the value of the first entry in **ARF**. Similarly, for times later than the last entry in **ARF**, **GetArte** returns the last value in **ARF**. The boolean variable **INCIDSW** is set to **TRUE** if the traffic incident is to be represented, otherwise **INCIDSW** is set to **FALSE**.

```
{------------------------------------------------------}
{> RushHour -- Calculations for fluid                  }
{>              approximation for a rush-hour.          }
{  Copyright Mike Tanner 1993                           }
{------------------------------------------------------}
Program RushHour;
{$N+,E+,R+ }
Uses Dos,Crt;
{$I \qthprog\qthtypes.pas  }
{$I \qthprog\qtherror.pas  }
{$I \qthprog\qthfile.pas   }
{-Define the arrival rate function----------------}
Const H=60;                         { minutes per hour  }
Type arvrte=record    { entry in arrival rate table }
              TIME:QTHreal;
              RATE:QTHreal;
           end;
Const ARFN=7;             { number of entries in ARF }
Const ARF:array[1..ARFN] of arvrte=(
    (TIME:7*H    ;RATE: 1), {  1 car/min  at 07:00 }
    (TIME:7*H+30;RATE: 3), {  3 cars/min at 07:30 }
    (TIME:8*H    ;RATE:10), { 10 cars/min at 08:00 }
    (TIME:8*H+30;RATE:15), { 15 cars/min at 08:30 }
    (TIME:8*H+45;RATE:25), { 25 cars/min at 08:45 }
    (TIME:8*H+55;RATE:15), { 15 cars/min at 08:55 }
    (TIME:9*H    ;RATE:10)); { 10 cars/min at 09:00 }
{-Subroutine to get arrival rate at a specific time}
Function GetArte(T:QTHreal):QTHreal;
Var I:integer;
begin
      {-Before first entry - use first value-}
      If T<=ARF[1].TIME
   then GetArte:=ARF[1].RATE
      {-After last entry - use last value----}
   else If T>=ARF[ARFN].TIME
   then GetArte:=ARF[ARFN].RATE
      {-Interpolate within the table---------}
   else begin
          For I:=2 to ARFN do
            If T<=ARF[I].TIME
            then begin
                  GetArte:=ARF[I-1].RATE+
                  (T-ARF[I-1].TIME)
                  *(ARF[I].RATE-ARF[I-1].RATE)
                  /(ARF[I].TIME-ARF[I-1].TIME);
                 Exit;
               end;
        end;
end;
{-Define the average service time-----------------}
Const INCIDSW=false;    { TRUE=traffic incident    }
Const THPUT:QTHreal=60/4.2;
{ Define the time interval and increments----------}
Const T0=8*H;             { start analysis at 08:00  }
Const DELTA=1;            { use intervals of 1 minute }
Const NINTVL=2*H;         { analyse 2 hours           }
```

```
{-Table in which to keep details for intervals-----}
Type interval=record
                    TIME,       { time                  }
                    THPT,       { throughput            }
                    RATE,       { arrival rate          }
                    NARV,       { number of arrivals    }
                    NCST,       { customers available   }
                    NDEP,       { number of departures  }
                    BKLG        { backlog at end        }
                          :QTHreal;
                    end;
Var INTVL:array[0..NINTVL] of interval;
Var I,J:integer;
Var T,RI:QTHReal;
Var CARV:QTHreal;           { cumulative arrivals   }
Var CDEP:QTHreal;           { cumulative departures }
Const TAB=chr(09);
Var MM:string[2];           { used to format time   }
{-Main routine of program------------------------}
begin
   QTHOpenFile('\qthprog\rushhour.prn');
   {-Throughput per time interval. If the switch---}
   {-INCIDSW is true, the throughput rate for a----}
   {-5 minute period is set to zero----------------}
   For I:=1 to NINTVL do INTVL[I].THPT:=DELTA*THPUT;
   If INCIDSW
   then For I:=31 to 35 do INTVL[I].THPT:=0;
   {-Set initial backlog, rest cumulative values---}
   INTVL[0].BKLG:=0;
   CARV:=0;
   CDEP:=0;
   {-Write headings to results file---------------}
   Writeln(OPF,'"TIME"',
                TAB,'"RATE"',TAB,'"THPT"',
                TAB,'"CARV"',TAB,'"CDEP"',
                TAB,'"BKLG"');
   {-Analyse each interval in turn----------------}
   For I:=1 to NINTVL do
   begin
       RI:=I;
       With INTVL[I] do
       begin
           {-Start of interval and average rate------}
           TIME:=T0+(RI-1)*DELTA;
           RATE:=GetArte(TIME+DELTA/2);
           {-Number of arrivals during interval------}
           NARV:=RATE*DELTA;
           CARV:=CARV+NARV;
           {-Number of customers available----------}
           NCST:=NARV+INTVL[I-1].BKLG;
           {-Customers processed--------------------}
           If NCST>THPT then NDEP:=THPT
                        else NDEP:=NCST;
           CDEP:=CDEP+NDEP;
           {-Backlog at end of this interval--------}
           BKLG:=NCST-NDEP;
```

```
{-Write results to file------------------}
If (I mod 15)=1
then begin
        Str((Trunc(TIME) mod 60):2,MM);
        If MM=' 0' then MM:='00';
        Write(OPF,'"',(Trunc(TIME) div 60):2,
              ':',MM,'"')
      end
    else Write(OPF,'""     ');
    Writeln(OPF,TAB,RATE:7:2,TAB,THPT:7:2,
              TAB,CARV:7:2,TAB,CDEP:7:2,
              TAB,BKLG:7:2);
  end;
 end;
 QTHWaitKey;
 QTHCloseFile;
end.
```

Further reading

Fluid approximations are discussed at length by Kleinrock (1976). Kleinrock also covers 'diffusion approximations', which are more sophisticated approximations taking some account of the randomness of arrivals and service-times. Newell (1982) covers the behaviour of a queue as server utilization increases from less than 100 per cent to more than 100 per cent.

5
Little's law

Introduction

In Fig. 5.1 a simple queueing system is shown, together with the basic quantities that describe the performance of this system. In this chapter we shall note the basic relationships between these quantities, which are obvious but none the less useful. We shall then look at Little's law, which is one of the few interesting and useful results that applies to practically all queueing systems.

First let us look at where customers spend their time. On average the time a customer waits before starting service is T_W, and the average time actually spent receiving service is T_S. The average time spent in the system as a whole, including both waiting and being served, is T_Q. We therefore have the obvious relationship

$$\text{time-in-system} = \text{waiting time} + \text{service-time} \tag{5.1}$$

$$T_Q = T_W + T_S \tag{5.2}$$

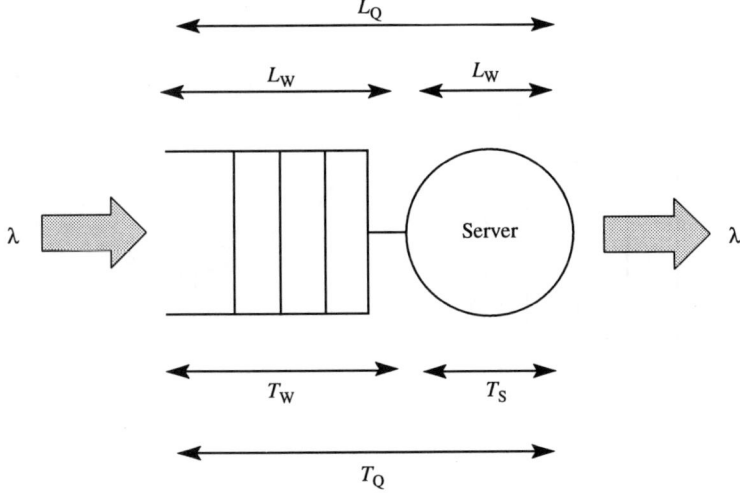

Figure 5.1. Basic quantities for a queueing system.

Equation 5.2 applies to each individual customer, and to the averages over all customers (or any subset of customers). In an exactly similar way we have a relationship between the number of customers waiting, L_W, the number of customers being served L_S, and the number of customers in the system as a whole L_Q.

$$\text{number in system} = \text{number waiting} + \text{number being served} \qquad (5.3)$$

$$L_Q = L_W + L_S \qquad (5.4)$$

Derivation of Little's law

Still referring to Fig. 5.1, we shall introduce Little's law by way of an intuitive justification. Consider individual customers. When a customer arrives at the system there will be on average a particular number of customers present in the system. When a customer leaves the system, he or she will on average leave behind a number of customers in the system. Assuming a steady state, these two averages must be the same. We know that the average time spent in the system is T_Q, and that customers arrive at an average rate λ. So the number of customers that will have arrived while a particular customer is present in the system will be λT_Q. This will also be the average number found by an arriving customer. Now if we further assume that the average number found by arriving customers is the same as the time-averaged number in the system (which is true for random arrivals), it follows that

$$L_Q = \lambda T_Q \qquad (5.5)$$

This is Little's law. The derivation given above is far from rigorous, but provides an intuitive justification. If we now look at Fig. 5.2, we have so far defined our

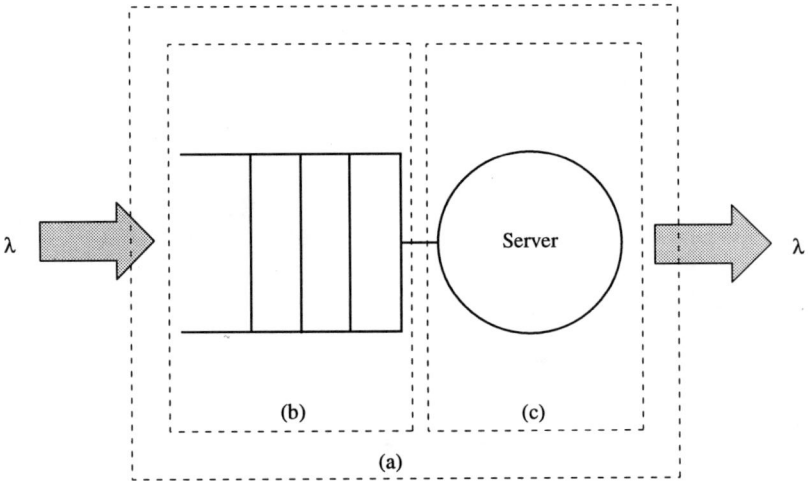

Figure 5.2. Little's law.

'system' to be the things inside the box labelled (a). We could alternatively define our system to be box (b) or box (c), and get different forms of Little's law.

In box (b) we are looking at just the waiting line, and we get

$$L_W = \lambda T_W \qquad (5.6)$$

In box (c) we consider just the server, and we get

$$L_S = \lambda T_S \qquad (5.7)$$

The quantity λT_S is also, for single-server systems, the server utilization. For multiple-server systems it is called the 'traffic intensity'. We have seen so far that the 'system' can be defined in several ways, resulting in different forms of Little's law. We could in fact divide the system up in almost any way and apply the same logic. For example, we could consider just a particular category of customers. More importantly, we can take a system of inter-linked queues and draw a box around any subset of the whole system in order to apply Little's law. Readers are referred to Lazowska *et al.* (1984) for an explanation of how this can be applied to computer systems.

Example 5A—The café

The owner of a small café knows that on average 18 customers per hour arrive, and there are typically 6 customers in the café. What is the average length of time each customers spends in the café?

We know that $\lambda = 18$ customers per hour, and that $L_Q = 6$, and we want to find T_Q. Using Eq. 5.5 we can write $6 = 18T_Q$, so that $T_Q = 1/3 \, \text{h} = 20 \, \text{min}$.

Example 5B—Transactions in a computer system

A computer system receives transactions at the rate of 8 per second. If each transaction is in the computer system for an average of 0.7 seconds, how many transactions, on average, are simultaneously present in the system?

Simply applying Eq. (5.5) gives $L_Q = \lambda T_Q = 8 \times 0.7 = 5.6$ transactions.

Part Two
The simple queue

6
Random arrivals and exponential service

Introduction

For a large proportion of queueing models customers are assumed to arrive 'at random'. To the layperson this may sound a rather vague statement. In fact, 'random arrivals' is a very carefully and precisely defined arrival pattern. In this chapter we shall look at what is meant by random arrivals. First, though, we shall investigate another common assumption in queueing theory—that service-times have an exponential distribution.

Exponential service-times

Let us look first at the shape of the exponential distribution, which is illustrated in Fig. 6.1. Small values are the most common, with progressively larger values

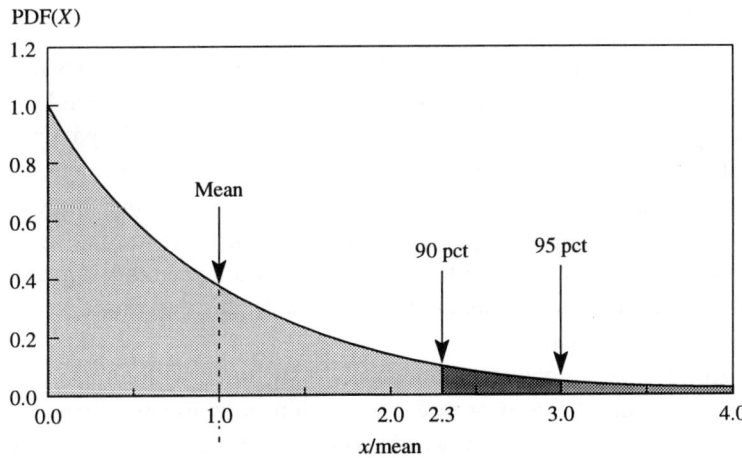

Figure 6.1. PDF of the exponential distribution.

becoming less and less frequent. The curve never gets down to zero for a finite value of x, so extremely large values of x do occur even if their probability is very small.

This distribution of service-times is assumed in many queueing-theory models. The reason for this is a very useful characteristic called the 'memoryless property' which will be discussed shortly. But how realistic is this distribution for representing real service-times? For some types of service, notably the length of a telephone conversation, the exponential distribution fits very well. This is a happy co-incidence, since the early developments and applications of queueing theory were concerned with the design of telephone systems! For other types of service we might object that the service-time has a minimum value, so the probability of calls less than this minimum is zero. Similarly, there might be a maximum value. The exponential distribution may seem a poor fit in many cases. Because of this, queueing systems with different service-time distributions have been analysed, and many of these results are in this book. However, as we look at queueing systems, we find ourselves always starting with exponential service-times to find out how a queueing system behaves, and then considering the effect of some other kind of service-time distribution. The assumption of exponential service-times is often pessimistic, since it allows for the possibility of occasional very long service-times.

Memoryless property of exponential distribution

Suppose a customer arrives at a service centre and finds another customer part-way through being served, and several other customers in the queue waiting to be served. How long will this newly-arrived customer have to wait before starting service? Say there are actually four customers waiting, plus the one being served. The wait-time for the newly-arrived customer will be the sum of

- remaining service-time for the customer already being served
- four service-times for the customers already waiting.

Assuming we know the distribution of service-time, the second element does not present a problem, but the first does. The remaining service-time will depend on how long service has already been going on. Of course, this depends on the particular time the new customer arrives. This is by no means an intractable problem, as we shall see in the chapters dealing with general service-time queueing models. However, if the service-time has an exponential distribution then the maths is much simplified. Why is this? The exponential distribution has a peculiar property that is very useful in the analysis of queues. We shall illustrate this property by discussing the length of a phone call.

Suppose the length of a phone call is exponentially distributed with an average of 3 min. If a call has already been in progress for 1 min, how much longer, on average, would we expect the call to last? It would seem reasonable that if calls last on average 3 min and a call has been in progress for 1 minute, the call might be expected to last a further 2 min. This seems intuitively correct but, as the reader will by now suspect, it is wrong. The correct answer is 3 min. This may seem odd. It may seem even more odd that, however long the call has been in progress, the average remaining length of the call is still 3 min. Mathematically this is easy to demonstrate, but that does not make it any easier to accept!

What do we mean by saying 'the length of a phone call is exponentially distributed with an average of 3 min'? Some calls will be short and some calls will be long. If we are concerned only with calls longer than 1 min, we are excluding or ignoring all calls lasting only a minute or less. Since we are excluding the 'short' calls, the average length of the remaining calls will obviously be greater than the overall average of 3 min. If we split the calls into two categories, those less than a minute long and those more, the properties of these categories are given in Table 6.1.

Table 6.1. Phone calls categorized by length

Category of calls	Percentage	Average call length
Calls less than 1 min	28.30	0.472 min
Calls greater than 1 min	71.70	4 min
All calls	100.00	3 min

So when we restrict ourselves to calls lasting over a minute, we are dealing with only a subset of calls, which will have a different distribution. This would happen with any probability distribution, but with the exponential, the distribution of calls longer than X minutes is exponential with a mean of X + (mean for all calls). The average time remaining is therefore always the average length of all calls. The 'shape' of the distribution remains the same, as shown in Fig. 6.2. This is called the 'memoryless' property of the exponential distribution, and it makes the maths of

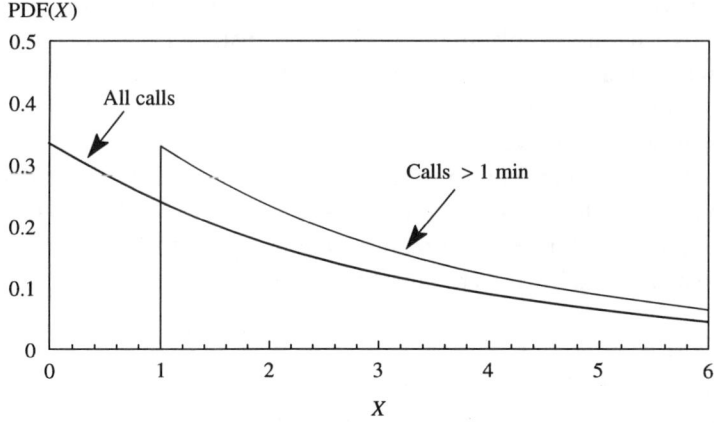

Figure 6.2. Exponential distribution.

queueing theory much easier. It is doubly fortunate that an exponential distribution is a reasonable assumption for many, though by no means all, service-times
found in practice.

Going back to our queue with four people waiting and one being served, we
need only specify the number of people in the system to describe the state of the
system completely. If service-time had a non-exponential distribution, then we
would also have to specify how much service the customer-in-service had already
received.

Random arrivals

What we mean by 'random' arrivals is precisely defined in queueing theory. The
definition, from first principles, divides time into equal small intervals, and makes
two key assumptions. The first is that the number of arrivals in a particular time-
interval is independent of how many arrivals there were in some other interval. A
higher than average number of arrivals in one interval implies nothing about
whether arrivals in another interval will be either higher or lower than the average.
Another way of thinking about this assumption is that it means each customer
arrives independently of all other customers.

The second assumption made is that if the intervals are made small enough,
then the probability of one arrival is proportional to the length of the interval, and
the probability of more than one arrival is negligible. Mathematically these
assumptions are very precise, and we shall look at what follows from this assumption presently. Intuitively, the reader should note that we have ruled out exactly
simultaneous arrivals, but still allowed a fair degree of 'bunching' of customers.

In describing arrival patterns, we have two complementary methods. We can
look at the number of arrivals in successive time periods (these periods will be
quite long ones, not to be confused with the small intervals referred to in defining
random arrivals). There will be a certain probability distribution for the number
of arrivals in each period. Alternatively we can look at the amount of time
between successive arrivals, and look at the probability distribution for 'inter-
arrival' times. The reader may not be surprised to learn that for random arrivals,
at an average rate λ, inter-arrival times have an exponential distribution with a
mean of $1/\lambda$. Suppose customers arrive on average 30 seconds apart, i.e. at a rate of
two per minute. If we ask at a particular instant how long on average before the
next customer arrives, then the answer is always 30 secs. This is something of
a paradox, but the important thing for the reader to note is how the exponential
distribution crops up again, and helps us by simplifying the maths.

Poisson distribution

Now let us turn to the number of arrivals in given-length time periods. For
random arrivals the distribution is called the Poisson distribution. If the time

periods are of length t, and the average arrival rate is λ per unit time, then the average number of arrivals per period is λt. The probability of a specific number of arrivals k is

$$\text{Probability}(k \text{ arrivals}) = \frac{e^{-\lambda t}(\lambda t)^k}{k!} \qquad (6.1)$$

The shape of this distribution is illustrated in Fig. 6.3, which shows the pdf for selected values of mean. For low means, such as 0.5, notice how unsymmetrical the

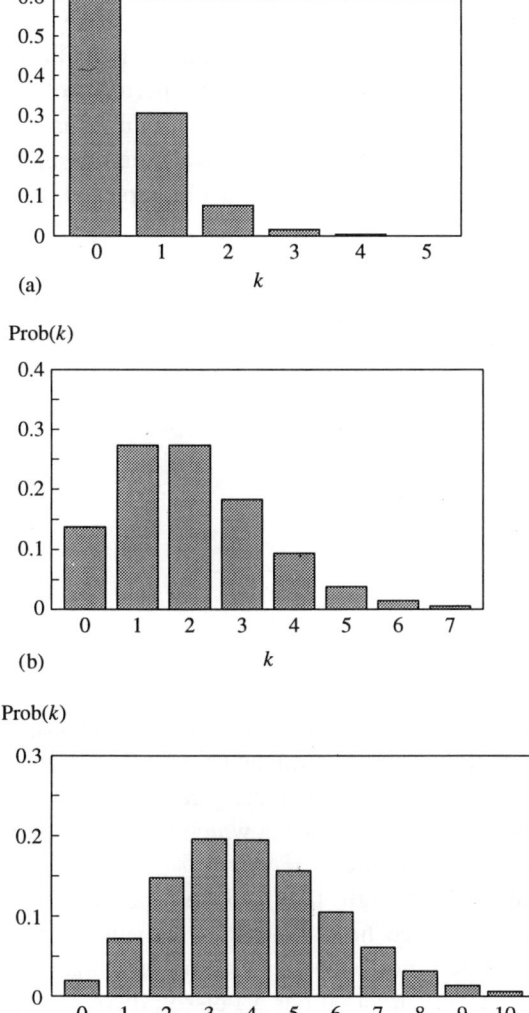

Figure 6.3. Poisson distribution with mean (a) 0.5, (b) 2, (c) 4.

shape is. As the mean increases, the shape becomes more symmetrical. For mean values of more than about 4, the most probable value (or one of the most probable values) is, in fact, the mean.

Another interesting characteristic of the Poisson distribution is revealed if we look at the 95th percentile of the number of arrivals in given-length time periods. Suppose we know the average number of customers that will arrive at a bank or shop during a particular hour such as 10:00 to 11:00 on Tuesdays. Assuming customers arrive at random, and therefore have a Poisson arrival pattern, we can say that on 95 per cent of Tuesdays, during that hour, no more than a particular number of customers will appear. That number is the 95th percentile, and is given in Table 6.2 for a range of values of mean. If the average number of arrivals is 4, then in 95 per cent of such time periods no more than 8 customers will arrive. This means that it is fairly likely that twice the average number might arrive. On the other hand, if the average number of customers is 50, then in 95 per cent of the time periods no more than 62 will arrive, so that one might say it is unusual for more than 1.24 times the average to arrive. The higher the arrival rate, the more 'steady' it is, while for low arrival rates marked fluctuations must be expected.

Table 6.2. Percentiles of the Poisson distribution

Mean	95th percentile	(95th percentile)/mean
1	3	3.00
2	5	2.50
3	6	2.00
5	9	1.80
10	15	1.50
30	39	1.30
50	62	1.24
100	117	1.17

Summary

We have indicated how the exponential distribution has a 'memoryless' property that helps a great deal in solving the mathematics of a queueing situation. Fortunately also, the exponential distribution is a not wholly unreasonable assumption for service-times in practice. In many cases it is pessimistic, which means we shall overestimate rather than underestimate waiting times.

Random arrivals have been defined, although the rigorous mathematical definition is more detailed than the one given here. 'Random arrivals' in fact means an exponential distribution for the times between customer arrivals. The number of customers arriving in a given time interval has a Poisson distribution, with a mean of

average arrival rate per unit time × length of interval   (6.2)

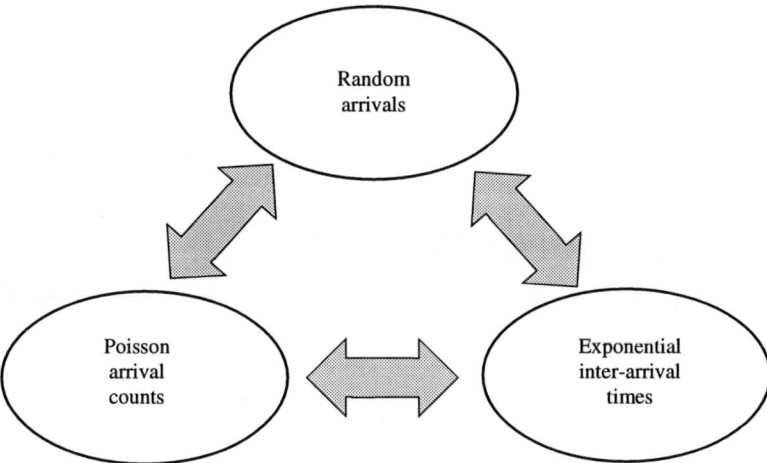

Figure 6.4. Random arrivals.

There is a 'duality' or correspondence between a Poisson distribution for the number of arrivals and an exponential distribution for times between arrivals, as depicted in Fig. 6.4.

7
The simple queue (M/M/1)

Introduction

As you would expect, we start our study of queues by looking at one of the simplest queueing situations possible. We have just a single server, and arriving customers form a single queue to be served on a first-come-first-served basis (Fig. 7.1). All customers have equal priority, and there is no limit on how many customers may be waiting. The length of the queue has no effect on the average rate at which customers arrive.

This queueing situation is also simple in another sense. Many queueing systems require difficult mathematics for their analysis, and many systems cannot be analysed in any great detail by mathematical techniques. The simple system we are looking at here, however, uses assumptions about the customer arrival pattern and the service time for each customer that make the mathematical analysis as tractable as possible. This does not mean the maths is trivial—unless you had graduate level probability and statistics you would have to work hard to follow the derivation of the results. Most of the results, are nice simple formulae that can easily be worked out by hand.

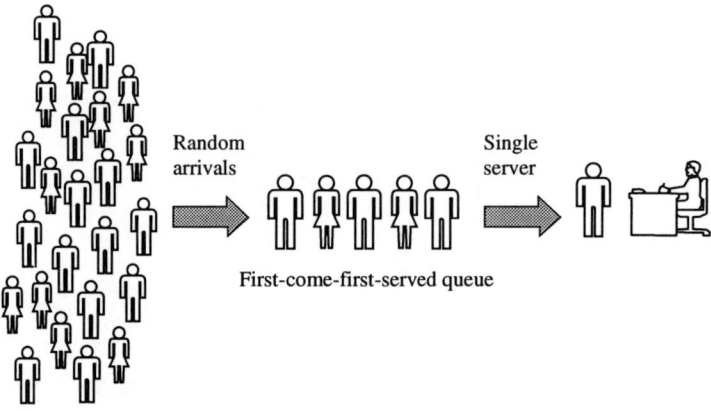

Figure 7.1. The simple queue.

What shall we get out of this simple model, and how useful is it in practice? First of all, we can get a number of valuable intuitive insights into how queues behave. Later in the book we shall see that these insights carry over into more complex situations, so let us start by just listing the intuitive insights that will be discussed at length in the rest of this chapter.

- The average time customers have to wait depends heavily on how busy the server is. This dependency is very non-linear, so the average wait when the server is 80 per cent busy is a lot more than double the average wait when the server is 40 per cent busy.
- There is a conflict between wanting to give good customer service and wanting to use the server efficiently. Keeping average customer waiting time down means having service capacity that is idle for a significant amount of time.
- A faster, or more powerful, server can handle proportionally more work than a slower server for the same level of customer service. A small number of powerful servers will, other things being equal (which, of course, they never are), give better service than a large number of less powerful servers.

The M/M/1 model is important also because it is the only bit of queueing theory that many system designers use. The authors hope is that this queueing-theory handbook will encourage and enable system designers to use more appropriate models. At least the limitations of the M/M/1 model need to be understood. A further important role of the M/M/1 model is as a yardstick for comparing other queueing systems. When we look at other queueing systems later in the book, their performance will be compared with M/M/1 because that is the familiar, or basic, model of queue behaviour.

Assumptions for the M/M/1 model

Random (or Poisson) arrivals are assumed, and service-times are exponentially distributed. The chapter on random arrivals/exponential service explains in detail what this means, and how it simplifies the mathematics.

The population of potential customers is assumed to be infinite, so the rate at which customers arrive is unaffected by how many are already either being served or waiting to be served.

The queue discipline is first-come-first-served.

Parameters and initial calculations for M/M/1

The parameters that need to be specified for M/M/1 are very few. They are:

the average customer arrival rate λ, e.g. customers per minute
the average service-time for a customer T_S, e.g. 2 minutes

The particular time unit used does not matter, as long as the same unit is used for arrival rate and service-time. It is convenient, when describing queueing behaviour, to assume that T_S is 1 time unit, so that the only variable is arrival rate or, equivalently, server utilization.

Having defined λ and T_S, the next thing to do is to calculate the utilization of the server. Server utilization ranges from 0, meaning the server is always idle, to 1, meaning the server is always busy. Often utilizations are quoted as percentages. As a general rule it is better to do all calculations with utilizations expressed in the range 0 to 1, and convert to percentages just for final presentations of results. Luckily, errors from mixing up proportions with percentages are likely to be dramatic and easily spotted. Server utilization is easily calculated as

$$\rho = \lambda T_S \tag{7.1}$$

An important check is that $\rho < 1$, in other words the server must not be overloaded. If the server is overloaded then the waiting line will grow remorselessly. The assumption of an infinite customer population, and an arrival pattern unaffected by the state of the queue, means that customers would continue to arrive at the same rate even though the server could not process customers at that rate. Note that the case of $\rho = 1$ may seem acceptable to the reader, but such a system would be unstable, and the formulae given below are not defined for $\rho = 1$.

Time-in-system (or queueing time)

Usually the best overall measure of the service that customers receive is the average time spent in the system, that is both waiting for service and actually being served. For the M/M/1 model this is easily calculated with the following formula, which is illustrated in Fig. 7.2:

$$T_Q = \frac{T_S}{1 - \rho} \tag{7.2}$$

We often want to know more about the distribution of the time in system than just the average. For the M/M/1 model we have the luxury of knowing explicitly what the distribution is. In fact, time-in-system has an exponential distribution, with a mean of T_Q, so that

$$\text{Prob(time in system} \le t) = 1 - \exp\left(-\frac{t}{T_Q}\right) \tag{7.3}$$

$$= 1 - \exp\left(-\frac{(1 - \rho)t}{T_S}\right) \tag{7.4}$$

From the properties of the exponential distribution, we know the variance of time-in-system is the square of the mean, or equivalently that the squared coefficient of

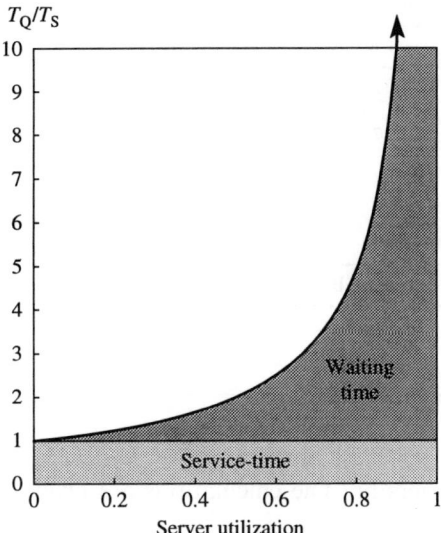

Figure 7.2. Time-in-system for M/M/1.

variation is 1, i.e. $\sigma^2_{T_Q} = T^2_Q$ and $C^2_{T_Q} = 1$. Also, we can easily find expressions for the percentiles.

$$\pi(90)_{T_Q} = 90\text{th percentile of time-in-system} = 2.3T_Q \tag{7.5}$$

$$\pi(95)_{T_Q} = 95\text{th percentile of time-in-system} = 3.0T_Q \tag{7.6}$$

$$\pi(r)_{T_Q} = r\text{th percentile of time-in-system} = \ln\left(\frac{100}{100 - r}\right)T_Q \tag{7.7}$$

Non-linearity of time-in-system

Figure 7.2 is probably the best-known graph in queueing theory. The most notable characteristic is that T_Q increases more and more rapidly as server utilization approaches 1. This is a fundamental characteristic of many types of queueing system, not just the very simple queue we are looking at in this chapter. As server utilization increases, queueing time increases faster and faster unless there is some mechanism at work to limit customer arrivals when the system is congested. We can illustrate the non-linearity of time spent in the system with a straightforward example. Suppose the average service-time is 2 min 24 secs (i.e. 144 secs, or 0.04 hrs). What happens if the arrival rate is 10 customers/hr and then increases to 12 customers/hr? Table 7.1 gives the calculations.

The increase from 10 to 12 customers/hr represents an increase of 20 per cent in the arrival rate. The corresponding increase in queueing time is from 240 to 277 secs, an increase of 15.4 per cent. Now let us look at what happens when the

Table 7.1. Non-linearity of queueing time

	Case A1	Case A2	Increase
λ	10/hr	12/hr	20%
ρ	0.40	0.48	
$T_Q = T_S/(1 - \rho)$	240 secs	277 secs	15.4%

Table 7.2. Non-linearity of queueing time

	Case B1	Case B2	Increase
λ	20/hr	22/hr	10%
ρ	0.80	0.88	
$T_Q = T_S/(1 - \rho)$	720 secs	1200 secs	67%

arrival rate increases from 20 to 22 customers/hr. The calculations are given in Table 7.2.

The increase in arrival rate from 20 to 22 customers/hr is a 10 per cent increase in arrival rate. The resulting effect on queueing time is a 67 per cent increase, a markedly sharper increase than the previous case.

Waiting time

Sometimes we focus on the waiting time rather than the time-in-system. The average waiting time is calculated with the formula

$$T_W = \text{average waiting time} = \frac{\rho T_S}{1 - \rho} \tag{7.8}$$

Average time-in-system, average waiting time and average service-time are obviously related by

$$T_Q = T_W + T_S \tag{7.9}$$

which can be written as

$$T_Q = \frac{T_S}{1 - \rho} = \frac{\rho T_S}{1 - \rho} + T_S \tag{7.10}$$

If we want to know more about waiting time, such as variance and percentiles, we can easily get these because the explicit distribution of waiting time is known. The cdf is

$$W(t) = \text{Prob(waiting time} \le t) = 1 - \rho \exp\left(-\frac{(1 - \rho)t}{T_S}\right) \tag{7.11}$$

which is illustrated in Fig. 7.3 for $\rho = 0.6$. This distribution is described as a modified exponential: it differs from the exponential by having a jump at $t = 0$, since

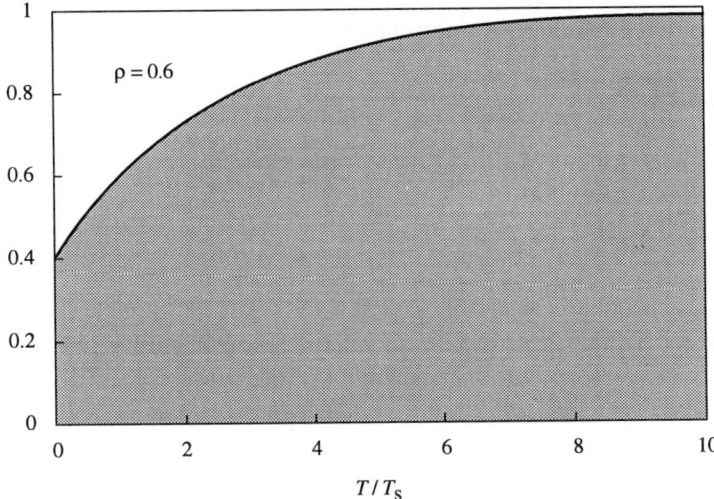

Figure 7.3. Cdf of waiting time for $M/M/1$.

there is a probability of zero waiting time of $1 - \rho$, i.e. the probability that the server is free.

The pdf of waiting time is given by

$$w(t) = \frac{\mathrm{d}W(t)}{\mathrm{d}t} = \frac{\rho(1-\rho)}{T_S} \exp\left(-\frac{(1-\rho)t}{T_S}\right) \tag{7.12}$$

From the pdf we can derive the variance of waiting time and squared coefficient of variation, i.e.

$$\sigma_{T_W}^2 = \frac{(2-\rho)\rho T_S^2}{(1-\rho)^2} \tag{7.13}$$

$$C_{T_W}^2 = \frac{\sigma_{T_W}^2}{T_W^2} = \frac{2-\rho}{\rho} \tag{7.14}$$

Since ρ never exceeds 1, C_W^2 is never less than 1 and increases rapidly as ρ gets near zero. On the face of it this is surprising, since we expect performance of the system to be worse and less stable as utilization increases. Let us think this through to see what it means. At low utilization waiting time T_W is very small; indeed, for a large proportion of customers it is zero. When customers do wait at low utilizations, they could be said to be unlucky. They will typically wait for one other customer to finish, so that their waiting time will be approximately T_S. This sort of distribution, where the average is nearly zero and the variance is small but non-zero, has a high coefficient of variation, but this is because the mean is very low, not because the variance is very high. At high utilizations both the mean and the variance of waiting time are high. What this means is that if you are trying to measure waiting time in a real system that approximates the M/M/1 model, a large sample of individual customer waiting times will be needed to get a reliable measurement.

From the cdf we can derive an expression for the percentiles of waiting time, i.e.

$$\pi_{T_W}(r) = r\text{th percentile of waiting time} = T_W \frac{1}{\rho} \ln\left(\frac{100\rho}{100 - r}\right) \tag{7.15}$$

$$= \frac{T_S}{1 - \rho} \ln\left(\frac{100\rho}{100 - r}\right) \tag{7.16}$$

When we looked at the percentile of time-in-system, the expressions were simple constant multiples of the average time in system, e.g. the 90th percentile of time in system is $2.3 T_Q$. Percentiles of waiting time depend in a slightly more complex way on server utilization. We can see this in Fig. 7.4, which plots $\pi_W(90)/T_W$ and $\pi_W(95)/T_W$ against utilization.

Pctile/T_W

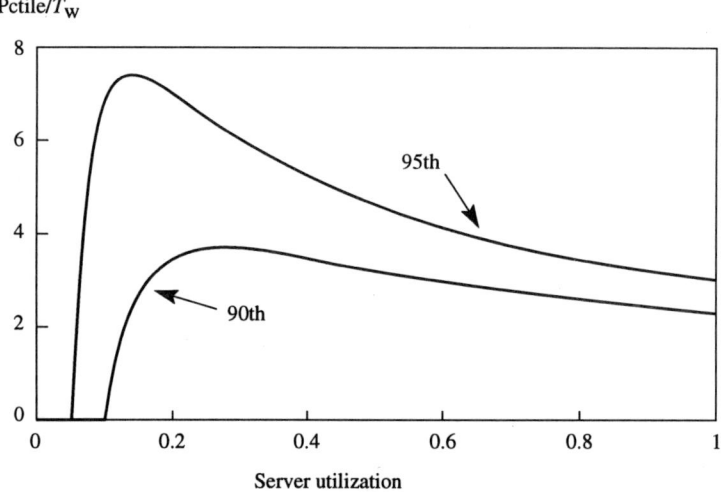

Figure 7.4. Waiting time percentiles for M/M/1.

Wait-time for delayed customers

A little care is required when talking about waiting times. The above formula for T_W (Eq. (7.8)) is the average waiting time for all customers, including those who arrive to find the server idle and so do not wait at all. Another way to express waiting time is to say how likely it is that a customer will have to wait, and the average waiting time for those who do wait. We can calculate these results as follows.

$$\text{Prob(customer has to wait)} = \rho \tag{7.17}$$

so that the proportion of customers that are delayed is the same as the server utilization. This is intuitively obvious, since an arriving customer will have to wait if the server is busy. We can now calculate the average waiting time for customers that do actually wait with the formula

$$T_{\text{WD}} = \text{average wait for delayed customers} \tag{7.18}$$

$$T_{\text{WD}} = \frac{T_{\text{W}}}{\text{proportion of customers that wait}} \tag{7.19}$$

$$T_{\text{WD}} = \frac{T_{\text{W}}}{\rho} = \frac{T_{\text{S}}}{1 - \rho} \tag{7.20}$$

In some applications it is important to make the distinction between delayed and non-delayed customers. For systems where the customers are human beings, we have to remember that a long wait will have much more impact on an individual's perception of the service received than a zero wait. We all remember and tell our friends about the terrible delay at the airport, the day the roads were hopelessly congested, and the long queue at the supermarket. Less often do we tell them about the times that everything went smoothly!

Number-in-system

Now we move on to the number of customers in the system, or queue size. For the M/M/1 model the number-in-system has a geometric distribution, which means

$$p_k = \text{Prob(exactly } k \text{ customers in system)} = (1 - \rho)\rho^k \tag{7.21}$$

Since we know the explicit distribution of number-in-system, we can easily calculate the mean and variance:

$$L_{\text{Q}} = \text{mean number-in-system} = \sum_{k=0}^{\infty} k p_k = \frac{\rho}{1 - \rho} \tag{7.22}$$

$$\sigma_{L_{\text{Q}}}^2 = \text{variance of number-in-system} = \frac{\rho}{(1 - \rho)^2} \tag{7.23}$$

It may be also be useful to be able to calculate the probability of k or more customers in the system, or alternatively no more than k customers present. The formulae for these probabilities are:

$$\text{Prob}(\geq k \text{ customers in system}) = \rho^k \tag{7.24}$$

$$\text{Prob}(\leq k \text{ customers in system}) = 1 - \rho^{k+1} \tag{7.25}$$

For example, if $\rho = 0.7$ then the average queue size is given by $L_{\text{Q}} = 0.7/0.3 = 2.3$. The distribution of the number of customers in the system is illustrated in Fig. 7.5, which shows the pdf, and Fig. 7.6, showing the cdf. The probability of none in the system is the probability that the server is idle, i.e. 0.7. If we want to

Prob(k)

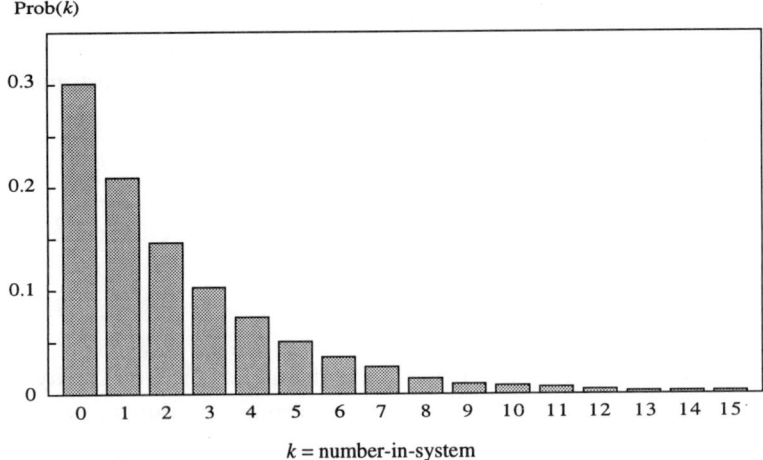

k = number-in-system

Figure 7.5. PDF of number of customers in system for $\rho = 0.7$.

Prob($\leq k$)

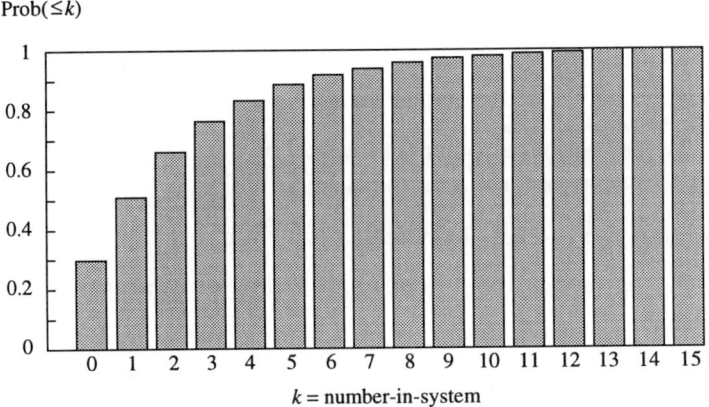

k = number-in-system

Figure 7.6. CDF of number-in-system for $\rho = 0.7$.

know the probability of three of more customers in the system, i.e. two or more customers waiting, the answer is $\rho^3 = (0.7)^3 = 0.343$.

Number waiting

The distribution of the number waiting is defined by Eq. (7.21), except that of course $k > 1$ customers in the system corresponds to $k - 1$ customers waiting, and 0 or 1 customers in the system means there are zero customers waiting. So we have

$$\text{Prob}(n \text{ customers waiting}) = 1 - \rho^2 \text{ for } n = 0 \tag{7.26}$$

$$= (1 - \rho)\rho^{n+1} \text{ for } n > 0 \tag{7.27}$$

From this distribution we can derive the mean and variance of the number of customers waiting:

$$L_W = \text{mean customers waiting} = \frac{\rho^2}{1 - \rho} \tag{7.28}$$

$$\sigma_{L_w}^2 = \text{variance of number waiting} = \frac{\rho^2(1 + \rho - \rho^2)}{(1 - \rho)^2} \tag{7.29}$$

We may also wish to calculate the probability that no more than a given number of customers are waiting.

$$\text{Prob}(\leq n \text{ customers waiting}) = 1 - \rho^{n+2} \tag{7.30}$$

The busy period

The server is alternately busy, idle, busy, idle, and so on. In this section we shall look at the distribution of the busy period, and the distribution of the number of customers served in a busy period. Common sense suggests that at low utilizations there will usually be just one customer served in a busy period, and the length of the busy period will have the same distribution as customer service times. On the other hand, when the server is working at high utilizations, many customers may be served before the queue will empty.

The non-mathematical reader may wish to take a deep breath here, since the next result is surprisingly complex compared with the generally straightforward formulae for M/M/1. Fortunately, as we shall see shortly, the formula for average busy-period length is much simpler than you might expect from looking at Eq. (7.31):

$$g(y) = \text{pdf of busy period} = \frac{1}{y\sqrt{\rho}} e^{-\left(\lambda + \frac{1}{T_S}\right)y} I_1\left[2y\sqrt{\frac{\lambda}{T_S}}\right] \tag{7.31}$$

where the I_1 means the 'modified Bessel function of the first kind or order one'. Bessel functions will have been encountered by readers who studied engineering or mathematics, since they arise in the solution of differential equations. They may be unfamiliar to many readers, but are no more unusual than the logarithm, sine and cosine functions. A routine for calculating $I_1(x)$ is given in Appendix 2, together with a graph illustrating the function.

The mean and variance of the busy-period duration are given by

$$g_1 = \text{mean length of busy period} = \frac{T_S}{1 - \rho} \tag{7.32}$$

$$\sigma_g^2 = \text{variance of length of busy period} = \frac{1 + \rho}{(1 - \rho)^3} T_S^2 \tag{7.33}$$

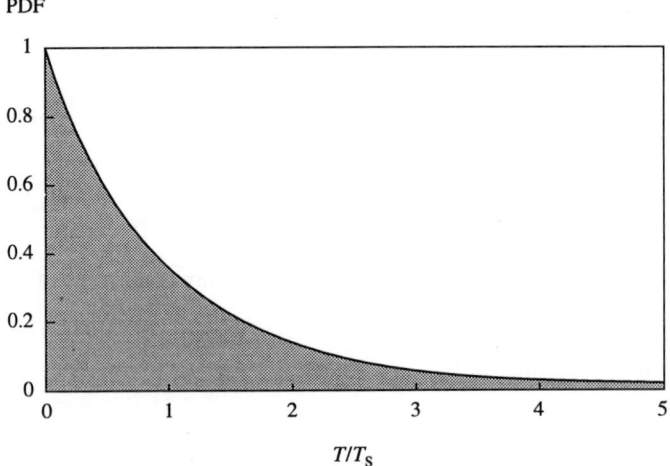

Figure 7.7. PDF of busy period of M/M/1 for $\rho = 0.1$.

It is worth noting the cubic power in the denominator of σ_g^2, which means that the variance gets extremely large at high utilizations. Figures 7.7 and 7.8 illustrate the busy-period distribution for low ($\rho = 0.1$) and high ($\rho = 0.9$) server utilizations.

For low utilizations, as we would expect, the distribution is very similar to the exponential. At high utilizations the graph looks quite similar, but in fact the probability of longer busy periods has increased significantly. It is perhaps easier to see this effect by looking at the distribution of the number of customers served during a busy period, which we shall now do.

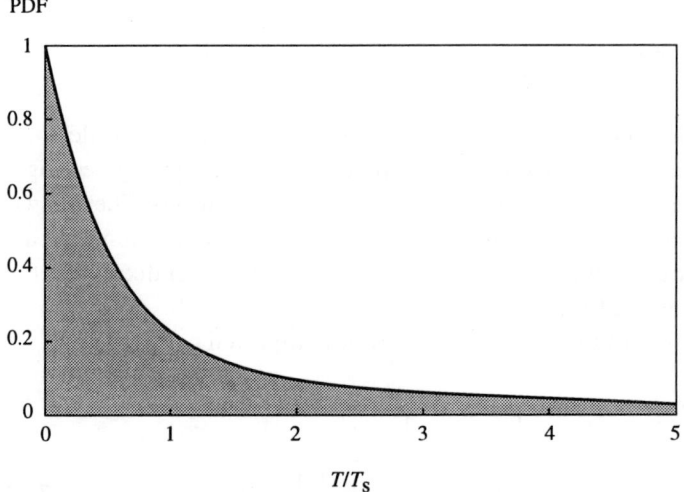

Figure 7.8. PDF of busy period of M/M/1 for $\rho = 0.9$.

Number served per busy period

We know explicitly the probability distribution for the number of customers served during a busy period. This is given by

$$f_n = \text{Prob(exactly } n \text{ customers served in a busy period)} \tag{7.34}$$

$$f_n = \frac{1}{n} \binom{2n-2}{n-1} \rho^{n-1} (1+\rho)^{1-2n} \tag{7.35}$$

The mean and variance of the number of customers served in a busy period are given by

$$h_1 = \text{mean number served per busy period} = \frac{1}{1-\rho} \tag{7.36}$$

$$\sigma_h^2 = \text{variance of number served per busy period} = \frac{\rho(1+\rho)}{(1-\rho)^3} \tag{7.37}$$

Figure 7.9 shows the 90th, 95th and 99th percentiles of the number of customers served in a busy period. If we consider the 90th percentile, what this means is that 10 per cent of busy periods (or runs of customers) involve serving more than that number of customers without a pause between customers. A server who is about 75 per cent utilized will in a typical busy period serve four customers (easily calculated using Eq. (7.36)), but not infrequently may serve about 50 customers without a pause. Interpretation of these graphs needs a little thought. A very long run of customers may occur in only a small percentage of busy periods, but that will represent a bigger percentage of customers served and work done by the server.

Number served

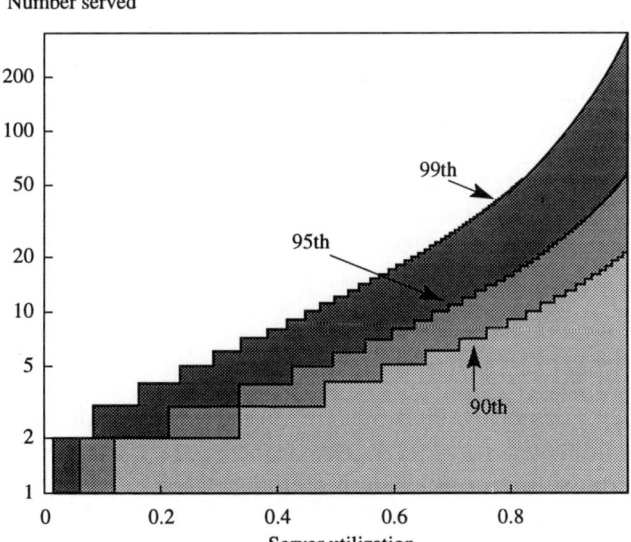

Figure 7.9. Percentiles of number served in busy period.

If the server is a human being, as opposed to a machine, the number of customers served in a busy period can be an important characteristic. People need a pause in their work now and again, and very long busy periods may not be productive, or even tolerable.

Server efficiency versus customer service

There is a clear conflict between wanting to make maximum use of the server and wishing to keep customer waiting and service-times to a minimum. It is an interesting exercise to try and think of answers to the questions 'why can I not load the server 100 per cent?'. This question does crop up sometimes from people not used to thinking about the statistical behaviour of systems, as opposed to deterministic simplifications. Answers that involve mathematical formulae, or that rely on statements such as 'the queue size would be infinite' are not very effective. The author's attempt at an answer is to point out that there will be times when the server is idle because there are no customers to serve. The irregular way in which customers arrive will mean there is some bunching of arrivals, and the converse of this is that quite long intervals may occur in which no customers arrive. The server idle time that occurs in these intervals can never be recovered, so the overall utilization of the server must be less than 100 per cent.

In some cases we could find out the optimum balance between server efficiency and customer waiting times. We shall look at a simple example here. A cost function needs to be defined, in terms of the performance statistics of the queueing system. This cost function is then minimized using some appropriate mathematical technique, and the optimum utilization is arrived at. Assume we want to minimize 'wasted' time, which we shall take to be the sum of server idle time and the waiting time for each customer. If customers arrive at λ per hour, and their average waiting time is T_W, then the total time wasted by customers is λT_W per hour. The server wasted time is $(1 - \rho)$ hours. So the function we need to minimize is

$$Y(\rho) = (1 - \rho) + \lambda T_W \tag{7.38}$$

$$Y(\rho) = (1 - \rho) + \lambda \frac{\rho T_S}{1 - \rho} \tag{7.39}$$

$$Y(\rho) = (1 - \rho) + \frac{\rho^2}{1 - \rho} \tag{7.40}$$

Applying some elementary calculus, we find that this function has a minimum value of $Y(\rho) = 0.83$ hours/hour when $\rho = 0.29$. The shape of the cost curve is shown in Fig. 7.10. The minimum is not very pronounced, but the idea of minimizing a cost function has been shown.

Another approach might be to say that the time wasted by the server is to equal the time wasted by the customers. The mathematics for this is very simple,

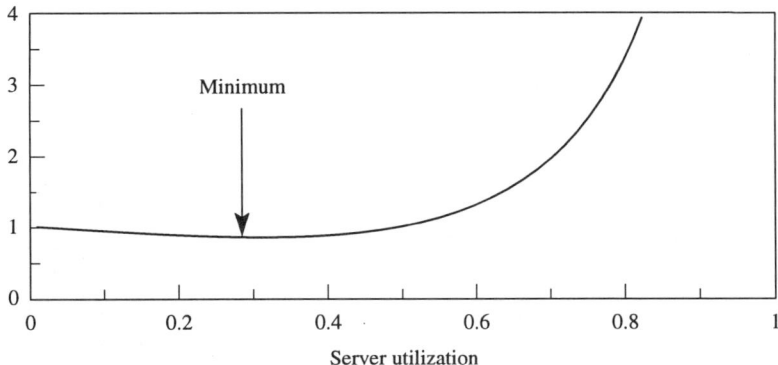

Figure 7.10. Optimum utilization.

since we want

$$\text{server wasted time} = \text{customer wasted time}$$

$$(1 - \rho) = \lambda T_{\text{W}} \tag{7.41}$$

$$1 - \rho = \lambda \frac{\rho T_{\text{S}}}{1 - \rho} = \frac{\rho^2}{1 - \rho} \tag{7.42}$$

so that

$$(1 - \rho)^2 = \rho^2 \tag{7.43}$$

and so $\rho = 0.5$. With the server 50 per cent loaded, and average of 1 hour per hour is wasted, made up of 30 min of server idle time and 30 min of waiting time shared between the customers that arrived during that hour.

The scaling effect

One of the most important generalizations that can be made about queueing systems is that systems dealing with large volumes of customers are more efficient than systems dealing with small volumes. We shall see this effect illustrated for many of the queueing systems described in this book. This phenomenon is called the scaling effect. We shall in a moment look at the scaling effect of M/M/1, but first it is important to note that in the real world there are many factors to be considered in designing a system, and a larger system is by no means always preferable to several smaller systems. The scaling effect is usually one factor among many.

Let us start with a simple example of an M/M/1 system, where the arrival rate λ is 5 customers per hour, and the average service-time T_{S} is 6 min. Therefore the server utilization is $\rho = 0.5$, and the average queueing time is $T_{\text{Q}} = 12$ min. These and the related results are given in Table 7.3. Now let us see what happens if the

Table 7.3. Scaling effect.

λ	T_S	ρ	T_Q
5/h	6 min	0.50	12 min
10/h	3 min	0.50	6 min
15/h	3 min	0.75	12 min

average customer arrival rate is doubled from 5 per hour to 10 per hour, but at the same time the average service-time is halved from 6 min to 3 min. The server utilization remains the same at $\rho = 0.5$, but the average time in the system, or queueing time, is reduced from $T_Q = 12$ min to $T_Q = 6$ min. So we have a situation where server utilization remains the same but the overall service given to customers has improved. This might lead us to ask how much the arrival rate could be increased while still providing the same level of service as the original system, i.e. an average time-in-system of 12 min. We can calculate this easily with some simple algebra, since we want to find λ such that

$$T_Q = 12 \text{ min} = \frac{T_S}{1 - \rho} = \frac{T_S}{1 - \lambda T_S} \text{ where } T_S = 3 \text{ min} \tag{7.44}$$

and the answer is $\lambda = 15$ customers per hour.

So by halving the service time we can triple the arrival rate while still providing the same service to customers. A bigger or faster server can, in general, be loaded to a higher utilization than a smaller or slower server for the same average time spent in the system by each customer. Another way of looking at the scaling effect is to ask ourselves what server capacity needs to be provided to cope with a given average customer arrival rate for a specified average queueing time. By 'capacity' we mean the rate at which the server could deal with customers if fully utilized. We shall, just for this section, use C to denote server capacity—in general C denotes a coefficient of variation. So we have

$$C = \text{server capacity} = \frac{1}{T_S} \text{customers per unit time} \tag{7.45}$$

and, since $\rho = \lambda T_S$

$$T_Q = \frac{T_S}{1 - \rho} = \frac{T_S}{1 - \lambda T_S} = \frac{1}{C - \lambda} \tag{7.46}$$

Rearranging this to express C in terms of λ and T_Q, we arrive at the expressions we need, i.e.

$$C = \lambda + \frac{1}{T_Q} \text{ and } \frac{C}{\lambda} = 1 + \frac{1}{\lambda T_Q} \tag{7.47}$$

The ratio C/λ is the capacity necessary in proportion to the capacity that will, on average, actually be utilized. This is a measure of the 'slack' or 'headroom' or 'over-configuration' required to maintain the performance, expressed by T_Q, that is deemed to be necessary. The extra capacity is necessary because of the irregular

Capacity/workload

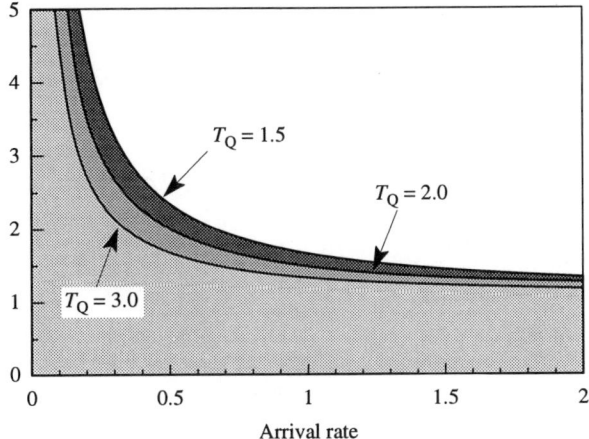

Figure 7.11. Scaling effect—capacity.

intervals at which customers arrive, and the irregular time taken to serve each customer. The capacity-to-workload ratio is plotted against arrival rate in Fig. 7.11. Three different values of T_Q are illustrated: obviously, the more stringent the target queueing time, the greater is the capacity that will be needed.

The message from Fig. 7.11 is that the greater the average rate at which customers arrive, the less 'excess' capacity will be needed to meet a given target queueing time. In this sense, systems dealing with high arrival rates are more efficient than systems dealing with low arrival rates. The ratio of capacity to workload is the reciprocal of utilization, so we could instead look at the maximum utilization that can be permitted to meet a target average queueing time as the average arrival rate varies. This is done in Fig. 7.12. The message is the same: larger systems can be driven harder, i.e. to higher utilizations, than smaller systems.

Example 7A—Public telephone

For this example, imagine a busy city street with crowds of people. On this street is a single public telephone. From time to time a passer-by decides to make a phone call. The average length of the phone call is 2 min, and on average people arrive to make calls once every 5 min. How long on average does each person have to wait? What is the 90th percentile of waiting time? How many people, on average, are waiting to make a call?

Using the M/M/1 queueing model, we have

$$T_S = \text{average service time} = 2 \text{ min}$$

$$\lambda = \text{average arrival rate} = 1 \text{ customer per 5 min}$$

$$= 0.2 \text{ customers per min}$$

Utilization

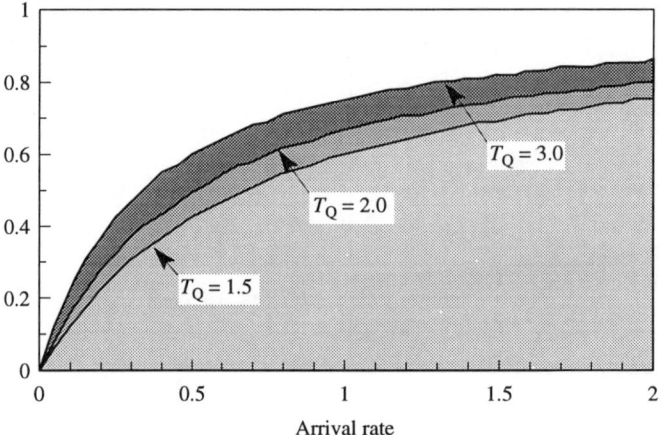

Figure 7.12. Scaling effect—utilization.

The server utilization is given by

$$\rho = \lambda T_S = (2)(0.2) = 0.4 \text{ or } 40 \text{ per cent} \tag{7.48}$$

The average waiting time is given by Eq. (7.8), so that

$$T_W = \text{average waiting time} = \frac{\rho T_S}{1 - \rho} = \frac{0.4 \times 2}{1 - 0.4} = \frac{4}{3} \text{ mins} = 80 \text{ secs} \tag{7.49}$$

Percentiles of waiting time are given by Eq. (7.25), so that

$$\pi_{T_W}(90) = 90\text{th percentile of waiting time} = T_W \frac{1}{\rho} \ln\left(\frac{100\rho}{100 - 90}\right) \tag{7.50}$$

$$= \frac{1.333}{0.4} \ln(10 \times 0.4) = 4.62 \text{ min} = 277 \text{ s} \tag{7.51}$$

The average number of people waiting is given by Eq. (7.28), so that

$$L_W = \text{mean customers waiting} = \frac{\rho^2}{1 - \rho} = \frac{(0.4)^2}{1 - 0.4} = 0.27 \tag{7.52}$$

It seems rather strange to say that there is on average about a quarter of a person waiting!

Example 7B—Newspaper seller

In a busy city-centre street a man is selling newspapers. Passers-by decide to buy a paper at the rate of one every 10 seconds. On average each customer takes 8 seconds to be served. On average, how many people are either waiting or being served? What is the average length of time that the newspaper seller is occupied dealing with customers before a pause in activity? What is the probability of there being more than three customers either waiting or being served?

The parameters of the system are

$$\lambda = 1/10 = 0.1 \text{ customers/sec}$$

$$T_S = 8 \text{ secs}$$

$$\rho = \lambda T_S = 0.1 \times 8 = 0.8 \tag{7.53}$$

The average number of people in the system is given by Eq. (7.22), so that

$$L_Q = \text{mean number-in-system} = \frac{\rho}{1-\rho} = \frac{0.8}{(1-0.8)} = 4 \tag{7.54}$$

The average length of a busy period is given by Eq. (7.32), so we have

$$g_1 = \text{mean length of busy period} = \frac{T_S}{1-\rho} = \frac{8}{1-0.8} = 40 \text{ secs} \tag{7.55}$$

Example 7C—Communications link

In a data communications network, messages arrive to be transmitted over a particular link. The average time required to transmit a message is 0.6 seconds, and messages arrive at an average rate of 1 message every second. How long does each message take, on average, to traverse the link? What is the 95th percentile of the number of messages either waiting to be sent or being transmitted?

The parameters of the system are

$$T_S = 0.6 \text{ secs}$$

$$\lambda = 1 \text{ message/sec}$$

$$\rho = \lambda T_S = 0.6 \tag{7.56}$$

The average time in the system is given by Eq. (7.2)

$$T_Q = \frac{T_S}{1-\rho} = \frac{0.6}{1-0.6} = 1.5 \text{ secs} \tag{7.57}$$

The cdf of the number of messages in the system is given by Eq. (7.25). For $\rho = 0.6$ we have to find k such that

$$\text{Prob}(\leq k \text{ customers in system}) = 0.95 = 1 - \rho^{k+1} \tag{7.58}$$

Simple trial and error will show that $k = 5$ is the answer. Alternatively, Eq. 7.58 can be solved in a straightforward manner as follows:

$$\rho^{k+1} = 1 - 0.95 = 0.05 \tag{7.59}$$

$$(k+1)\log\rho = \log(0.05) \tag{7.60}$$

$$k = \frac{\log(0.05)}{\log\rho} - 1 = 4.86 \tag{7.61}$$

Obviously, k must be an integer to be meaningful, so we take the next highest
integer, and $k = 5$.

The programs

The calculations for M/M/1 do not require complicated intermediate functions to
be implemented, and most of the characteristics usually of interest can be calculated in a straightforward way. The routine MM1Calc provides most of the things
that the reader will need, and is illustrated below.

```
{----------------------------------------------------}
{ MM1Calc - Calculations for M/M/1 queueing model  }
{ Inputs:                                            }
{      LAMBDA      customer arrival rate             }
{      TS          mean service time                 }
{ Outputs:                                           }
{      RHO         server utilisation                }
{      TQ/SDVTQ    mean/std dev time in system       }
{      P90TQ/P95TQ 90/95th pctile of time in system }
{      TW/SDVTW    mean/std dev waiting time         }
{      P90TW/P95TW  90/95th pctile of waiting time   }
{      PZW         probability of zero wait          }
{      LQ/SDVLQ    mean/std dev number in system     }
{      LW/SDVLW    mean/std dev number waiting       }
{      TB/SDVTB    mean/std dev of busy period length }
{      NB/SDVNB    mean/std dv no.served in bsy period}
{      VALID       T=results valid,F=not             }
{ Copyright Mike Tanner 1993                         }
{----------------------------------------------------}
Procedure MM1Calc(LAMBDA,TS:QTHreal;Var RHO,TQ,
                  SDVTQ,P90TQ,P95TQ,TW,SDVTW,P90TW,
                  P95TW,PZW,LQ,SDVLQ,LW,SDVLW,
                  TB,SDVTB,NB,SDVNB:QTHreal;
                  Var VALID:boolean);
begin
   RHO:=LAMBDA*TS;
   VALID:=(RHO<1.0); If not VALID then Exit;
   TQ:=TS/(1-RHO); SDVTQ:=TQ;
   P90TQ:=2.3*TQ;  P95TQ:=3.0*TQ;
   TW :=TQ-TS;  SDVTW:=TQ*Sqrt(RHO*(2-RHO));
   PZW:=1-RHO;
   If PZW>=0.90 then P90TW:=0.0
                else P90TW:=TQ*Ln(10*RHO);
   If PZW>=0.95 then P95TW:=0.0
                else P95TW:=TQ*Ln(20*RHO);
   LQ:=RHO/(1-RHO); SDVLQ:=Sqrt(RHO)/(1-RHO);
   LW:=RHO*RHO/(1-RHO);
   SDVLW:=Sqrt(1+RHO-RHO*RHO)*RHO/(1-RHO);
   TB:=TQ; SDVTB:=TQ*Sqrt((1+RHO)/(1-RHO));
   NB:=1/(1-RHO);
   SDVNB:=Sqrt(RHO*(1+RHO)/(1-RHO))/(1-RHO);
end;
```

In addition to MM1Calc, there are some routines that may sometimes be needed when dealing with M/M/1. These are contained in the file MM1AUXY.PAS. The first of these routines is MM1WaitCdf, which calculates the cumulative distribution function for wait-time, and is illustrated below. The value returned by MM1WaitCdf is the probability that wait-time will be less than or equal to the argument T. Other arguments are λ and T_S.

```
{-------------------------------------------------------}
{ MM1WaitCdf   -- cdf of waiting time                   }
{ Inputs: T        time for which cdf required          }
{         LAMBDA   customer arrival rate                }
{         TS       mean service time                    }
{ Copyright Mike Tanner 1993                            }
{-------------------------------------------------------}
Function MM1WaitCdf(T,LAMBDA,TS:QTHreal):QTHreal;
Var RHO:QTHreal;
begin
   RHO:=LAMBDA*TS;
   If RHO>=1.0 then QTHError('MM1WaitCdf error');
   MM1WaitCdf:=1-RHO*Exp(-(1-RHO)*T/TS);
end;
```

The next two routines in MM1AUXY.PAS are for calculating the probability and cumulative distribution functions of queue size, i.e. the number of customers in the system. MM1QszePdf returns the probability that the instantaneous queue size will be a particular value MM1QszeCdf returns the probability that the instantaneous queue size will not exceed a specified value.

```
{-------------------------------------------------------}
{ MM1QszePdf   -- pdf of queue size                     }
{ MM1QszeCdf   -- cdf of queue size                     }
{ Inputs: K         no. for which pdf/cdf required      }
{         LAMBDA   customer arrival rate                }
{         TS       mean service time                    }
{ Copyright Mike Tanner 1993                            }
{-------------------------------------------------------}
Function MM1QszePdf(K:integer;
                    LAMBDA,TS:QTHreal):QTHreal;
Var RHO,RK:QTHreal;
begin
   RHO:=LAMBDA*TS;
   If RHO>=1.0 then QTHError('MM1QszePdf error');
   RK:=K;
   MM1QszePdf:=Exp(Ln(1-RHO)+RK*Ln(RHO));
end;
{-------------------------------------------------------}
Function MM1QszeCdf(K:integer;
                    LAMBDA,TS:QTHreal):QTHreal;
Var RHO,RK:QTHreal;
```

```
begin
   RHO:=LAMBDA*TS;
   If RHO>=1.0 then QTHError('MM1QszePdf error');
   RK:=K;
   MM1QszeCdf:=1.0-Exp((RK+1)*Ln(RHO));
end;
```

8
Modelling

Introduction

When we use queueing theory we are building a 'model' of the system we are interested in. What exactly do we mean by a 'model'? A model is a representation of something in the real world, an abstraction or subset of the characteristics of whatever it is we are seeking to understand. Models take many different forms, depending on what aspect of reality we are concerned with. Take a new car. A full-size replica, made from clay or wood, will be constructed to show the appearance of the new car. A computer program will be used to 'simulate' or model the behaviour of the car's suspension system to ensure it fulfils its task. An accountant will no doubt create a financial model in the form of a spreadsheet to investigate the investment and profit associated with the car. So models are used for many different purposes, and take many different forms.

One of the most common models we use is a map. In fact we use many different kinds of map. An atlas shows us where the countries of the world are in relation to each other, but would be little use for finding our way around a strange city, for which we use a street map, i.e. a different kind of model for a different purpose. A road map is good for finding our way on a long journey, but may not tell us much about the detailed scenery or topography in a particular area. Flat maps are convenient to handle, but for some purposes a globe is necessary. Maps for seafarers tend to show land areas as empty, rather uninteresting spaces, whereas the opposite is true for the majority of maps. We take all this for granted, but the principles of building and using a model are the same whether we are making a map or constructing a sophisticated mathematical entity to predict how fast a computer system will perform the work we give it.

The modelling cycle

The process of building a model is illustrated in Fig. 8.1. We enter the diagram at the top, and start by trying to understand the real-world problem. Once we believe we have a good enough understanding, we proceed to build a model. We decide

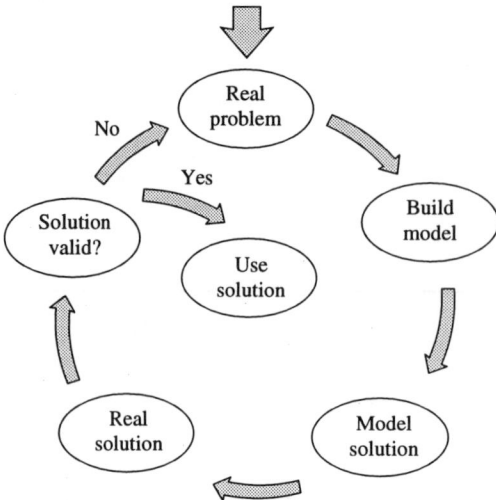

Figure 8.1. The modelling cycle.

which aspects of the real problem should be included in the model, and what can be left out. We also decide the nature of the model (queueing-theory, simulation, deterministic formulae, etc.), and the level of detail that is necessary. This first step, extracting from the real problem the salient factors, and only the salient factors, is the art and skill of modelling. Technical skills are necessary in the next step, but getting this first step right is where the foundations of a good model are laid, and where experience is of most value.

Having constructed the model, we have to solve it. By this we mean, for queueing theory, to derive the mathematical formulae that will tell us about waiting times and queue lengths, plus other characteristics we may be interested in. In fact, instead of deriving the formulae, we may solve the model's equations by some numerical or approximation technique. This is the step where all the specialized knowledge of queueing theory, or whatever other method we may use, is needed. This is often the technically difficult and academically interesting part of a modelling study. It is vital, however, not to lose sight of the fact that this step is just one step, and not the total modelling exercise.

The next two steps are sometimes not clearly distinguishable. Once we have a model and a solution method that can produce answers, we have to translate the model's answers back into real-world terms, and check that the model's recommendations are sensible. The model may be highly abstract, and will almost certainly have omitted some real-world constraints. The model may be recommending that we use more servers than is physically possible, or organize the working patterns of human servers in unacceptable ways. We must also check that the model behaves in a similar way to the real world! Calibration, or checking that our model can explain the performance of an existing system, is an essential step in building a model that is credible. It is quite possible that in building the model we decided to omit some factor that actually has a significant effect on the performance of the system being modelled. Conversely, we may discover that we have included something that could be safely left out.

It is almost always necessary to go round the modelling cycle more than once for any particular modelling study. The first attempt at building a model may do nothing more than reveal our lack of understanding of the real system being studied. Model building is a powerful educational method, and only when the analyst has progressed to a good understanding of the real system will the model become an effective predictor of how the system, or modified system, will behave.

Collecting data for a model

Whatever type of model is to be built, data must be collected about arrival rates and service-times. The data available is often incomplete, so that, for example, only average service-times are known and nothing at all is known about how much service-times vary. Similarly, customer arrival rates may be known only approximately, and it is rare indeed to be presented with reliable evidence about whether customers arrive randomly or regularly, or in groups. Sometimes a lot of data is available, but it has not been collected with a view to predicting the waiting times of customers and the effect of organizational changes on waiting times. An example of this would be where data has been collected to work out the cost of various types of service. Invariably this is based only on averages, e.g. an average of so many 'service transactions' a day of each kind are performed, with no information about what combinations of 'service transactions' actual customers demand.

It would be very easy to say simply that an extensive data-gathering exercise must be done before modelling can proceed. However, it is important to stand back and think about the purpose of the modelling exercise. If the objective is to provide an accurate prediction of actual waiting times, then the expense and time of collecting data may indeed be inevitable.

Another approach might be to use whatever data is to hand, together with opinions about service-times and arrival patterns from a small number of people with experience of the system to be modelled. This may have two benefits. First, putting 'guesstimates' into an explicit model may demonstrate whether the perceptions of people involved in the system are consistent with each other. For example, opinions about arrival rates and service-times may be inconsistent with the opinions of the same people or others about waiting times. This can be a valuable exercise because it starts a process of the people involved developing their intuitive understanding, so that their views of what is achievable in terms of customer service, and the resources required, become more realistic and consistent.

A second benefit is that the sensitivity of the model outputs to different sets of input assumptions can be explored. If only crude information about service-times is available, but large changes to the service-time assumptions have little effect on how the queueing system will be operated, then there is little point in an expensive data-gathering exercise. On the other hand, a preliminary model based on poor-quality data may demonstrate that small changes to the input assumptions will have a dramatic effect on management decisions. In this case the expense of thorough data-gathering will be more easily justified.

The effort to be expended on data-gathering will obviously also depend on how detailed a model is required. The range of possible models is discussed in the next section.

Analysis methods

There is, of course, a range of technical approaches to assessing the performance of a system, be it a computer system, a telecommunication network, a super-market, bank branch, etc. Figure 8.2 illustrates how the possible techniques relate to each other. Which technique should be used depends, of course, on the accuracy needed, which in turn depends on the decision that is to be made based on the results of the analysis. The guiding principle in all mathematical modelling is to use the simplest possible model that enables whatever management, design, or operational decision is required to be made with confidence. A more sophisticated approach might consider the cost of a 'better' model against the risk and con-sequences of an incorrect decision. The key point is that building a model is not an end in itself, but is part of a decision or design process.

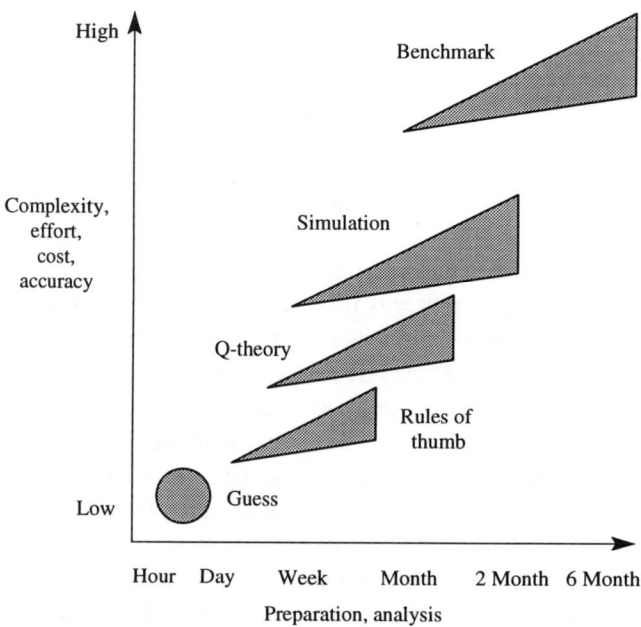

Figure 8.2. Analysis methods.

Queueing theory is shown as somewhere near the middle of the range in terms of the effort needed, the complexity, and the accuracy provided. Simulation tends to take a little more effort, if only because a number of different simulation runs are needed to get statistically reliable results and to plot a performance curve. A benchmark means taking a real system and subjecting it to an artificially generated workload. Benchmarks for computer systems are of mixed value. While they are expensive and time-consuming to set up, especially to ensure comparable tests on different systems, they are at least superficially credible because in one sense the 'model' is the real system. On the other hand, it is difficult to explore the effect of a workload or system configuration different from that actually used in the bench-mark. Benchmarks provide most value when they are part of a larger modelling exercise using queueing theory or simulation. A benchmark can provide the calibration, while the other models can provide the insights into why a system performs in a particular way.

Figure 8.2 also shows 'rules of thumb'. A rule of thumb (ROT) is a simple rule such as 'do not load computer systems more than 60 per cent'. An ROT is valuable if it represents the distillation of practical experience with similar systems, or the conclusion of a modelling exercise. Unfortunately, it is too easy to misapply an ROT that is valid for one type of system to a completely different type of system. Figure 8.3 shows the wrong and the right way to establish an ROT. Readers should beware of any ROT whose justification and origin cannot be explained.

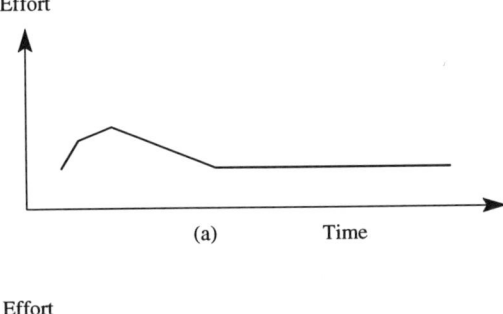

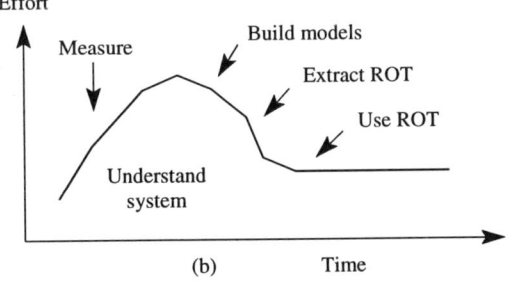

Figure 8.3. Rules of thumb: (a) wrong; (b) right.

Art versus science

Modelling is often said to be more art than science. The science comes in the technicalities of queueing theory, simulation, or other relevant mathematical or operational research methods. These are specialized skills, requiring significant effort to master, but that still leaves the 'art'. The art of modelling lies in being able to understand diverse technologies and systems sufficiently well to identify how the parts of the system interact. Experience, and confidence, play a big role in this.

Part Three
Single server with general service-times

9
Queues with general service-times (M/G/1)

Introduction

So far we have looked only at the simple, classic, queueing system with random arrivals and exponential service-times. In Kendall's notation this is the M/M/1 model. Everything about an M/M/1 queue that we are likely to be interested in can be explicitly calculated with straightforward formulae.

One of the most intuitively obvious lessons from the M/M/1 model is that the average time a customer spends in the system, or waiting, depends heavily on the server utilization. The busier the server, the longer customers have to wait. In this chapter we shall look at the effect of service-time variability. We shall see that the more variable service-time is, the longer customers must wait. Let us try and get an intuitive handle on why waiting time should be affected by whether service-times are all similar or very different. Remember we are assuming 'random' arrivals, so a customer may arrive soon after several other customers arrive, or may arrive after a long interval when no other customers have appeared. This, we may consider, is a matter of luck. If the customer arrives to find the server idle, then waiting time is of course zero, and is unaffected by the variation in other customers' service-times. What has to occur for a particular customer to suffer a long wait? First of all, that customer must be unlucky enough to arrive when there are already some other customers waiting, so let us assume that this has happened. If service-times are fairly constant, then long waits will be due to bad luck as regards arrivals. If some service-times are very long, as with exponentially distributed service-times, then our customer may be even more unlucky and find that someone ahead of him or her in the queue requires one of these very long service-times. So not only will our particular customer have a long wait, but there will be an increased chance of subsequent arrivals also having a long wait. This is not a watertight explanation, as readers will have realized, but perhaps it will be of use to some. For watertight demonstrations of how service-time variation affects queueing we must turn to mathematical techniques. Most of the standard academic texts on queueing theory referenced in the bibliography contain rigorous derivations of the M/G/1 results. In this book we shall concentrate on the usable results.

Review of variance and coefficient of variation

This is a good place to review what we mean by variance of service-time, and the ways in which variance can be expressed. In the chapter on probability and statistics, variance of a random variable was defined as

$$\text{variance} = \sigma^2 = \frac{1}{n}\sum_{i=1}^{i=n}(x_i - \mu)^2\, \text{Prob}(X = x_i) \tag{2.22}$$

for discrete variables, with equivalent definitions for continuous variables. For descriptive purposes, the standard deviation σ is often better than variance, since it is in the same 'units' as the mean and is more easily interpreted informally, e.g. for many distributions most values lie within plus or minus two standard deviations of the mean. A statistician usually works with variance, because variance is more convenient when transforming or combining random variables, and formal statistical tests usually use variance rather than standard deviation. What matters in queueing theory is the magnitude of the variance relative to the average, and we measure this with the 'squared coefficient of variation', which is a fearsome-sounding name for the variance divided by the square of the mean, i.e.

$$\text{coeff. of variation squared} = C_S^2 = \frac{\text{variance}}{\text{mean}^2} = \frac{\sigma_{T_S}^2}{T_S^2} \tag{2.24}$$

Most queueing theory formulae use C^2 rather than C, and it is a good idea to be carefully explicit about whether the coeff of variation, or the squared coeff of variation is being quoted.

So what values of C^2 are commonly encountered? If $C^2 = 0$ then variance is zero, and so the distribution is actually constant. When $C^2 = 1$, we have an exponential distribution. This is usually considered to be a moderately high variance, and so some books on queueing theory state that most service-time distributions that will be encountered will have a coefficient of variation somewhere between 0 and 1. It is true that in many cases the coefficient is between 0 and 1, but it is also not at all uncommon to find C^2 higher than this. If you encountered C^2 greater than, say, 5, that would warrrant re-examination of the service-time data.

Parameters and initial calculations for M/G/1

In this chapter we shall give the results that can be obtained for M/G/1 when just the three parameters listed below are specified. In Chapter 10 we shall look at additional results that can be obtained when more information about the service-time distribution is available.

> λ, the average customer arrival rate, e.g. customers per minute
> T_S, the average service time for a customer, e.g. 2 minutes
> C_S^2, the squared coefficient of variation of service-time

The time units used for arrival rate and service-time must of course be the same, and it is often convenient to 'normalize' the parameters so that $T_S = 1$. Instead of the squared coefficient of variation for service-time, the variance or standard deviation may be given, or the distribution may be stated to be a particular type. If so, then C_S^2 should be calculated as a first step.

The next thing to do is to calculate the utilization of the server, $\rho = \lambda T_S$, and to check that ρ is less than 1.

Time-in-system (queueing time)

The formula for average time-in-system for the M/G/1 model is

$$T_Q = \text{average time-in-system} = \left(1 + \frac{\rho(1 + C_S^2)}{2(1 - \rho)}\right) T_S \qquad (9.1)$$

This formula is a bit more complicated than the corresponding result for M/M/1, but the factor $(1 - \rho)$ still appears in the denominator. This means that T_Q increases faster and faster as server utilization approaches 100 per cent, just as with M/M/1. Without more information about the service-time distribution, the variance and percentiles of time-in-system cannot be calculated. As long as $C_S^2 \leqslant 1$, the percentiles of queuing time from the M/M/1 model can be used as pessimistic estimates, i.e.

$$\pi(90)_{T_Q} = \text{90th percentile of time-in-system} \approx 2.3 T_Q \qquad (9.2)$$

$$\pi(95)_{T_Q} = \text{95th percentile of time-in-system} \approx 3.0 T_Q \qquad (9.3)$$

If $C_S^2 > 1$, then these estimates may be optimistic, but they are still useful in the absence of more information on service-times. In fact, it can be shown that at high utilizations these formulae for percentiles are good approximations.

Waiting time

The formula for average waiting time for the M/G/1 model is

$$T_W = \text{average waiting time} = \frac{\rho(1 + C_S^2)}{2(1 - \rho)} T_S \qquad (9.4)$$

Number-in-system and waiting-line size

The average number of customers in the system is given by the following formula, known as the Pollacek–Khinchine formula:

$$L_Q = \text{mean queue size} = \rho + \rho^2 \frac{(1 + C_S^2)}{2(1 - \rho)} \tag{9.5}$$

and the mean waiting-line size is given by

$$L_W = \text{mean waiting-line size} = L_Q - \rho \tag{9.6}$$

$$= \rho^2 \frac{(1 + C_S^2)}{2(1 - \rho)} \tag{9.7}$$

Effect of service-time variation

Referring back to the M/M/1 model, the formula for T_Q with exponential service–time is Eq. (7.2). This can be rewritten as

$$T_{Q(M/M/1)} = \text{mean time-in-system for M/M/1} = T_S + \frac{\rho T_S}{1 - \rho} \tag{9.8}$$

where the first term is the average service-time and the second term is the average waiting time. It is instructive to rewrite the M/G/1 formula in a similar way (Eq. (9.9)), and compare it with the M/M/1 version.

$$T_Q = \text{mean time-in-system for M/G/1} = T_S + \frac{\rho T_S}{1 - \rho} \left(\frac{1 + C_S^2}{2} \right) \tag{9.9}$$

When the service-time has a general, rather than an exponential, distribution, the waiting time becomes multiplied by the factor $(1 + C_S^2)/2$. For the exponential distribution, $C_S^2 = 1$, and this factor reduces to 1. Equation (9.8) shows directly the effect of variation in service-time.

Now let us compare queueing systems with the same average, but different variance of, service-time. As our base case we take M/M/1 where $C = 1$. We know now that service-time variance matters, so let us also look at the ideal case where we can make the service-time constant, i.e. $C = 0$. This would be referred to as an M/D/1 system.

As another yardstick for comparison we can use the D/D/1 system, where customers arrive at exactly regular intervals as well as having constant service-times. We do not need any sophisticated, or even simple, mathematics to tell us the characteristics of D/D/1 system, an example of which would be a conveyor belt carrying items at equally spaced positions, with the server being an assembly worker who has to carry out some operation on the items. The operation is always the same and always takes the same time. As long as the speed of the conveyor belt means the assembly worker has sufficient time to deal with each item, the 'customers' (the items on the conveyor belt) will not have to wait at all. The time-in-system for customers will be exactly the time required for the assembly operation to be carried out. As the speed of the conveyor belt, i.e. the customer arrival rate, is increased, waiting time will continue to be zero up to the point at which items are arriving slightly too fast for one item to be dealt with before the next item arrives. At that point the items will accumulate remorselessly and the assembly worker will never stop the backlog growing. The performance characteristics are shown in Fig. 9.1.

An interesting observation can be made about the different performance characteristics for M/M/1, M/D/1 and D/D/1. We have shown that variability, either in the arrival pattern or in the service-time, of itself increases the average waiting time. If in a real system there are things we can do to make service-times less variable and/or get more regular customer arrivals, the effect is to pull the waiting-time curve into the bottom right-hand corner. This is indicated by the arrow on the graph. This is good, because customers are having to wait less. On the other hand, the more regular the system is, the sharper the deterioration in waiting time as utilization approaches 100 per cent. In the extreme for the D/D/1 system, a small increase in customer arrival rate causes waiting time to go from zero to infinity!

Another observation is a purely presentational one. Figure 9.1 is a line graph rather than a bar chart, since for a continuous variable such as average time-in-system a simple line graph seems the natural choice. However, the human eye is likely to perceive the M/M/1 and M/D/1 curves as very similar, because they run

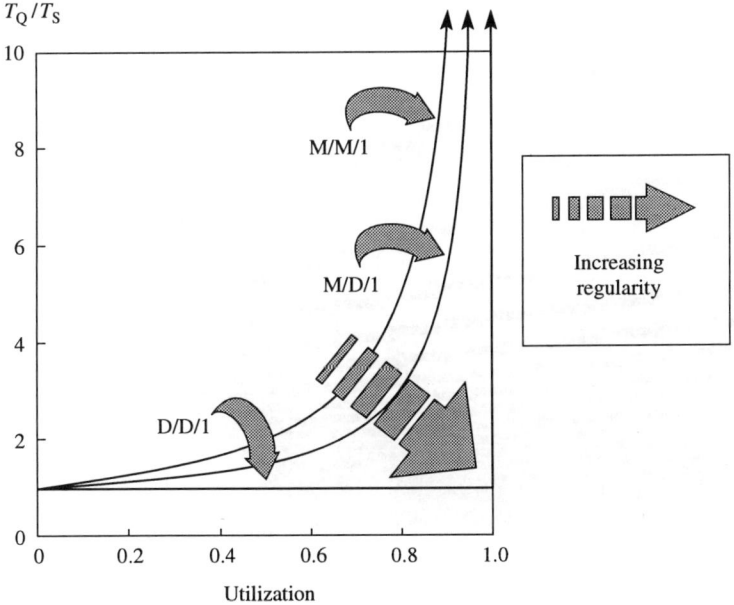

Figure 9.1. Effect of service-time variance.

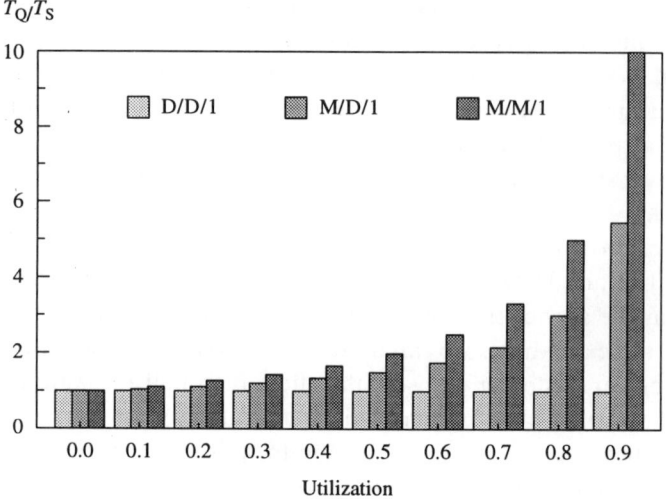

Figure 9.2. Bar chart showing effect of service-time variance.

parallel to each other. This is contrary to the point the graph is intended to make, which is that the curves are really quite different. This point is better made visually with a bar chart, as in Fig. 9.2.

A more general way of showing the combined effects on average waiting time of server utilization and service-time variability is with the 3-D chart in Fig. 9.3. At

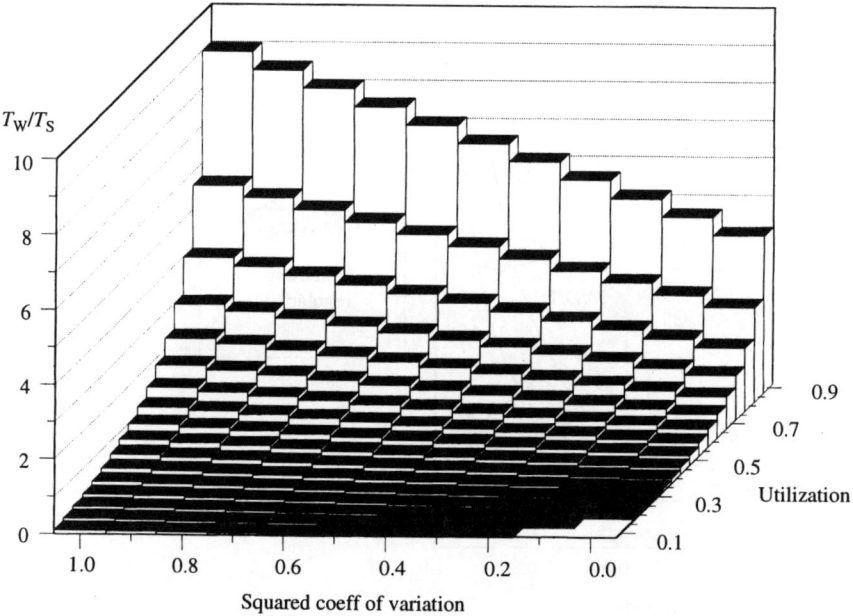

Figure 9.3. Combined effects of utilization and service-time variance.

low utilizations the variance of service-time does not matter much, but at high utilizations the variance has a significant effect.

Programs for M/G/1

In order to use the routines for M/G/1, a program must contain the following statements.

```
{$N+,E+ }
{$I \qthprog\QTHTypes.pas }
{$I \qthprog\QTHError.pas }
{$I \qthprog\MG1Calc.pas  }
```

MG1Calc—general calculations for M/G/1

This routine performs the basic calculations for M/G/1.

```
{------------------------------------------------------}
{> MG1Calc - Calculations for basic M/G/1 model      }
{ Inputs:                                            }
{    LAMBDA   customer arrival rate                  }
{    TS       mean service time                      }
{    COEFF2   sqd coeff of variation service time    }
{ Outputs:                                           }
{    RHO         server utilisation                  }
{    TQ          mean time in system                 }
{    TQ90        approx 90th pctile of time in system }
{    TQ95        approx 95th pctile of time in system }
{    TW          mean waiting time                   }
{    LQ          mean customers in system            }
{    LW          mean customers waiting              }
{    VALID       T=results valid, F=not valid        }
{  Copyright Mike Tanner 1993                        }
{------------------------------------------------------}
Procedure MG1Calc(LAMBDA,TS,COEFF2:QTHreal;
             Var RHO,TQ,TQ90,TQ95,TW,LQ,LW:QTHreal;
             Var VALID:boolean);
Var ZA,ZB:QTHreal;
begin
   RHO:=LAMBDA*TS;
   VALID:=(RHO<1.0);   If not VALID then Exit;
   ZA:=RHO/(1-RHO);   ZB:=(1+COEFF2)/2;
   TW:=TS*ZA*ZB;
   TQ :=TS+TW; TQ90:=2.3*TQ;   TQ95:=3.0*TQ;
   LW :=RHO*ZA*ZB; LQ :=LW+RHO;
 end;
```

Example 9A—Communications link

Messages arrive at random to be sent across a communications link. The link speed is 9600 bits per second. Assuming the link is 70 per cent utilized, and the average message length is 1000 bytes (i.e. 8000 bits), calculate the average waiting time for constant-length messages, and for exponentially distributed message lengths.

We shall apply Eq. (9.4) to each case. We have $T_S = 8000/9600 = 0.833$ secs, and $\rho = 0.7$. For constant-length messages $C_S^2 = 0$, so that Eq. 9.4 gives

$$T_W = \text{average waiting time} = \frac{\rho(1 + C_S^2)}{2(1 - \rho)} T_S = \frac{0.7}{2(0.3)} 0.833 = 0.972 \text{ secs}$$

For exponentially distributed message lengths $C_S^2 = 1$, so that Eq. 9.4 gives

$$T_W = \text{average waiting time} = \frac{\rho(1 + C_S^2)}{2(1 - \rho)} T_S = \frac{0.7(2)}{2(0.3)} 0.833 = 1.944 \text{ secs}$$

Example 9B—Small post office

In a small post office there is a single person serving. Seventy per cent of customers take 1 min to serve, 20 per cent take 3 min and 10 per cent take 10 min. Calculate the average time spent in the post office, and the average number of people in the post office, when customers arrive at an average rate of (a) one per 3 min, (b) one per 4 min, (c) one per 5 min.

Using Eqs. (2.17), (2.22) and (2.24) we can calculate

$$T_S = \text{average service-time} = 2.3 \text{ min}$$

$$\sigma_{T_S}^2 = \text{variance of service-time} = 7.21$$

$$C_S^2 = \frac{\sigma_{T_S}^2}{T_S^2} = \frac{7.21}{(2.3)^2} = 1.36$$

Applying Eqs 9.1 and 9.5 to each arrival rate in turn we get the requested results. The working is shown for (a), the others are left for the reader.

$$\rho = \text{server utilization} = \lambda T_S = (1/3) \times 2.3 = 0.767$$

$$T_Q = \text{average time-in-system} = \left(1 + \frac{\rho(1 + C_S^2)}{2(1 - \rho)}\right) T_S$$

$$= \left(1 + \frac{0.767(1 + 1.36)}{2(1 - 0.767)}\right) 2.3 = 11.23 \text{ min}$$

$$L_Q = \text{mean queue size} = \rho + \rho^2 \frac{(1 + C_S^2)}{2(1 - \rho)}$$

$$= 0.767 + (0.767)^2 \frac{(1 + 1.36)}{2(1 - 0.767)}$$

$$= 3.75 \text{ customers}$$

10
More about queues with general service-times (M/GX/1)

Introduction

In the last chapter we looked at the M/G/1 queueing model when only the mean and standard deviation (or equivalently coeff of variation) of service-time are specified. We found that only averages of the queue's characteristics could be found. If we want more information about queueing time, etc., such as variance and percentiles, then we have to specify the service-time distribution in more detail. We may have an empirical distribution of service-time, or we may be willing to assume a standard distribution. We shall look at general service-times in this chapter, and at specific service-time distributions other than exponential in following chapters.

Parameters and initial calculations for M/GX/1

We assume a single server, with an infinite population of potential customers, and customers served in first-come-first-served order. The parameters to be specified are as follows:

λ = average arrival rate of customers
$s_1, s_2, s_3, \ldots$ = moments of service-time about the origin

The number of moments of service-time that are known will vary. If we are using a standard theoretical distribution then we can calculate as many moments as we want. With empirical data about service-time, we can in theory calculate a large number of moments, but more than the first few moments may be unreliable. This is because we are, in effect, estimating the moments from a sample, and the higher the moments we calculate, the lower will be the accuracy of the estimates.

In terms of our previous notation for describing the service-time distribution, we have

$$T_S = s_1 \tag{10.1}$$

$$\sigma^2_{T_S} = s_2 - s_1^2 \tag{10.2}$$

$$C^2_S = \frac{s_2 - s_1^2}{s_1} \tag{10.3}$$

As usual, the first thing we do is to calculate $\rho = \lambda T_S$ and check that $\rho < 1$ so that the system is stable.

Waiting time

The moments of waiting time (about zero) are denoted by $w_1, w_2, w_3, \ldots$, and their values can be calculated from the following recurrence relation, due to Takacs. This is not a very convenient thing to have to do by hand, but is straightforward enough as a computer program.

$$w_k = \frac{\lambda}{1-\rho} \sum_{i=1}^{k} \binom{k}{i} \frac{s_{i+1}}{i+1} w_{k-i} \text{ where } w_0 = 1 \tag{10.4}$$

If we expand Eq. (10.4) for the first and second moments of waiting time we get

$$w_1 = \frac{\lambda s_2}{2(1-\rho)} \tag{10.5}$$

$$w_2 = \frac{\lambda}{1-\rho} \left(\frac{\lambda s_2^2}{2(1-\rho)} + \frac{s_3}{3} \right) \tag{10.6}$$

An important thing to observe is that to get the second moment of waiting time, we have to know up to the third moment of service-time. In general, we need to know up to the $(i+1)$th moment of service-time to calculate the ith moment of queueing or waiting time. Readers may recall that for the simple analysis of M/G/1, we needed to know the second moment of service-time (or equivalently the coefficient of variation) in order to calculate the mean (first moment) of waiting or queueing time. It seems we have to put a little more information in than we are able to get out!

Using our previous notation, the mean and variance of waiting time can be expressed as

$$T_W = w_1 \tag{10.7}$$

$$\sigma^2_{T_W} = w_2 - w_1^2 \tag{10.8}$$

Time-in-system (or queueing time)

Queueing time is waiting time plus service-time. This means that the distribution of queueing time is what is known as a 'convolution' of two distributions, i.e. the distribution of waiting time and the distribution of service-time. It follows from this that the moments of queueing time about zero, which we shall denote by q_1, q_2, q_3,..., can be calculated using another recurrence relation due to Takacs, i.e.

$$q_k = \sum_{i=0}^{k} \binom{k}{i} w_{k-i} s_i \tag{10.9}$$

Expanding Eq. (10.9) for the first couple of moments of queueing time we get

$$q_1 = w_1 + s_1 \tag{10.10}$$

$$q_2 = w_2 + 2w_1 s_1 + s_2 \tag{10.11}$$

Equation (10.10) is saying no more than $T_Q = T_W + T_S$, while Eq. (10.11) can be used to calculate the variance of queueing time, i.e.

$$\sigma_{T_Q}^2 = q_2 - q_1^2 \tag{10.12}$$

Waiting-line size and number-in-system

There are no straightforward results for p_k, the probability of there being a particular number of customers in the system. It is, however, possible to obtain the moments of the distribution of the number of customers waiting or the number in the system. Since even these are rather unwieldy formulae, we shall confine ourselves to the mean and variance of these quantities. For the waiting-line size we have

$$L_W = \text{average customers waiting} = \frac{\rho^2}{1-\rho}\left(\frac{1+C_S^2}{2}\right) \tag{10.13}$$

$$\sigma_{L_W}^2 = \text{variance of no. waiting}$$

$$= \frac{\lambda^3 s_3}{3(1-\rho)} + \left(\frac{\lambda^2 s_2}{2(1-\rho)}\right)^2 + \frac{\lambda^2(3-2\rho)s_2}{2(1-\rho)} \tag{10.14}$$

For the number of customers in the system, either waiting or being served, we can find the mean and variance using results for number waiting and the following relationships

$$L_Q = \text{average customers in system} = L_W + \rho \tag{10.15}$$

$$\sigma_{L_Q}^2 = \text{variance of customers in system} = \sigma_{L_W}^2 + \rho(1-\rho) \tag{10.16}$$

The busy period

Kleinrock (1975) gives an extensive analysis of the busy-period distribution for M/G/1. We shall denote the moments of the busy-period length about zero by g_1, g_2, g_3,.... The first result is the average length of the busy period, which is given by

$$g_1 = \text{average length of busy period} = \frac{s_1}{1 - \rho} \equiv \frac{T_S}{1 - \rho} \qquad (10.17)$$

The interesting thing to observe here is that the average length of the busy period does not depend on the 'shape' of the service-time distribution, it just depends on T_S, the average service-time. Looking now at the higher moments of busy-period length, the particular service-time distribution does have an effect. The second moment of the busy period is

$$g_2 = \frac{s_2}{(1 - \rho)^3} \qquad (10.18)$$

The factor $(1 - \rho)^3$ in the denominator tells us immediately that for high utilizations the length of the busy period is extremely variable. From g_1 and g_2 we can construct expressions for the variance of the busy period and its coefficient of variation:

$$\sigma_g^2 = \text{variance of busy period} = g_2 - g_1^2 = \frac{\sigma_{T_S}^2 + \rho T_S^2}{(1 - \rho)^3} \qquad (10.19)$$

$$C_g^2 = \text{squared coeff of variation of busy period} = \frac{\sigma_g^2}{g_1^2} = \frac{C_S^2 + \rho}{1 - \rho} \qquad (10.20)$$

Equation 10.20 for C_g^2 gives some scale to the busy-period variation. Taking, as an example, a 90 per cent loaded M/M/1 system, we would have $C_g^2 = 19$, and even for constant service-times, i.e. $C_S^2 = 0$, we would have $C_g^2 = 9$ which is still very high. The third and fourth moments of the busy-period duration are given by

$$g_3 = \frac{s_3}{(1 - \rho)^4} + \frac{3\lambda s_2^2}{(1 - \rho)^5} \qquad (10.21)$$

$$g_4 = \frac{s_4}{(1 - \rho)^5} + \frac{10\lambda s_2 s_3}{(1 - \rho)^6} + \frac{15\lambda^2 s_2^3}{(1 - \rho)^7} \qquad (10.22)$$

Number served in a busy period

The number of customers served in a busy period is the number served 'in a row' or without a break between one customer and the next. The moments of the distribution of the number served are $h_1, h_2, \ldots$, where

$$h_1 = \text{average number served in a busy period} = \frac{1}{1 - \rho} \qquad (10.23)$$

It is interesting to note from Eq. 10.23 that the average depends only on the server utilization. The variance and higher moments of the number served per busy period do, however, depend on the characteristics of the service-time distribution.

$$h_2 = \frac{2\rho(1-\rho) + \lambda^2 s_2}{(1-\rho)^3} + \frac{1}{1-\rho} \tag{10.24}$$

$$\sigma_h^2 = \text{variance of no. served in busy period} = h_2 - h_1^2 = \frac{\rho(1-\rho) + \lambda^2 s_2}{(1-\rho)^3} \tag{10.25}$$

Programs for M/GX/1

In order to use the routines for M/GX/1, a program must contain the following statements.

```
{$N+,E+ }
{$I \qthprog\QTHTypes.pas  }
{$I \qthprog\QTHError.pas  }
{$I \qthprog\MG1XCalc.pas  }
{$I \qthprog\bicoeff.pas   }
```

MG1XCalc—general calculations for M/G/1

All the calculations for M/GX/1 are done by the single routine MG1XCalc. MG1XCalc calculates as many moments of queueing and waiting time as possible, given the number of service-time moments that are supplied. At least the first three moments of service-time must be supplied—if only the first two are available then use MG1Calc.

```
{--------------------------------------------------------}
{> MG1XCalc - Extended calculations for M/G/1,           }
{>            when service time distribution is          }
{>            specified by its moments.                  }
{ Inputs:                                                }
{     LAMBDA       customer arrival rate                 }
{     S            moments of service time about         }
{                  the origin                            }
{ Outputs:                                               }
{     RHO          server utilisation                    }
{     TQ           mean time in system                   }
{     TW           mean waiting time                     }
{     LW           mean no. waiting                      }
{     SDVLW        std dev of no. waiting                }
{     LQ           mean no. in system                    }
{     SDVLQ        std dev of no. in system              }
{     TQM[1..10]   moments of time in system             }
{     TWM[1..10]   moments of waiting time               }
{     BSY          mean busy-period length               }
{     SDVBSY       std dev of busy-period length         }
{     VALID        true if results valid, false if       }
{                  system not stable                     }
{   Copyright Mike Tanner 1993                           }
{--------------------------------------------------------}
```

```
Procedure MG1XCalc(LAMBDA:QTHreal;S:Moments;
                   Var RHO,TQ,TW,LW,SDVLW:QTHreal;
                   Var LQ,SDVLQ:QTHreal;
                   Var TQM,TWM:Moments;
                   Var BSY,SDVBSY:QTHreal;
                   Var VALID:boolean);
Var TS,C2S,VS,RI,VLW,VLQ,G2:QTHreal;I,K:integer;
begin
    {-Check at least first 3 service-time moments---}
    If S.MVM<3 then QTHError('MG1XCalc error');
    {-Preliminary calculations---------------------}
    TS:=S.MOM[1];
    RHO:=LAMBDA*TS;
    VALID:=(RHO<1.0);
    If not VALID then Exit;
    VS:=S.MOM[2]-TS*TS;
    C2S:=VS/(TS*TS);
    {-Waiting-line---------------------------------}
    LW:=Sqr(RHO)*((1+C2S)/2)/(1-RHO);
    VLW:=Sqr(LAMBDA)*S.MOM[2]/(2*(1-RHO));
    VLW:=Sqr(VLW)
          +LAMBDA*Sqr(LAMBDA)*S.MOM[3]/(3*(1-RHO));
    VLW:=VLW+Sqr(LAMBDA)*(3-2*RHO)*S.MOM[2]/(2*(1-RHO));
    SDVLW:=Sqrt(VLW);
    {-Queue size-----------------------------------}
    LQ:=LW+RHO;
    VLQ:=VLW+RHO*(1-RHO);
    SDVLQ:=Sqrt(VLQ);
    {-Moments of waiting-time----------------------}
    TWM.MOM[0]:=1;
    TWM.MVM:=S.MVM-1;
    For K:=1 to (S.MVM-1) do
    begin
        TWM.MOM[K]:=0;
        For I:=1 to K do
        begin
           RI:=I;
           TWM.MOM[K]:=TWM.MOM[K]
                       +BiCoeff(K,I)*S.MOM[I+1]
                        *TWM.MOM[K-I]/(RI+1);
        end;
        TWM.MOM[K]:=TWM.MOM[K]*LAMBDA/(1-RHO);
    end;
    TW:=TWM.MOM[1];
    {-Moments of queueing time---------------------}
    TQM.MOM[0]:=1;
    TQM.MVM:=S.MVM-1;
    For K:=1 to (S.MVM-1) do
    begin
        TQM.MOM[K]:=0;
        For I:=0 to K do
          TQM.MOM[K]:=TQM.MOM[K]
                    +BiCoeff(K,I)*TWM.MOM[K-I]*S.MOM[I];
    end;
```

```
      TQ:=TQM.MOM[1];
      {-Busy period----------------------------------}
      BSY:=TS/(1-RHO);
      G2:=S.MOM[2]/((1-RHO)*Sqr(1-RHO));
      SDVBSY:=Sqrt(G2-Sqr(BSY));
end;
```

11
Erlang-k distribution of service-times (M/E$_k$/1)

Introduction

The Erlang distribution is a special case of the gamma distribution, so the reader may wonder why the M/E$_k$/1 is included in its own right. The Erlang distribution is named after A.K. Erlang, who pioneered queueing theory for its application to congestion in telephone networks. As well as the Erlang distribution, a number of useful formulae are named after Erlang, and we shall come to these in later chapters. The Erlang distribution was an early, and very successful, way of constructing a queueing model without assuming an exponential distribution for service-times, but still keeping most of the mathematical tractability of the M/M/1 model. So for its historical value, and its usefulness, the M/E$_k$/1 model needs to be included in any book on queueing theory.

The idea is that the service facility consists of a number of stages (Fig. 11.1). A customer has to pass through each stage of service. The next customer cannot start service until the previous customer has completed all the stages. The stages may in some cases correspond to how service is actually provided, or they may be purely conceptual. The time taken for each stage has an exponential distribution, and the mean time for each stage is the same. The distribution of service-time for the whole facility is said to be Erlang-k, where k is the number of stages. The mathematical

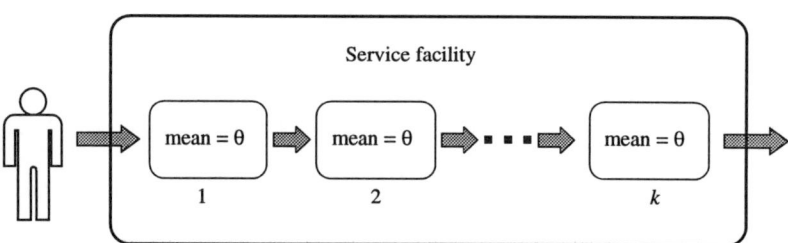

Figure 11.1. Erlang-k service facility.

analysis is still greatly helped by the use of the exponential distribution for each stage, but the service-time has an Erlang-k distribution, whose properties we shall look at next.

The Erlang-k distribution

The Erlang-k distribution can have a coefficient of variation squared ranging from $C^2 = 1$, when it is in fact the exponential distribution, to $C^2 = 0$ when it is a constant distribution. We choose the coeff of variation we want by selecting the value of k, since

$$C^2 = \text{squared coeff of variation} = \frac{1}{k} \tag{11.1}$$

We do not have complete freedom in setting C^2 since k must be an integer, but a useful selection of values can be obtained. The probability and cumulative distribution functions are given in Appendix 1. Figure 11.2 illustrates the shape of the Erlang-k distribution for several values of k. The minimum value for k is 1, which gives the exponential distribution. As k increases, the distribution becomes more symmetrical and also more tightly concentrated around the mean. For very large k the distribution is effectively a constant.

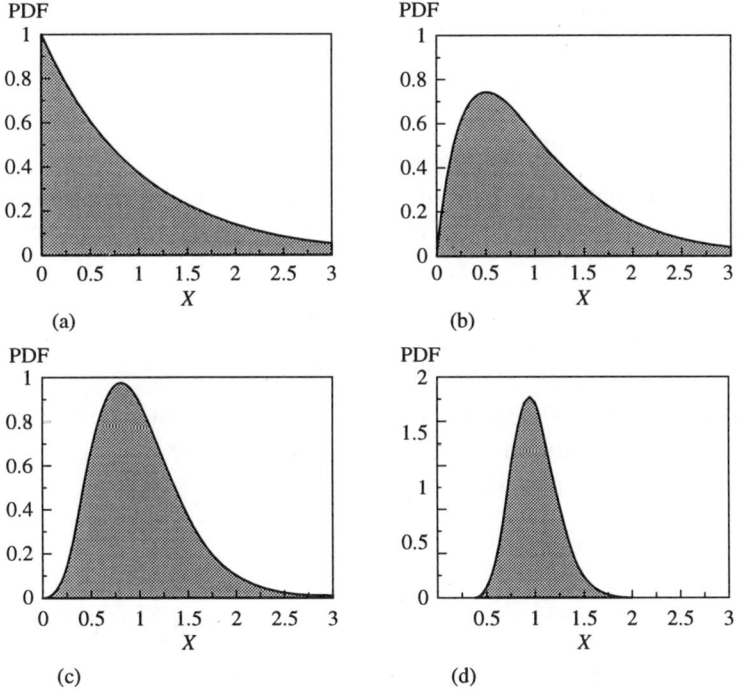

Figure 11.2. PDF Erlang-k for mean $= 1$: (a) $k = 1$; (b) $k = 2$; (c) $k = 5$; (d) $k = 20$.

Assumptions, parameters, and initial calculations for the M/E_k1 model

We assume a single server with random arrivals from an infinite population of customers. Customers are served in FIFO order. The parameters that need to be specified for the $M/E_k/1$ model are

> λ, the average customer arrival rate
> T_S, the average service-time per customer
> k, the number of stages in the Erlang distribution of service-time, or equivalently the squared coeff of variation of service-time C_S^2, or the variance of service-time σ_S^2

We first calculate the server utilization $\rho = \lambda T_s$ and check that $\rho < 1$. If this condition is not met then the system is not stable. If C_S^2, the squared coeff of variation of service-time is given, then k should be chosen to be the largest value of k such that

$$k = \text{Erlang parameter} \leq \frac{1}{C_S^2} \tag{11.2}$$

The effect of increasing k is non-linear, so that the difference between $k = 1$ and $k = 2$ is about the same as between $k = 2$ and $k = 5$, or between $k = 5$ and $k = \infty$. Unless k is quite small its exact value is not very important. If $\sigma_{T_s}^2$ is given, then first calculate $C_S^2 = \sigma_{T_s}^2 / T_S^2$ and then choose k as described above.

Time-in-system (or queueing time)

The formula for the mean time-in-system for $M/G/1$ is Eq. (9.01). If we substitute $C_S^2 = 1/k$ into Eq. (9.1) and rearrange slightly, then we get the result specific to $M/E_k/1$:

$$T_Q = \text{average time-in-system} = \frac{T_S}{1-\rho}\left[1 - \frac{\rho}{2}\left(1 - \frac{1}{k}\right)\right] \tag{11.3}$$

Turning now to the variance of the time-in-system, we know the moments of the Erlang-k distribution, and we can put these into the appropriate $M/GX/1$ results to obtain a formula for the variance, which is

$$\sigma_{T_Q}^2 = \text{variance of time-in-system}$$

$$\sigma_{T_Q}^2 = \frac{T_S^2}{(1-\rho)^2}\left\{\left[1 - \frac{\rho(4-\rho)}{6}\left(1 - \frac{1}{k}\right)\right]\left[1 + \frac{1}{k}\right] - \left[1 - \frac{\rho}{2}\left(1 - \frac{1}{k}\right)\right]^2\right\} \tag{11.4}$$

Waiting time

The M/G/1 formula for the average waiting time is Eq. (9.4), which with $C_S^2 = 1/k$ gives the M/E$_k$/1 form of this result

$$T_W = \text{average waiting time} = \frac{\rho T_S}{2(1-\rho)}\left(1 + \frac{1}{k}\right)$$

(11.5)

Similarly to queueing time, by putting the moments of the Erlang-k distribution into the appropriate M/GX/1 formula, we get the following expression for the variance of waiting time:

$$\sigma_{T_W}^2 = \text{variance of waiting time} = \frac{T_S^2\rho(k+1)}{12(1-\rho)^2 k^2}[4(k+2) - \rho(k+5)]$$

(11.6)

Number-in-system

The M/G/1 formula for the average number of customers in the system is Eq. (9.5), which with $C_S^2 = 1/k$ gives the M/E$_k$/1 formula for mean number-in-system:

$$L_Q = \text{average number-in-system} = \frac{\rho}{1-\rho}\left[1 - \frac{\rho}{2}\left(1 - \frac{1}{k}\right)\right]$$

(11.7)

Using the moments of the Erlang-k distribution in the appropriate M/GX/1 results to obtain a formula for the variance of the number-in-system, we get

$$\sigma_{L_Q}^2 = \text{variance of number-in-system}$$

$$\sigma_{L_Q}^2 = \frac{\rho}{(1-\rho)^2}\left\{1 - \frac{\rho}{2}\left[3 - \frac{\rho(10-\rho)}{6} - \frac{(3 - 3\rho + \rho^2)}{k} - \frac{\rho(8-5\rho)}{6k^2}\right]\right\}$$

(11.8)

Waiting-line size

The M/G/1 formula for the average number of customers waiting is Eq. (9.7). If we use $C_S^2 = 1/k$, then for M/E$_k$/1 the mean number waiting is given by

$$L_W = \text{average number waiting} = \frac{\rho^2(k+1)}{2k(1-\rho)}$$

(11.9)

Using the moments of the Erlang-k distribution in the appropriate M/GX/1 results to obtain a formula for the variance of the number waiting, we get

$$\sigma_{L_W}^2 = \text{variance of number waiting}$$

$$\sigma_{L_W}^2 = \frac{\rho^3(k+2)(k+1)}{3(1-\rho)k^2} + \left(\frac{\rho^2(k+1)}{2(1-\rho)k}\right)^2 + \frac{\rho^2(3-2\rho)(k+1)}{2(1-\rho)k}$$

(11.10)

The busy period

We know that the average length of the busy period is independent of the precise shape of the service-time distribution and depends only on the average service-time, i.e.

$$g_1 = \text{average length of busy period} = \frac{T_S}{1 - \rho} \qquad (11.11)$$

The variance of the busy period does, however, depend on the shape of the service-time distribution. For Erlang-k service the variance is given by

$$\sigma_g^2 = \text{variance of busy period} = \frac{T_S^2}{(1 - \rho)^3} \left(\rho + \frac{1}{k} \right) \qquad (11.12)$$

Number served in busy period

Just as the average length of the busy period does not depend on the precise service-time distribution, neither does the average number of customers served during a busy period. This average is

$$h_1 = \text{average number served per busy period} = \frac{1}{1 - \rho} \qquad (11.13)$$

The variance of the number served per busy period does, however, depend on the nature of the service-time distribution, although this quantity depends more markedly on the server utilization:

$$\sigma_h^2 = \text{variance of number served per busy period} = \frac{\rho(k + \rho)}{k(1 - \rho)^3} \qquad (11.14)$$

Programs for M/E$_k$/1

Prerequisite routines

There are no prerequisite routines for the $M/E_k/1$ calculation routines, except for type definitions, so to use the routine for this chapter the following set of 'include file' statements would be used in Turbo-Pascal.

```
{$N+,E+ }
{$I \qthprog\qthtypes.pas    }
{$I \qthprog\mek1calc.pas    }
```

MEk1Calc—general calculations for M/E$_k$/1

The routine ME_k1Calc calculates most of the statistics of interest for $M/E_k/1$. The variable VALID is set to TRUE if the results are meaningful, and to FALSE if server utilization is not less than 1.

```
{------------------------------------------------}
{> MEk1Calc - Calculations for M/Ek/1           }
{ Inputs:                                        }
{     LAMBDA      customer arrival rate          }
{     TS          mean service time              }
{     K           Erlang parameter               }
{ Outputs:                                       }
{     RHO         server utilisation             }
{     TQ/SDVTQ    mean/std dev time in system    }
{     TW/SDVTW    mean/std dev waiting time       }
{     LQ/SDVLQ    mean/std dev number in system  }
{     LW/SDVLW    mean/std dev number waiting     }
{     TB/SDVTB    mean/std dev of busy period    }
{     NB/SDVNB    mean/stdv no.served in busy period }
{     VALID       T=results valid, F=not valid    }
{  Copyright Mike Tanner 1993                    }
{------------------------------------------------}
Procedure MEk1Calc(LAMBDA,TS:QTHreal;K:integer;
                   Var RHO,TQ,SDVTQ,TW,SDVTW:QTHreal;
                   Var LQ,SDVLQ,LW,SDVLW:QTHreal;
                   Var TB,SDVTB,NB,SDVNB:QTHreal;
                   Var VALID:boolean);
Var RK:QTHreal;
begin
  RK:=K;  RHO:=LAMBDA*TS;
  VALID:=(RHO<1.0);  If not VALID then Exit;
  {-Time in system-------------------------------}
  TQ:=TS*(1-RHO*(1-1/RK)/2)/(1-RHO);
  SDVTQ:=(TS/(1-RHO))
         *Sqrt((1-RHO*(4-RHO)*(1-1/RK)/6)*(1+1/RK)
              -Sqr(1-RHO*(1-1/RK)/2));
  {-Waiting time---------------------------------}
  TW:=RHO*TS*(1+1/RK)/(2*(1-RHO));
  SDVTW:=RHO*(RK+1)*(4*(RK+2)-RHO*(RK+5))/12;
  SDVTW:=Sqrt(SDVTW)*TS/((1-RHO)*RK);
  {-Number in system-----------------------------}
  LQ:=RHO*(1-RHO*(1-1/RK)/2)/(1-RHO);
  SDVLQ:=1-(RHO/2)*(3-RHO*(10-RHO)/6
  -(3-3*RHO+Sqr(RHO))/RK-RHO*(8-5*RHO)/(6*Sqr(RK)));
  SDVLQ:=Sqrt(RHO*SDVLQ)/(1-RHO);
  {-Number waiting-------------------------------}
  LW:=RHO*RHO*(RK+1)/(2*RK*(1-RHO));
  SDVLW:=RHO*Sqr(RHO)*(RK+2)*(RK+1)/(3*(1-RHO)*Sqr(RK));
  SDVLW:=SDVLW+Sqr(Sqr(RHO)*(RK+1)/(2*(1-RHO)*RK));
  SDVLW:=SDVLW+Sqr(RHO)*(3-2*RHO)*(RK+1)/(2*(1-RHO)*RK);
  SDVLW:=Sqrt(SDVLW);
  {-Busy period----------------------------------}
  TB:=TS/(1-RHO);
  SDVTB:=Sqr(TS)*(RHO+1/RK)/((1-RHO)*Sqr(1-RHO));
  SDVTB:=Sqrt(SDVTB);
  {-Number served in busy period-----------------}
  NB:=1/(1-RHO);
  SDVNB:=Sqrt(RHO*(RK+RHO)/(RK*(1-RHO)))/(1-RHO);
  end;
```

Example 11A—Communications link

Look again at Example 7C, where a message took on average 1.5 secs to traverse the link. Now suppose that instead of the message lengths being exponentially distributed, they have an Erlang-3 distribution. What now is the average time taken by a message to get across the link? What arrival rate can now be allowed while still meeting the previous performance of $T_Q = 1.5$ secs?

Using Eq. (11.3) we get, with $T_S = 0.6$ secs, $\lambda = 1$ message/sec, $\rho = 0.6$

$$T_Q = \frac{T_S}{1-\rho}\left[1 - \frac{\rho}{2}\left(1 - \frac{1}{k}\right)\right] = \frac{0.6}{1-0.6}\left[1 - \frac{0.6}{2}\left(1 - \frac{1}{3}\right)\right] = 1.2 \text{ secs}$$

A little trial and error with Q-Calc shows that arrival rate can be increased from 1 to 1.15 messages/sec while still keeping $T_Q \leqslant 1.5$ secs.

12
M/G/1 with gamma distribution of service-times (M/Ga/1)

Introduction

As a special case of M/G/1 we shall assume that service-times have a gamma distribution. The gamma distribution is quite a general one: the shape is plausible for many types of service-time and the coefficient of variation can take on any value except zero. In the chapter on M/G/1 we saw how variation in service-time affects the *average* time-in-system. Later in this chapter we shall use the gamma distribution to investigate how the variation in service-time affects the *variation* of time-in-system.

The gamma distribution

The formula for the probability distribution function of the gamma distribution is

$$\mathrm{pdf}(x) = \left(\frac{x}{\alpha}\right)^{\beta-1} \frac{e^{-x/\alpha}}{\alpha\Gamma(\beta)} \tag{12.1}$$

where α is called the scale factor, and β is called the shape factor. Mathematically, the pdf is easier to manipulate when expressed using the scale and shape factors. On the other hand, in practice we want to use T_S and C_S^2 to describe the service-time distribution. Conversion between the two ways of specifying a gamma distribution is quite easy, since

$$T_S = \alpha\beta \text{ and } C_S^2 = \frac{1}{\beta} \tag{12.2}$$

and, in the reverse direction

$$\alpha = T_S C_S^2 \text{ and } \beta = \frac{1}{C_S^2} \tag{12.3}$$

There is no convenient analytical expression for the cumulative distribution function of the gamma distribution. Numerical approximations have been

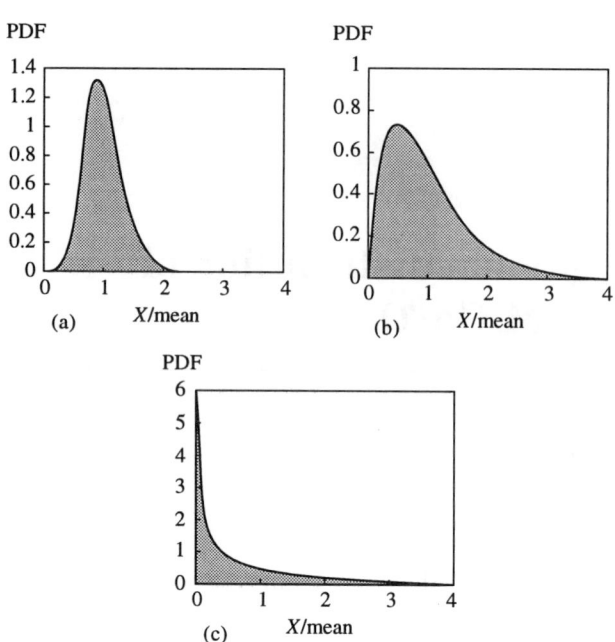

Figure 12.1. Pdf of gamma distribution: (a) $C^2 = 0.1$, (b) $C^2 = 0.5$, (c) $C^2 = 2.0$.

devised, and are given in Appendix 1. Let us take a look at the shape of the gamma distribution, which is illustrated in Fig. 12.1 for a selection of values of C_S^2. For small values of C_S^2 the shape is nearly symmetrical. As C_S^2 increases the shape becomes more skewed, and for $C_S^2 = 1$ the gamma distribution is in fact the exponential distribution. (Several important statistical distributions are special cases of the gamma distribution. For queueing theory the special cases of interest are the exponential and Erlang distributions.) As C_S^2 increases beyond 1, most of the distribution clusters around small values, with a long narrow tail of very large values. For service times with large values of C_S^2 the gamma distribution can be used, but the hyperexponential can also be used.

Parameters and initial calculations for M/Ga/1

The parameters that need to be specified for the M/Ga/1 model are

       γ, the average customer arrival rate
       T_S, the average service-time per customer
       C_S^2, the squared coefficient of variation for service-time

We first calculate the server utilization $\rho = \lambda T_S$ and check that $\rho < 1$. If this condition is not met then the system is not stable. The formulae in this chapter are all in terms of the gamma distribution parameters α and β. The formulae are slightly easier to manipulate using these parameters rather than T_S and C_S^2. We must therefore calculate α and β using Eq. (12.3).

Waiting time

The M/G/1 formula for average waiting time is Eq. (9.4), which for gamma service-times becomes

$$T_W = \text{average waiting time} = \frac{\rho\alpha(1+\beta)}{2(1-\rho)} \tag{12.4}$$

To get variance of the waiting time we substitute moments of the gamma distribution in the appropriate M/GX/1 formula to obtain

$$\sigma_{T_W}^2 = \text{variance of waiting time} = T_W^2 + \frac{2}{3}\alpha(2+\beta)T_W \tag{12.5}$$

Time-in-system, or queueing time

The M/G/1 formula for average time-in-system is Eq. (9.1). Using the gamma distribution parameters this becomes

$$T_Q = \text{average time-in-system} = \frac{\alpha}{2(1-\rho)}(\rho + 2\beta - \rho\beta) \tag{12.6}$$

To get variance of the waiting time we substitute moments of the gamma distribution in the appropriate M/GX/1 formula to obtain

$$\sigma_{T_Q}^2 = \text{variance of time-in-system} = \sigma_{T_W}^2 + \alpha^2\beta \tag{12.7}$$

Number waiting

The M/G/1 formula for the average number of customers waiting is Eq. (9.7). Using the gamma distribution parameters this becomes

$$L_W = \text{average number waiting} = \frac{\rho^2}{2(1-\rho)}\left(1 + \frac{1}{\beta}\right) \tag{12.8}$$

To get variance of the number waiting we substitute moments of the gamma distribution in the appropriate M/GX/1 formula to obtain

$$\sigma_{L_W}^2 = \rho\alpha(1+\beta)\left(\frac{\lambda^2\alpha(2+\beta)}{3(1-\rho)} + \frac{\rho\lambda^2\alpha(1+\beta)}{4(1-\rho)^2} + \frac{\lambda(3-2\rho)}{2(1-\rho)}\right) \tag{12.9}$$

Number-in-system

The M/G/1 formula for the average number of customers in the system is Eq. (9.5). Using the gamma distribution parameters this becomes

$$L_Q = \text{average number-in-system} = \frac{\rho}{2(1-\rho)}\left(\frac{\rho}{\beta} + 2 - \rho\right) \tag{12.10}$$

To get variance of the number of customers in the system we substitute moments of the gamma distribution in the appropriate $M/GX/1$ formula to obtain

$$\sigma_{L_Q}^2 = \text{variance of number-in-system} = \sigma_{L_w}^2 + \rho(1 - \rho) \tag{12.11}$$

Busy period

We know that the average length of the busy period is independent of the precise shape of the service-time distribution and depends only on average service-time, i.e.

$$g_1 = \text{average length of busy period} = \frac{T_S}{1 - \rho} \tag{12.12}$$

The variance of the busy period does, however, depend on the shape of the service-time distribution. For gamma service-times the variance is given by

$$\sigma_g^2 = \text{variance of busy period} = \frac{\alpha^2 \beta(1 + \rho\beta)}{(1 - \rho)^3} \tag{12.13}$$

Number served in busy period

Just as the average length of the busy period does not depend on the precise service-time distribution, neither does the average number of customers served during a busy period. This average is

$$h_1 = \text{average number served per busy period} = \frac{1}{1 - \rho} \tag{12.14}$$

The variance of the number served per busy period does, however, depend on the nature of the service-time distribution, although this quantity depends more markedly on the server utilization:

$$\sigma_h^2 = \text{variance of number served in busy period} = \frac{\rho(1 + \lambda\alpha)}{(1 - \rho)^3} \tag{12.15}$$

Variance of queueing times versus variance of service-time

The gamma distribution is a fairly general distribution. Several important statistical distributions are in fact special cases of the gamma distribution, in particular the exponential and the Erlang distributions. So if we use the Gamma distribution for service-time, and look at how the variation in queueing time changes as variance in service-time changes, we get some interesting insights.

In Fig. 12.2 the squared coefficient of variation for service-time is the horizontal axis. For a particular value of C_S^2 we can calculate the higher moments of service-time, and then apply the $M/Gx/1$ results to get $C_{T_Q}^2$, the squared coefficient of variation for queueing time. Notice first that separate lines are plotted for different

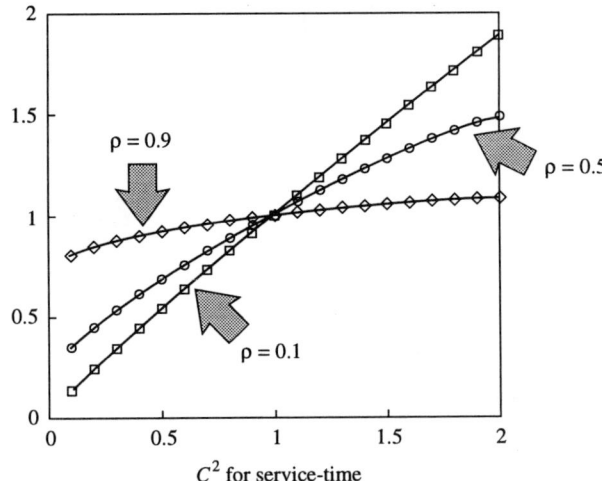

Figure 12.2. Variance of queueing time versus variance of service-time.

values of server utilization, and that the shape of the line differs markedly with utilization. Take first the curve for $\rho = 0.1$. At low utilization there will be very little queueing, and the time that most customers spend in the system will be their service-time. It follows that the distribution of queueing time will follow closely the distribution of service-time, so at low utilizations we have $C_{T_Q}^2 \approx C_S^2$.

Now look at the curve for $\rho = 0.9$. It is apparent that at high utilizations $C_{T_Q}^2 \approx 1$, in other words the distribution of queueing time tends towards an exponential distribution. An intuitive, or common-sense, explanation of this characteristic is not obvious. In some way the arrival pattern dominates the queueing time at high utilizations, whereas the service pattern dominates at low utilizations. The arrival pattern is random, by which we mean that the intervals between arrivals have an exponential distribution, and this distribution shows itself in the pattern of queueing time. This tendency of the queueing time to an exponential distribution at high utilization is of great importance, and approximations and bounds for queueing time have been developed using this property. For utilizations that are neither high nor low, the graph naturally falls between the two extremes, as shown by the graph for $\rho = 0.5$. For all values of ρ, the graph passes through the point $(C_S^2 = 1, C_{T_Q}^2 = 1)$, recalling the result that queueing time in an M/M/1 system has an exponential distribution.

It is as well to emphasize here that we are looking at how the variation in queueing time is affected by the variation in service-time. The average queueing time is also affected by variation in service-time, as we saw in Chapter 9. As a reminder of that effect, Fig. 12.3 shows average queueing time versus variation in service-time.

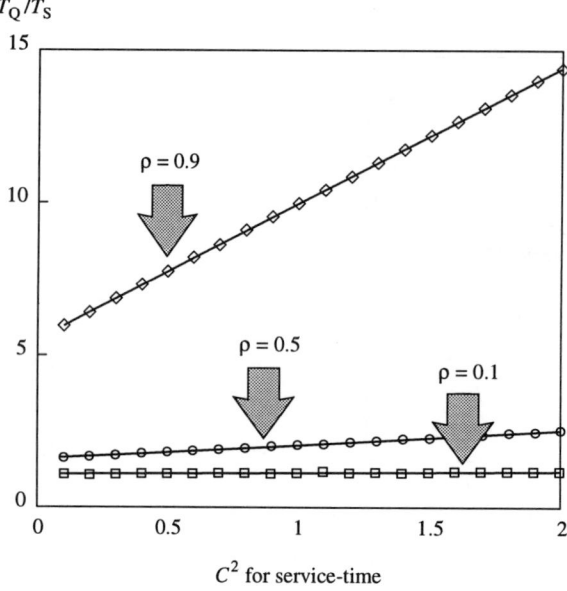

Figure 12.3. Average queueing time versus variance of service-time.

Programs for M/Ga/1

Prerequisite routines

There are no prerequisite routines for the M/Ga/1 calculation routines, except for
type definitions, so to use the routine for this chapter the following set of 'include
file' statement would be used in Turbo-Pascal.

```
{$N+,E+ }
{$I \qthprog\qthtypes.pas      }
{$I \qthprog\mga1calc.pas      }
```

MGa1Calc—general calculations for M/Ga/1

The routine MGa1Calc calculates most of the statistics of interest for M/Ga/1.
The variable VALID is set to TRUE if the results are meaningful, and to FALSE
if server utilization is not less than 1.

```
{-----------------------------------------------------------}
{> MGa1Calc - Calculations for M/Gamma/1                    }
{ Inputs:                                                   }
{      LAMBDA     customer arrival rate                     }
{      TS         mean service time                         }
{      COEFF2     sqd coeff of var service time             }
{ Outputs:                                                  }
{      RHO        server utilisation                        }
```

```
{       TQ/SDVTQ   mean/std dev time in system         }
{       TW/SDVTW   mean/std dev waiting time           }
{       LQ/SDVLQ   mean/std dev number in system       }
{       LW/SDVLW   mean/std dev number waiting         }
{       TB/SDVTB   mean/std dev of busy period         }
{       NB/SDVNB   mean/stdv no served in bsy period   }
{       VALID      T=results valid, F=not valid        }
{ Copyright Mike Tanner 1993                           }
{------------------------------------------------------}
Procedure MGa1Calc(LAMBDA,TS,COEFF2:QTHreal;
                   Var RHO,TQ,SDVTQ,TW,SDVTW,LQ,
                       SDVLQ,LW,SDVLW,TB,SDVTB,
                       NB,SDVNB:QTHreal;
                   Var VALID:boolean);
Var ALPHA,BETA:QTHreal;
begin
   {-Calculate utilisation and check stability-----}
   RHO:=LAMBDA*TS;
   VALID:=(RHO<1.0);  If not VALID then Exit;
   {-Calculate parameters of gamma distribution----}
   ALPHA:=TS*COEFF2;  BETA:=1/COEFF2;
   {-Waiting time----------------------------------}
   TW:=RHO*ALPHA*(1+BETA)/(2*(1-RHO));
   SDVTW:=Sqrt(Sqr(TW)+TW*2*ALPHA*(2+BETA)/3);
   {-Time in system--------------------------------}
   TQ:=ALPHA*(RHO+2*BETA-RHO*BETA)/(2*(1-RHO));
   SDVTQ:=Sqrt(Sqr(SDVTW)+Sqr(ALPHA)*BETA);
   {-Number waiting--------------------------------}
   LW:=Sqr(RHO)*(1+1/BETA)/(2*(1-RHO));
   SDVLW:=Sqr(LAMBDA)*ALPHA*(2+BETA)/(3*(1-RHO))
         +Sqr(LAMBDA/(2*(1-RHO)))*RHO*ALPHA*(1+BETA)
         +LAMBDA*(3-2*RHO)/(2*(1-RHO));
   SDVLW:=Sqrt(SDVLW*RHO*ALPHA*(1+BETA));
   {-Number in system------------------------------}
   LQ:=(RHO/BETA+2-RHO)*RHO/(2*(1-RHO));
   SDVLQ:=Sqrt(Sqr(SDVLW)+RHO*(1-RHO));
   {-Busy period-----------------------------------}
   TB:=TS/(1-RHO);
   SDVTB:=Sqrt(BETA*(1+RHO*BETA)/(1-RHO));
   SDVTB:=SDVTB*ALPHA/(1-RHO);
   {-Number served in busy period-----------------}
   NB:=1/(1-RHO);
   SDVNB:=Sqrt(RHO*(1+LAMBDA*ALPHA)/(1-RHO));
   SDVNB:=SDVNB/(1-RHO);
 end;
```

Example 12A—The Kayos Confectionery Kiosk

At the Kayos Confectionery Kiosk, customers arrive at an average rate of one every 5 min. The average service-time is 3 min, but service-time is very variable, with $C_S^2 = 1.7$. What is the average waiting time and the average number of customers waiting? The kiosk owner wants to improve service, and (having read a book on queueing theory once) feels he can do this by making service-times more consistent. What would be the average waiting time and average customers waiting if C_S^2 could be reduced to 1, and then further to 0.33?

In each of the three cases $\lambda = 0.2$ customers min and $T_S = 3$ min. We use Eq. (12.3) to find the parameters of the gamma distribution, Eq. (12.4) to find the average waiting time, and Eq. (12.8) to find the average number waiting. The results are given in Table 12.1.

Table 12.1. Results for Example 12A

C_S^2	Average waiting time (min)	Average number waiting
1.70	6.08	1.22
1.00	4.50	1.13
0.33	3.00	0.60

13
Hyperexponential service-times (M/H$_k$/1)

Introduction

The hyperexponential is the 'mirror image' of the Erlang distribution. While the Erlang distribution is a combination of exponential distributions resulting in a coefficient of variation less than 1, the hyperexponential combines exponential distributions to get a coefficient of variation greater than 1. A hyperexponential distribution can be represented as a number of exponential distributions in parallel. Here we shall look only at the hyperexponential of degree 2, which means only 2 exponential distributions are combined, as depicted in Fig. 13.1.

The exponentials are selected randomly, with the probability of selecting each exponential being α_1 and α_2. The mean values of the exponentials are θ_1 and θ_2. Unlike the Erlang-k distribution, where the component exponential distributions all have the same mean, for the hyperexponential the component means must be different or the structure would simply reduce to a single exponential distribution.

An interesting discussion of Erlang and hyperexponential distributions can be

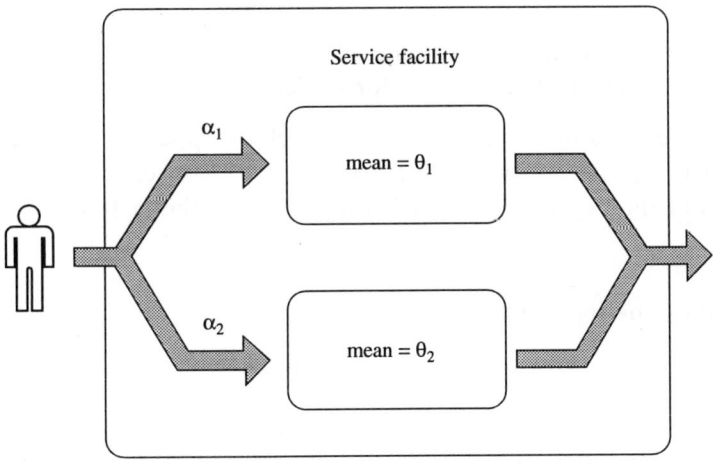

Figure 13.1. Structure of hyperexponential-2 distribution.

found in Kleinrock (1975), including more complicated combinations of exponential distributions in series and parallel to produce a very wide range of service-time distributions. Theoretically, these complicated distributions still allow the maths of the queueing model to be solved, exploiting the memoryless property of the exponential distribution. However, the solutions are far from easy.

The hyperexponential distribution

The hyperexponential is mainly used in simulation. The ease and efficiency of generating random numbers with an exponential distribution make the hyperexponential a convenient way of generating random values having a coeff of variation squared greater than 1.

If we are given the mean and squared coefficient of variation for service-time, we need to calculate α_1, α_2, θ_1 and θ_2. A convenient method of doing this is suggested by Allen (1978). The calculations are

$$\alpha_1 = \frac{1}{2}\left(1 - \sqrt{\frac{C_S^2 - 1}{C_S^2 + 1}}\right) \text{ and } \alpha_2 = 1 - \alpha_1 \tag{13.1}$$

$$\theta_1 = \frac{T_S}{2\alpha_1} \text{ and } \theta_2 = \frac{T_S}{2\alpha_2} \tag{13.2}$$

Figure 13.2 shows the shape of the distribution for a selection of values for the squared coeff of variation.

Assumptions, parameters, and initial calculations for M/H2/1

We assume a single server with random arrivals from an infinite population of customers. Customers are served in FIFO order. The parameters that need to be specified for the M/H2/1 model are

> λ, the average customer arrival rate
> T_S, the average service-time per customer
> C_S^2, the squared coefficient of variation for service-time

We first calculate the server utilization $\rho = \lambda T_S$ and check that $\rho < 1$. If this condition is not met then the system is not stable. Next we must convert T_S and C_S^2 into the parameters of the hyperexponential-2 distribution by the method given above.

Time-in-system (or queueing time)

The formula for the mean time-in-system for M/G/1 is Eq. (9.1), which for the hyperexponential service becomes

$$T_Q = \text{average time-in-system} = T_S\left[1 + \frac{\rho}{4(1 - \rho)}\left(\frac{1}{\alpha_1} + \frac{1}{\alpha_2}\right)\right] \tag{13.3}$$

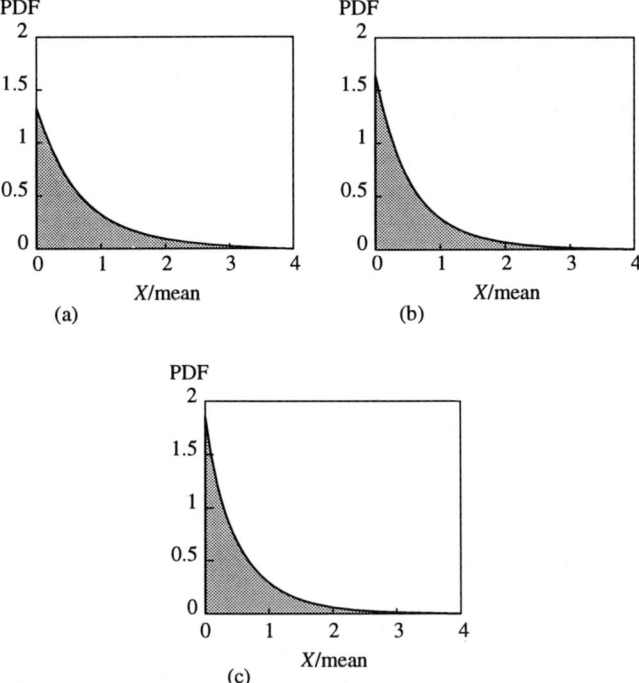

Figure 13.2. Shape of the hyperexponential distribution for: (a) $C^2 = 2$, (b) $C^2 = 5$, (c) $C^2 = 10$.

Substituting appropriate expressions for moments of service-time in the M/GX/1 formula, we get an expression for the variance of time-in-system.

$$\sigma^2_{T_Q} =$$

$$\left\{ \left[\frac{\rho}{4(1-\rho)} \left(\frac{1}{\alpha_1} + \frac{1}{\alpha_2} \right) \right]^2 + \frac{\rho}{4(1-\rho)} \left(\frac{1}{\alpha_1^2} + \frac{1}{\alpha_2^2} \right) + \frac{1}{2} \left(\frac{1}{\alpha_1} + \frac{1}{\alpha_2} \right) - 1 \right\} T_S^2$$

(13.4)

Waiting time

The M/G/1 formula for the average waiting time is Eq. (9.4), which for hyperexponential service-times becomes

$$T_W = \text{average waiting time} = \frac{\rho T_S}{4(1-\rho)} \left(\frac{1}{\alpha_1} + \frac{1}{\alpha_2} \right)$$

(13.5)

Substituting appropriate expressions for moments of service-time in the M/GX/1 formula we get an expression for the variance of waiting time:

$$\sigma^2_{T_W} = \text{variance of waiting time}$$

$$\sigma^2_{T_W} = \left\{ \left[\frac{\rho}{4(1-\rho)} \left(\frac{1}{\alpha_1} + \frac{1}{\alpha_2} \right) \right]^2 + \frac{\rho}{4(1-\rho)} \left(\frac{1}{\alpha_1^2} + \frac{1}{\alpha_2^2} \right) \right\} T_S^2$$

(13.6)

Waiting-line size

The M/G/1 formula for the average number of customers waiting is Eq. (9.7), which for hyperexponential service becomes

$$L_W = \text{average number waiting} = \frac{\rho^2}{4(1-\rho)}\left(\frac{1}{\alpha_1}+\frac{1}{\alpha_2}\right) \tag{13.7}$$

Substituting appropriate expressions for moments of service-time in the M/GX/1 formula, we get an expression for the variance of waiting-line size:

$$\sigma^2_{L_W} = \text{variance of number waiting}$$

$$\sigma^2_{L_W} = \frac{\rho^3}{4(1-\rho)}\left(\frac{1}{\alpha_1^2}+\frac{1}{\alpha_2^2}\right) + \left[\frac{\rho^2}{4(1-\rho)}\left(\frac{1}{\alpha_1}+\frac{1}{\alpha_2}\right)\right]^2$$

$$+ \frac{\rho^2(3-2\rho)}{4(1-\rho)}\left(\frac{1}{\alpha_1}+\frac{1}{\alpha_2}\right) \tag{13.8}$$

Number-in-system

The M/G/1 formula for the average number of customers in the system is Eq. (9.5). For hyperexponential service, rather than using Eq. (9.5) with C_S^2 expressed in terms of the parameters of the hyperexponential distribution, it is easier to use the relationship

$$L_Q = \text{average number-in-system} = L_W + \rho \tag{13.9}$$

Similarly, the variance of the number of customers in the system is most easily calculated by

$$\sigma^2_{L_Q} = \text{variance of number-in-system} = \sigma^2_{L_W} + \rho(1-\rho) \tag{13.10}$$

The busy period

We know that the average length of the busy period is independent of the precise shape of the service-time distribution, and depends only on the average service-time, i.e.

$$g_1 = \text{mean length of busy period} = \frac{T_S}{1-\rho} \tag{13.11}$$

The variance of the busy-period duration does depend on the service-time distribution, and for hyperexponential service we get

$$\sigma^2_g = \text{variance of busy period} = \frac{T_S^2}{(1-\rho)^2}\left[\frac{1}{2(1-\rho)}\left(\frac{1}{\alpha_1}+\frac{1}{\alpha_2}\right)-1\right] \tag{13.12}$$

Number served in busy period

Just as the average length of the busy period does not depend on the precise service-time distribution, neither does the average number of customers served during a busy period. This average is

$$h_1 = \text{average number served} = \frac{1}{1 - \rho} \tag{13.13}$$

The variance of the number served does depend on the particular distribution of service-time, and is given by

$$\sigma_h^2 = \text{variance of number served} = \frac{1}{(1 - \rho)^2}\left[\rho + \frac{\rho^2}{2(1 - \rho)}\left(\frac{1}{\alpha_1} + \frac{1}{\alpha_2}\right)\right]$$

$$\tag{13.14}$$

Programs for M/H2/1

Prerequisite routines

There are no prerequisite routines for the M/H2/1 calculation routines, except for type definitions, so to use the routine for this chapter the following set of 'include file' statements would be used in Turbo-Pascal.

```
{$N+,E+ }
{$I \qthprog\qthtypes.pas    }
{$I \qthprog\mh21calc.pas    }
```

MH21Calc—general calculations for M/H2/1

The routine MH21Calc calculates most of the statistics of interest for M/H2/1. The variable VALID is set to TRUE if the results are meaningful, and to FALSE is server utilization is not less than 1.

```
{------------------------------------------------------}
{> MH21Calc - Calculations for M/H2/1                  }
{ Inputs:                                              }
{     LAMBDA      customer arrival rate                }
{     TS          mean service time                    }
{     COEFF2      squared coeff of variation for       }
{                 service time                         }
{ Outputs:                                             }
{     RHO         server utilisation                   }
{     TQ/SDVTQ    mean/std dev time in system          }
{     TW/SDVTW    mean/std dev waiting time            }
{     LQ/SDVLQ    mean/std dev number in system        }
{     LW/SDVLW    mean/std dev number waiting          }
{     TB/SDVTB    mean/std dev busy period length      }
{     NB/SDVNB    mean/stdv no.served in bsy period    }
{     VALID       T=results valid, F=not valid         }
{   Copyright Mike Tanner 1993                         }
{------------------------------------------------------}
```

```
Procedure MH21Calc(LAMBDA,TS,COEFF2:QTHreal;
                   Var RHO,TQ,SDVTQ,TW,SDVTW:QTHreal;
                   Var LQ,SDVLQ,LW,SDVLW:QTHreal;
                   Var TB,SDVTB,NB,SDVNB:QTHreal;
                   Var VALID:boolean);
Var ALPHA1,ALPHA2,F1,F2:QTHreal;
begin
   If COEFF2<1 then QTHError('MH21Calc invalid COEFF2');
   ALPHA1:=(1-Sqrt((COEFF2-1)/(COEFF2+1)))/2;
   ALPHA2:=1-ALPHA1;
   F1:=1/ALPHA1+1/ALPHA2;
   F2:=1/Sqr(ALPHA1)+1/Sqr(ALPHA2);
   {-Calculate utilisation and check stability-----}
   RHO:=LAMBDA*TS;
   VALID:=(RHO<1.0);  If not VALID then Exit;
   {-Time in system-------------------------------}
   TQ:=TS*(1+RHO*F1/(4*(1-RHO)));
   SDVTQ:=Sqr(RHO*F1/(4*(1-RHO)))
          +RHO*F2/(4*(1-RHO))+F1/2-1;
   SDVTQ:=Sqrt(SDVTQ)*TS;
   {-Waiting time---------------------------------}
   TW:=TQ-TS;
   SDVTW:=Sqr(RHO*F1/(4*(1-RHO)))+RHO*F2/(4*(1-RHO));
   SDVTW:=Sqrt(SDVTW)*TS;
   {-Number waiting-------------------------------}
   LW:=Sqr(RHO)*F1/(4*(1-RHO));
   SDVLW:=RHO*Sqr(RHO)*F2/(4*(1-RHO));
   SDVLW:=SDVLW+Sqr(F1*Sqr(RHO)/(4*(1-RHO)));
   SDVLW:=SDVLW+Sqr(RHO)*(3-2*RHO)*F1/(4*(1-RHO));
   SDVLW:=Sqrt(SDVLW);
   {-Number in system-----------------------------}
   LQ:=LW+RHO;
   SDVLQ:=Sqrt(Sqr(SDVLW)+RHO*(1-RHO));
   {-Busy period----------------------------------}
   TB:=TS/(1-RHO);
   SDVTB:=F1/(2*(1-RHO))-1;
   SDVTB:=Sqrt(SDVTB)*TS/(1-RHO);
   {-Number served in busy period-----------------}
   NB:=1/(1-RHO);
   SDVNB:=RHO+Sqr(RHO)*F1/(2*(1-RHO));
   SDVNB:=Sqrt(SDVNB)/(1-RHO);
end;
```

Example 13A—Communications link

Let us revisit the communications link of Examples 7C and 11A. Suppose now
that the messages are a mixture of short and long messages. Short messages have
an exponentially distributed transmission time with an average of 0.1 s. The transmission time for long messages is also exponentially distributed, but with a mean
of 1.5 s. 65 per cent of messages are short and 35 per cent long.

Using the formulae in Appendix 1 for the hyperexponential distribution, we have

$$\alpha_1 = 0.65, \alpha_2 = 0.35, \theta_1 = 0.1 \text{ secs}, \theta_2 = 1.5 \text{ secs}$$

$$T_S = 0.59 \text{ secs}, C_S^2 = 3.562$$

Now we can use either the basic M/G/1 formula Eq. (9.1) or Eq. (13.3). Either way we get $T_Q = 2.53$ secs, compared to 1.5 secs in Example 7C. Simple trial and error, perhaps with Q-Calc, easily shows that to get the average time-in-system back down to 1.5 secs the arrival rate must be reduced to 0.68 messages/sec.

Part Four
Single server with general arrival pattern

14
General arrivals and exponential service (G/M/1)

Introduction

With the G/M/1 model, we keep the helpful assumption of exponential service times, but allow the arrival pattern to be quite general. The arrival pattern is defined by the distribution of inter-arrival times. Suppose customers arrive in bursts, meaning that for short periods of time customers arrive randomly, then there is an extended period during which no customers arrive and then another burst occurs. The first customer of a burst will almost certainly find an empty system. Subsequent customers in a burst will have a higher probability of waiting than the long-term average server utilization would suggest. Conversely, if customers arrive more regularly then we get something more like a conveyor-belt system, and customers are much less likely to have to wait than with, say, random arrivals. So we can see that high variability of inter-arrival times will make waiting times worse, and more regular inter-arrival times will decrease waiting times. This parallels the effect of more or less variability of service-times as seen with M/G/1, and later in this chapter we shall compare the effect of inter-arrival time variance with service-time variance.

One of the characteristics that we often calculate for a queueing system is the average number of customers in the system. This is the average than an external observer would see if the system were inspected at times chosen randomly and unrelated to the state of the system. Looking at things from the point of view of the customers, we could ask what is the average number of customers in the system as seen by arriving customers. For this we would inspect the system at arrival instants.

The relationship between average over time and average at arrival instants is discussed later in this chapter. These two averages are not, in general, the same. However, the two averages are the same if the arrival pattern is random, i.e. if the inter-arrival times have an exponential distribution. Many queueing models assume random arrivals, and many of those that do not are very difficult if not impossible to analyse mathematically. So the G/M/1 model is a good vehicle for exploring the effects of different arrival patterns, and drawing some intuitive conclusions.

Assumptions for the G/M/1 model

- Single server, FIFO queue, infinite population of customers.
- Average customer arrival rate is λ, with a general distribution of inter-arrival times.
- Service-times are exponentially distributed with mean T_S.

Parameters and initial calculations for G/M/1

As usual, we first calculate $\rho = \lambda T_S$, and check that $\rho < 1$ so that the system is stable. Next we need to calculate

$$\theta = \text{Prob(arriving customer finds the server busy)} \qquad (14.1)$$

which for G/M/1, because of the general pattern of arrivals, is not the same as ρ, the server utilization. In fact, it is only for random arrivals that $\theta = \rho$. The value of θ is the solution to the equation.

$$\theta = A^* \left(\frac{1 - \theta}{T_S} \right) \text{ such that } 0 \le \theta < 1 \qquad (14.2)$$

where $A^*(s)$ is the Laplace–Stieltjes transform of the pdf of inter-arrival times. The author has tried to avoid asking readers to deal with Laplace transforms and other mathematical tools they either were once familiar with but are no longer, or never studied. Luckily, all the mathematically timorous reader needs to know is that every pdf has a thing called the Laplace–Stieltjes transform. A useful selection of inter-arrival time pdfs, their transforms, and the resulting particular forms of Eq. (14.2) are given below. In some cases Eq. (14.2) can be solved analytically, but in general a numerical procedure is required. Note that $\theta = 1$ is always a solution to Eq. (14.2), but that is not the root we need. We must strictly have $0 \le \theta < 1$.

It is interesting to compare the way the mathematical analysis proceeds for G/M/1 and M/G/1. For M/G/1, the details of the service-time distribution, in terms of its moments, enter into almost all the formulae for characteristics of interest, and we can obtain only moments of queueing and waiting time rather than explicit distributions. In contrast, for G/M/1 all the information about 'G' distribution is used to calculate the parameter θ, and we can then get explicit distributions for queueing and waiting times, and more straightforward formulae for other characteristics.

Calculating θ for specific distributions

In this section we present solutions to Eq. (14.2) for specific inter-arrival time distributions. For each of the equations a corresponding subroutine is provided for use in solving the equation numerically. Table 14.1 gives solutions for a range of ρ.

Table 14.1. Values of θ for inter-arrival time distributions

r	Regular (constant)	Uniform	Erlang-3	Erlang-2	Hyper-exponential-2 ($C^2 = 5$)	Gamma ($C = 5$)
0.100	0.000045	0.052786	0.012656	0.029180	0.164110	0.526732
0.200	0.006977	0.112684	0.059030	0.093774	0.321406	0.626906
0.300	0.040882	0.182753	0.131885	0.178046	0.468338	0.698087
0.400	0.107355	0.265313	0.223949	0.275305	0.600000	0.755958
0.500	0.203188	0.360768	0.330533	0.381966	0.711325	0.805981
0.600	0.324243	0.468403	0.448549	0.495841	0.800000	0.850767
0.700	0.466996	0.587161	0.575863	0.615477	0.868338	0.891784
0.800	0.628630	0.715972	0.710934	0.739853	0.921406	0.929949
0.900	0.806900	0.853867	0.852610	0.868218	0.964110	0.965877
0.950	0.901695	0.925952	0.925638	0.933712	0.982738	0.983142
0.980	0.960268	0.970151	0.970101	0.973393	0.993242	0.993303

The uniform distribution has $C^2 = 1/3$, as does the Erlang-3 distribution: the corresponding values of θ are similar except at low utilizations. For the hyper-exponential and gamma distributions, C^2 has been set to 5, and again the corresponding values of θ are close at moderate to high utilizations.

Regular arrivals D/M/1

For regular arrivals the inter-arrival times are constant. The Laplace–Stieltjes transform of a constant distribution is

$$A^*(s) = e^{-s/\lambda}, \quad \text{where } \lambda = \frac{1}{T_A} \tag{14.3}$$

so that Eq. (14.2) becomes

$$\theta = e^{-(1-\theta)/\rho} \tag{14.4}$$

No analytic solution is available for this, but numerical solution is straightforward.

Uniform arrivals U/M/1

For uniformly distributed arrival times we assume that the arrival times are evenly spread between 0 and $2/\lambda$, so as to give the arrival rate λ. With these limits the Laplace–Stieltjes transform of the pdf is

$$A^*(s) = \frac{\lambda}{2s}\left(1 - e^{-2s/\lambda}\right) \tag{14.5}$$

so that Eq. (14.02) becomes

$$\theta = \frac{\rho(1 - e^{-2(1-\theta)/\rho})}{2(1 - \theta)} \tag{14.6}$$

This equation must be solved numerically.

Erlang-3 arrivals E3/M/1

Erlang-3 inter-arrival times have a coefficient of variation of 1/3, and the Laplace–Stieltjes transform of the pdf is

$$A^*(s) = \left(\frac{3\lambda}{3\lambda + s}\right)^3 \tag{14.7}$$

so that Eq. (14.2) becomes

$$\theta = \left(\frac{3\rho}{3\rho + (1 - \theta)}\right)^3 \tag{14.8}$$

which is the form used for numerical solution. After some algebra and making the substitution $\phi = 1 - \theta$, we get the equation

$$0 = \phi^3 + (9\rho - 1)\phi^2 + 9\rho(3\rho - 1)\phi + 27\rho^2(\rho - 1) \tag{14.9}$$

Erlang-3 is a special case of the gamma distribution, but is given separately since it is possible to solve the above cubic equation analytically if required.

Erlang-2 arrivals E2/M/1

Erlang-2 inter-arrival times have a coefficient of variation of 1/2, and the Laplace–Stieltjes transform of the pdf is

$$A^*(s) = \left(\frac{2\lambda}{2\lambda + s}\right)^2 \tag{14.10}$$

so that Eq. (14.2) becomes

$$\theta = \left(\frac{2\rho}{2\rho + (1 - \theta)}\right)^2 \tag{14.11}$$

which is the form used for numerical solution. After a little algebra we get

$$0 = (\theta - 1)[\theta^2 - (4\rho + 1)\theta + 4\rho^2] \tag{14.12}$$

Erlang-2 is a special case of the gamma distribution, but is given separately since the above equation can be solved straightforwardly. We know that $\theta = 1$ is a solution of Eq. (14.2) that is of no interest to us, so we need to solve the quadratic equation in square brackets. Using the usual method of solving a quadratic, and taking note of the constraint $0 \leqslant \theta < 1$, we get

$$\theta = 2\rho + \frac{1}{2} - \sqrt{2\rho - \frac{1}{4}} \tag{14.13}$$

Gamma arrivals Ga/M/1

For a gamma distribution of inter-arrival times, we can choose C_A^2, the squared coefficient of variation, to have any value greater than zero. The mean inter-arrival time if $T_A = 1/\lambda$. For the gamma distribution it is convenient to work with the parameters

$$\alpha = T_A C_A^2 \text{ and } \beta = \frac{1}{C_A^2} \tag{14.14}$$

The Laplace–Stieltjes transform of the pdf is then

$$A^*(s) = \left(\frac{1}{1 + \alpha s}\right)^\beta \tag{14.15}$$

so that Eq. (14.2) becomes

$$\theta = \left(\frac{T_S}{T_S + \alpha(1 - \theta)}\right)^\beta \tag{14.16}$$

Hyperexponential-2 arrivals H2/M/1

A two-component hyperexponential distribution can be constructed to have any squared coefficient of variation greater than or equal to 1. The distribution is made up from two exponential distributions with means $1/\lambda_1$ and $1/\lambda_2$, in the proportions α and $(1 - \alpha)$ respectively. See chapter 13 for details on calculating these values to give a specific T_A and C_A^2. The Laplace–Stieltjes transform of the pdf is

$$A^*(s) = \frac{\alpha \lambda_1}{\lambda_1 + s} + \frac{(1 - \alpha)\lambda_2}{\lambda_2 + s} \tag{14.17}$$

so that Eq. (14.2) becomes

$$\theta = \frac{\alpha \lambda_1}{\lambda_1 + \mu(1 - \theta)} + \frac{(1 - \alpha)\lambda_2}{\lambda_2 + \mu(1 - \theta)}, \text{ where } \mu = \frac{1}{T_S} \tag{14.18}$$

This is the form used for numerical solution in the routines later in this chapter. A tidier form can be derived by making the substitution $\phi = 1 - \theta$, after which a little algebra results in

$$\theta = \mu^2 \phi^2 + \mu[\lambda_1 + \lambda_2 - \mu]\phi - \mu[(1 - \alpha)\lambda_1 + \alpha\lambda_2] + \lambda_1\lambda_2 \tag{14.19}$$

where

$$\mu = 1/T_S$$

Time-in-system (or queueing time)

The average time-in-system is given by

$$T_Q = \text{average time-in-system} = \frac{T_S}{1 - \theta} \tag{14.20}$$

We can compare this with the M/M/1 system, where

$$T_{Q(M/M/1)} = \text{average time-in-system for M/M/1} = \frac{T_S}{(1 - \rho)} \tag{14.20a}$$

so that G/M/1 has a longer queueing time than M/M/1 if

$$\frac{T_S}{1 - \theta} > \frac{T_S}{1 - \rho} \tag{14.21}$$

which means $\theta > \rho$. This is intuitively sensible, since θ is the probability that an arriving customer will have to wait. The corresponding probability for M/M/1 is ρ, so if $\theta > \rho$ an arriving customer to a G/M/1 system is more likely to have to wait than for an M/M/1 system, and will also on average have a longer waiting time.

The time-in-system has an exponential distribution, so that

$$\text{Prob(time in system} < t) = 1 - e^{-t/T_Q} \tag{14.22}$$

and we can easily calculate variance and percentiles of queueing time, i.e.

$$\sigma_{T_Q}^2 = \text{variance of time-in-system} = T_Q^2 \tag{14.23}$$

$$\pi_{T_Q}(r) = r\text{th percentile of time-in-system} = \ln\left(\frac{100}{100 - r}\right) T_Q \tag{14.24}$$

$$\pi_{T_Q}(90) = 90\text{th percentile of time-in-system} = 2.3 T_Q \tag{14.25}$$

$$\pi_{T_Q}(95) = 95\text{th percentile of time-in-system} = 3 T_Q \tag{14.26}$$

Waiting time

The average waiting time is given by

$$T_W = \text{average waiting time} = \frac{\theta T_S}{(1 - \theta)} \tag{14.27}$$

Comparing this with M/M/1, the formula for T_W has the same form but for G/M/1 depends on the parameter θ rather than the server utilization ρ. It is not hard to see that waiting time will be longer for G/M/1 than for M/M/1 if $\theta > \rho$. The waiting time has a modified exponential distribution, so that

$$\text{Prob(waiting time} < t) = 1 - \theta e^{-(1-\theta)t/T_S} = 1 - \theta e^{-t/T_Q} \tag{14.28}$$

Knowing the exact form of the distribution for waiting time, we can easily calculate variance and percentiles. These are given by

$$\sigma^2_{T_{\rm w}} = \text{variance of waiting time} = \frac{\theta(2-\theta)}{(1-\theta)^2} T_{\rm S}^2 \tag{14.29}$$

$$\pi_{T_{\rm w}}(r) = r\text{th percentile of waiting time} = \ln\left(\frac{100\theta}{100-r}\right) T_{\rm Q} \tag{14.30}$$

$$\pi_{T_{\rm w}}(90) = 90\text{th percentile of waiting time} = \ln(10\theta) T_{\rm Q} \tag{14.31}$$

$$\pi_{T_{\rm w}}(95) = 95\text{th percentile of waiting time} = \ln(20\theta) T_{\rm Q} \tag{14.32}$$

Number-in-system

The probability of there being k customers in the system is p_k, where

$$p_k = \text{Prob}(k \text{ customers in system}) = \rho(1-\theta)\theta^{k-1} \text{ for } k = 1, 2, 3, \ldots \tag{14.33}$$

$$p_0 = 1 - \rho \tag{14.34}$$

It should be emphasized that these are the probabilities as seen by an external observer of the system. Later in this chapter we shall look at the state of the system as seen at the instants at which customers arrive. When analysing a particular queueing system we have to think about why we want to know p_k or L_Q, or the corresponding values for arrival instants. Are we interested in the system behaviour as seen from outside? Or are we concerned with how arriving customers perceive the system? It all depends, of course, on what we are trying to do! Proceeding with the external observer's viewpoint, the mean and variance of the number of customers in the system are given by:

$$L_Q = \text{average number in system} = \frac{\rho}{1-\theta} \tag{14.35}$$

$$\sigma^2_{L_Q} = \text{variance of number in system} = \frac{\rho(1+\theta-\rho)}{(1-\theta)^2} \tag{14.36}$$

Number waiting

Equations (14.33) and (14.34) also define the distribution of the number of customers waiting, from which we can deduce that

$$L_{\rm W} = \text{average number waiting} = \frac{\theta\rho}{1-\theta} \tag{14.37}$$

$$\sigma^2_{L_{\rm W}} = \text{variance of number waiting} = \frac{\rho\theta[1+\theta(1-\rho)]}{(1-\theta)^2} \tag{14.38}$$

Number-in-system at arrival instants

It has already been mentioned in this chapter that the probability of there being a specified number of customers in the system and the mean and variance of the number of customers in the system are different depending on whether we inspect the system at times chosen randomly by an external observer or at the instants at which customers arrive. In this section we deal with how the system appears at the instants at which customers arrive. To start with, let us look at the probability of there being k customers in the system when a customer arrives:

$$r_k = \text{Prob}(k \text{ customers in system at arrival instant}) \qquad (14.38a)$$

$$r_k = (1 - \theta)\theta^k \text{ for } k = 0, 1, 2, \ldots \qquad (14.39)$$

The mean and variance of the number of customers present at arrival instants are denoted by L_A and $\sigma^2_{L_A}$ respectively, and are given by:

$$L_A = \text{average number-in-system at arrival instants} = \frac{\theta}{1 - \theta} \qquad (14.40)$$

$$\sigma^2_{L_A} = \text{variance of number-in-system at arrival instants} = \frac{\theta}{(1 - \theta)^2}$$

$$(14.41)$$

Figure 14.1 compares L_A, the average number-in-system at arrival instants, with L_B, the average number-in-system seen by an external observer. These two quantities are plotted against a range of values of C^2_A, the squared coefficient of inter-arrival times, which have been assumed to have a gamma distribution.

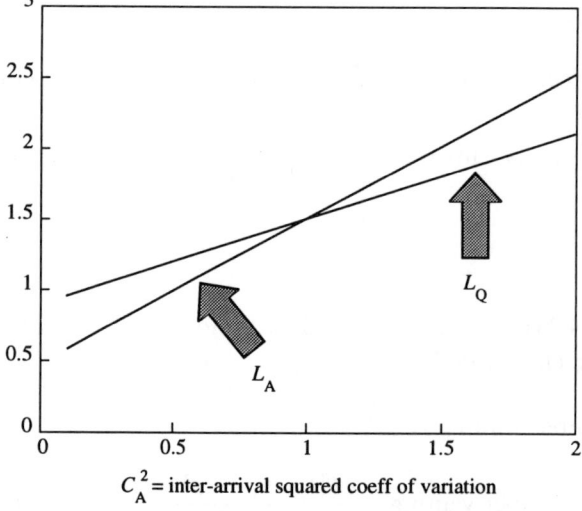

Average number-in-system

C^2_A = inter-arrival squared coeff of variation

Figure 14.1. Number-in-system and number at arrival instants (for $\rho = 0.6$).

Notice first that the lines coincide when $C_A^2 = 1$, which corresponds to an M/M/1 system. Smaller variance of inter-arrival times means that arrivals are more regular, and we would expect to find, on average, fewer customers in the system because arrivals are less bunched together. This is apparent from the graph, which also shows that arriving customers see fewer in the system than the externally-viewed average when $C_A^2 < 1$. For $C_A^2 > 1$ we get the opposite effects. As C_A^2 increases the average number in the system also increases, whether we measure this by L_A or L_Q. But now L_A is greater than L_Q, which means that arriving customers see a more congested system than an external observer.

Effect of arrival variance on waiting time

We have already suggested that, other things being equal, average waiting time increases as the arrival pattern becomes more irregular. In this section we look at this effect. Figure 14.2 shows average waiting time versus server utilization for three separate values of C_A^2. The effect of increased variance in the inter-arrival time is apparent, and is very marked at high utilizations.

An alternative way of showing the combined effects of server utilization and inter-arrival time variance is shown in Fig. 14.3, which can be compared with Figure 9.3 for M/G/1.

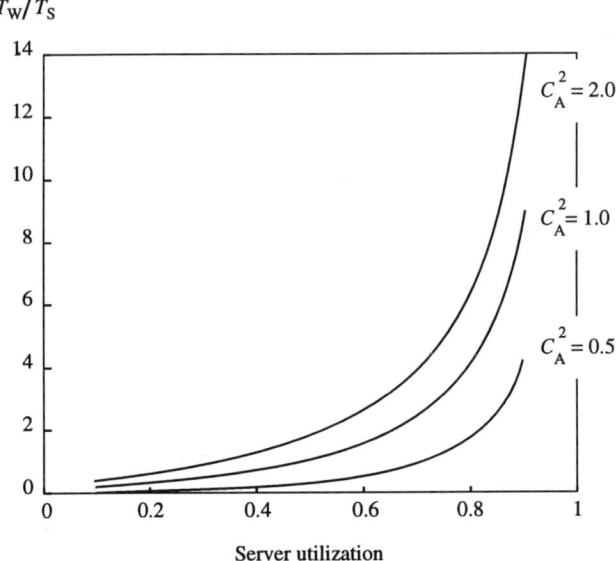

Figure 14.2. Average waiting time versus server utilization for different inter-arrival time variances.

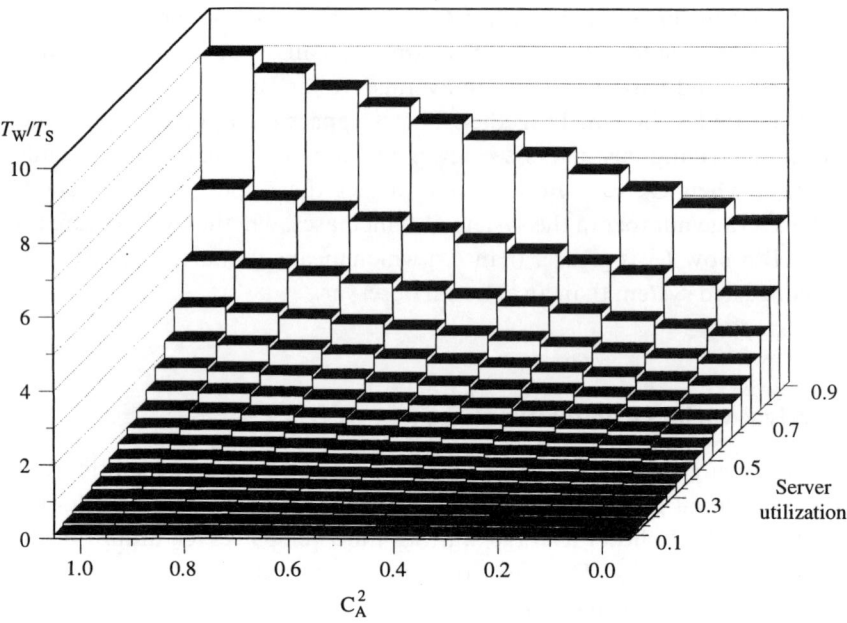

Figure 14.3. G/M/1: combined effect of server utilization and inter-arrival time variance.

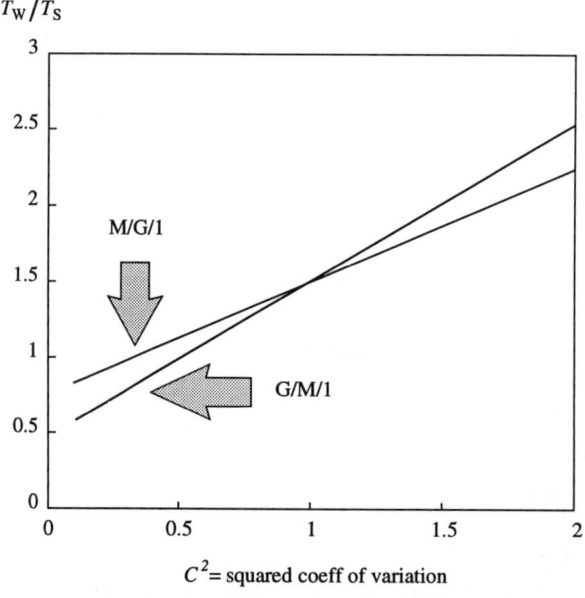

Figure 14.4. Comparison of average waiting time for G/M/1 and M/G/1.

Effect of arrival variance compared with service variance

Both the M/G/1 and G/M/1 models show that greater variability in a system results in longer average waiting times. The M/G/1 model demonstrates this for service-times, while the G/M/1 model demonstrates this for inter-arrival times. In this section we ask which is worse for waiting time, service-time variance or inter-arrival time variance? In Chapter 15 the reader will see that, unfortunately, there is no exact analysis available for G/G/1, so we shall answer our question by comparing M/Gamma/1 with Gamma/G/1. Figure 14.4 (opposite) plots average waiting time against squared coefficient of variation. The M/G/1 line is plotted against C_S^2, and the G/M/1 line against C_A^2. The similarity between the two curves is more significant than the differences.

Programs for G/M/1

Prerequisite routines

Apart from the type definitions, subroutines dealing with the hyperexponential distribution are used, so that the set of 'include' statements required to make use of the routines from this chapter is as follows.

```
{$I \qthprog\qthtypes.pas    }
{$I \qthprog\hypexdst.pas    }
{$I \qthprog\gm1theta.pas    }
{$I \qthprog\gm1calc.pas     }
{$I \qthprog\gm1auxy.pas     }
{$I \qthprog\gm1qsze.pas     }
```

GM1Theta—calculate θ for G/M/1

This routine provides the numerical method for solving Eq. (14.2). The parameter EQN is a function that calculates the particular equation to be solved for θ. A range of functions is provided, corresponding to the specific cases of G/M/1 discussed earlier in this chapter. The method used is bisection. The allowable range $0 \leqslant \theta < 1$ is progressively divided to narrow down the required value of θ. The accuracy sought by the routine is hard-coded as 10^{-7}, which is good enough for our purposes. The Newton–Raphson method could have been used as the method of solving Eq. (14.2), and has the advantage of converging to the solution faster than bisection. However, Newton–Raphson can produce erroneous results if the initial estimate of θ is very different from the actual solution. Bisection, on the other hand, is very robust, and we usually do not need the fastest possible method.

```
{-------------------------------------------------------}
{> GM1Theta - Calculate THETA for G/M/1               }
{ Inputs:                                             }
{     EQN       function to calculate eqn to solve    }
{     LAMBDA    customer arrival rate                 }
{     CA2       sqd coeff of var inter-arrival times  }
{     TS        mean service time                     }
{ Outputs:                                            }
{     VALID     false if utilisation >= 1             }
{     THETA     parameter for G/M/1 calculations      }
{  Copyright Mike Tanner 1993                         }
{-------------------------------------------------------}
Type GM1eqn=function(X,LAMBDA,CA2,TS:QTHreal):Qthreal;
Procedure GM1Theta(EQN:GM1eqn;LAMBDA,CA2,TS:QTHreal;
             Var THETA:QTHreal;Var VALID:boolean);
Var RHO,LOW,HGH,FLOW,FHGH,F:QTHreal; Label L1;
begin
   RHO:=LAMBDA*TS;
   VALID:=(RHO<1.0);  If not VALID then Exit;
   {-Solve equation for THETA by bisection--------}
   LOW:=0.0;             FLOW:=EQN(LOW,LAMBDA,CA2,TS);
   HGH:=0.9999999;       FHGH:=EQN(HGH,LAMBDA,CA2,TS);
L1:THETA:=(LOW+HGH)/2;   F:=EQN(THETA,LAMBDA,CA2,TS);
   If (HGH-LOW)<0.0000001 then Exit;
   If F<0 then LOW:=THETA
          else If F>0 then HGH:=THETA
                       else Exit;
   Goto L1;
end;
```

GM1D–calculate G/M/1 equation for deterministic

This routine evaluates Eq. (14.4).

```
{-------------------------------------------------------}
{> GM1D  - G/M/1 equation for D/M/1                   }
{ Inputs:                                             }
{     X           value at which eqn to be evaluated  }
{     LAMBDA      customer arrival rate               }
{     CA2         dummy parameter for consistency     }
{     TS          mean service time                   }
{ Returns:        parameter for G/M/1 calculations    }
{  Copyright Mike Tanner 1993                         }
{-------------------------------------------------------}
Function GM1D(X,LAMBDA,CA2,TS:QTHreal):QTHreal;
Var RHO:QTHreal;
begin
   RHO:=LAMBDA*TS;
   GM1D:=X-Exp(-(1-X)/RHO);
end;
```

GM1U—calculate G/M/1 equation for uniform

This routine evaluates Eq. (14.6).

```
{------------------------------------------------}
{> GM1U  - G/M/1 equation for U/M/1              }
{ Inputs:        as for function GM1D            }
{ Returns:       parameter for G/M/1 calculations }
{ Copyright Mike Tanner 1993                     }
{------------------------------------------------}
Function GM1U(X,LAMBDA,CA2,TS:QTHreal):QTHreal;
Var RHO,W:QTHreal;
begin
   RHO:=LAMBDA*TS;
   W:=2*(1-X)/RHO;
   GM1U:=X-(1-Exp(-W))/W;
end;
```

GM1E3–calculate G/M/1 equation for Erlang-3

This routine evaluates Eq. (14.9).

```
{------------------------------------------------}
{> GM1E3 - G/M/1 equation for E3/M/1             }
{ Inputs:        as for function GM1D            }
{ Returns:       parameter for G/M/1 calculations }
{ Copyright Mike Tanner 1993                     }
{------------------------------------------------}
Function GM1E3(X,LAMBDA,CA2,TS:QTHreal):QTHreal;
Var RHO,RHO2,X2:QTHreal;
begin
   X2:=Sqr(X);
   RHO:=LAMBDA*TS;  RHO2:=Sqr(RHO);
   GM1E3:=X*X2-(9*RHO+2)*X2+(1+9*RHO+27*RHO2)*X
          -27*RHO*RHO2;
end;
```

GM1E2–calculate G/M/1 equation for Erlang-2

This routine evaluates Eq. (14.12).

```
{------------------------------------------------}
{> GM1E2 - G/M/1 equation for E2/M/1             }
{ Inputs:        as for function GM1D            }
{ Returns:       parameter for G/M/1 calculations }
{ Copyright Mike Tanner 1993                     }
{------------------------------------------------}
Function GM1E2(X,LAMBDA,CA2,TS:QTHreal):QTHreal;
Var RHO,PHI:QTHreal;
begin
   RHO:=LAMBDA*TS;
   GM1E2:=(X-1)*(Sqr(X)-(4*RHO+1)*X+4*Sqr(RHO));
end;
```

GM1Ga—calculate G/M/1 equation for Gamma

This routine evaluates Eq. (14.16).

```
{--------------------------------------------------}
{> GM1Ga - G/M/1 equation for Ga/M/1               }
{  Inputs:       as for function GM1D              }
{  Returns:      parameter for G/M/1 calculations  }
{  Copyright Mike Tanner 1993                      }
{--------------------------------------------------}
Function GM1Ga(X,LAMBDA,CA2,TS:QTHreal):QTHreal;
Var ALPHA,BETA,W:QTHreal;
begin
   ALPHA:=CA2/LAMBDA;  BETA:=1/CA2;
   W:=1/(1+ALPHA*(1-X)/TS);
   GM1Ga:=X-Exp(BETA*Ln(W));
end;
```

GM1H2—calculate G/M/1 equation for hyperexponential

```
{---------------------------------------------------}
{> GM1H2 - G/M/1 equation for H2/M/1                }
{  Inputs:       as for function GM1Ga             }
{  Returns:      parameter for G/M/1 calculations  }
{  Copyright Mike Tanner 1993                      }
{---------------------------------------------------}
Function GM1H2(X,LAMBDA,CA2,TS:QTHreal):QTHreal;
Var ALPHA,MEAN1,MEAN2,LAMBDA1,LAMBDA2,W:QTHreal;
begin
   HypexFit(1/LAMBDA,CA2,ALPHA,MEAN1,MEAN2);
   LAMBDA1:=1/MEAN1; LAMBDA2:=1/MEAN2;
   W:=(1-X)/TS;
   GM1H2:=X-ALPHA*LAMBDA1/(LAMBDA1+W)
          -(1-ALPHA)*LAMBDA2/(LAMBDA2+W);
end;
```

GM1Calc—general calculations for G/M/1

This routine calculates, for G/M/1, the queueing characteristics usually of
interest.

```
{--------------------------------------------------}
{> GM1Calc - Calculations for G/M/1 queueing model }
{  Inputs:                                         }
{      LAMBDA    customer arrival rate             }
{      THETA     distribution-dependent parameter  }
{      TS        mean service time                 }
{  Outputs:                                        }
{      RHO       server utilisation                }
{   TQ,SDVTQ  mean & stdev of time in system       }
{   TW,SDVTW  mean & stdev of waiting time         }
{   LQ,SDVLQ  mean & stdev of number in system     }
```

```
{  LW,SDVLW   mean & stdev of number waiting          }
{  LA,SDVLA   mean & stdev of number in system at     }
{            arrival instants                         }
{  VALID      true if results valid, false if not     }
{  Copyright Mike Tanner 1993                         }
{-----------------------------------------------------}
Procedure GM1Calc(LAMBDA,TS,THETA:QTHreal;
             Var RHO,TQ,SDVTQ,TW,SDVTW:QTHreal;

             Var LQ,SDVLQ,LW,SDVLW,LA,SDVLA:QTHreal;
             Var VALID:boolean);
begin
   RHO:=LAMBDA*TS;
   VALID:=(RHO<1.0); If not VALID then Exit;
   {-Time in system-------------------------------}
   TQ:=TS/(1-THETA); SDVTQ:=TQ;
   {-Waiting time---------------------------------}
   TW:=THETA*TS/(1-THETA);
   SDVTW:=Sqrt(THETA*(2-THETA))*TS/(1-THETA);
   {-Number in system-----------------------------}
   LQ:=RHO/(1-THETA);
   SDVLQ:=Sqrt(RHO*(1+THETA-RHO))/(1-THETA);
   {-Number waiting-------------------------------}
   LW:=THETA*RHO/(1-THETA);
   SDVLW:=Sqrt(RHO*THETA*(1+THETA*(1-RHO)))/(1-THETA);
   {-Number in system at arrival instants---------}
   LA:=THETA/(1-THETA); SDVLA:=Sqrt(THETA)/(1-THETA);
end;
```

GM1QtmeCdf—cdf of queueing time (time-in-system) for G/M/1

This routine calculates the cdf of the time-in-system, using Eq. (14.22).

```
{-----------------------------------------------------}
{> GM1QtmeCdf  -- Cdf of time in system for G/M/1  }
{  Inputs: T       time for which cdf required     }
{          THETA   arrival distribution parameter  }
{          TS      mean service time               }
{  Copyright Mike Tanner 1993                      }
{-----------------------------------------------------}
Function GM1QtmeCdf(T,THETA,TS:QTHreal):QTHreal;
Var TQ:QTHreal;
begin
   TQ:=TS/(1-THETA);
   GM1QtmeCdf:=1-Exp(-T/TQ);
end;
```

GM1WaitCdf—cdf of waiting time for G/M/1

This routine calculates the cdf of waiting time, using Eq. (14.28).

```
{-----------------------------------------------------------}
{> GM1WaitCdf   -- Cdf of waiting time for G/M/1          }
{   Inputs: T           time for which cdf required         }
{           THETA       arrival distribution parameter     }
{           TS          mean service time                  }
{   Copyright Mike Tanner 1993                             }
{-----------------------------------------------------------}
Function GM1WaitCdf(T,THETA,TS:QTHreal):QTHreal;
Var TQ:QTHreal;
begin
    TQ:=TS/(1-THETA);
    GM1WaitCdf:=1-THETA*Exp(-T/TQ);
end;
```

GM1QtmePct—percentiles of queueing time (time-in-system) for G/M/1

```
{-----------------------------------------------------------}
{> GM1QtmePct - Pctile of time in system for G/M/1 }
{   Inputs: PCT         percentile required              }
{           THETA       arrival distribution parameter    }
{           TS          mean service time                 }
{   Copyright Mike Tanner 1993                            }
{-----------------------------------------------------------}
Function GM1QtmePct(PCT,THETA,TS:QTHreal):QTHreal;
Var TQ:QTHreal;
begin
    TQ:=TS/(1-THETA);
    GM1QtmePct:=Ln(100/(100-PCT))*TQ;
end;
```

GM1WaitPct—percentiles of waiting time for G/M/1

This routine calculates the percentiles of waiting time, using Eq. (14.30)

```
{-----------------------------------------------------------}
{> GM1WaitPct -- Percentile of wait time for G/M/1 }
{   Inputs: PCT         percentile required              }
{           THETA       arrival distribution parameter    }
{           TS          mean service time                 }
{   Copyright Mike Tanner 1993                            }
{-----------------------------------------------------------}
Function GM1WaitPct(PCT,THETA,TS:QTHreal):QTHreal;
Var TQ:QTHreal;
begin
    TQ:=TS/(1-THETA);
    GM1WaitPct:=Ln(100*THETA/(100-PCT))*TQ;
end;
```

GM1QszePdf/GM1QszeCdf—pdf/cdf of number-in-system for G/M/1

This routine calculates the pdf of the number of customers in the system, using Eqs (14.33) and (14.34).

```
{-----------------------------------------------------}
{> GM1QszePdf  -- Pdf of queue size for G/M/1         }
{> GM1QszeCdf  -- Cdf of queue size for G/M/1         }
{  Inputs:                                            }
{      K            no. for which pdf/cdf required    }
{      LAMBDA       customer arrival rate             }
{      THETA        arrival distribution parameter    }
{      TS           mean service time                 }
{  Copyright Mike Tanner 1993                         }
{-----------------------------------------------------}
Function GM1QszePdf(K:integer;
            LAMBDA,THETA,TS:QTHreal):QTHreal;
Var RHO,RK:QTHreal;
begin
   RHO:=LAMBDA*TS;
   If RHO>=1.0 then QTHError('GM1QszePdf error');
   If K=0 then GM1QszePdf:=1-RHO
          else begin
                   RK:=K;
                   GM1QszePdf:=RHO*(1-THETA)
                           *Exp((RK-1)*Ln(THETA));
               end;
end;
{------------------------}
Function GM1QszeCdf(K:integer;
            LAMBDA,THETA,TS:QTHreal):QTHreal;
Var RHO,RK:QTHreal;
begin
   RHO:=LAMBDA*TS;
   If RHO>=1.0 then QTHError('GM1QszeCdf error');
   RK:=K;
   GM1QszeCdf:=1.0-RHO*Exp(RK*Ln(THETA));
end;
```

GM1NarvPdf/GM1NarvCdf—pdf/cdf of number-in-system on arrival for G/M/1

These routines calculate the pdf and cdf of the numbers in the system at instants when customers arrive, using Eq. 14.39.

```
{-----------------------------------------------------}
{> GM1NarvPdf  -- Pdf of no. in system on arrival     }
{> GM1NarvCdf  -- Cdf of no. in system on arrival     }
{  Inputs:                                            }
{      K            no. for which pdf/cdf required    }
{      THETA        arrival distribution parameter    }
{  Copyright Mike Tanner 1993                         }
{-----------------------------------------------------}
Function GM1NarvPdf(K:integer;THETA:QTHreal):QTHreal;
Var RK:QTHreal;
```

```
begin
   RK:=K;
   GM1NarvPdf:=(1-THETA)*Exp(RK*Ln(THETA));
\end;
{--------------------}
Function GM1NarvCdf(K:integer;THETA:QTHreal):QTHreal;
Var RK:QTHreal;
begin
   RK:=K;
   GM1NarvCdf:=1-Exp((RK+1)*Ln(THETA));
end;
```

Example 14A—The airport tourist desk

The city airport has a tourist information desk, staffed by a single person. A proportion of arriving passengers decide to make enquiries about places to visit. The time taken to help each enquirer is on average 5 min, exponentially distributed. Customers arrive on average once every 10 min, but the nature of the arrival pattern is unknown. Assess the difference in waiting times for the following cases:

(i)   Erlang-2 inter-arrival times
(ii) random arrivals
(iii) hyperexponential inter-arrival times with $C_A^2 = 1.5$.

Preliminary statements and calculations are

$$\lambda = \text{arrival rate} = \frac{1}{10} = 0.1 \text{ customers/min}$$

$$T_S = \text{average service-time} = 5 \text{ min}$$

$$\rho = \lambda T_S = 0.5$$

Case (i): Erlang-2

From Table 14.1 we have $\theta = 0.381966$. Using Eq. (14.27) we get

$$T_W = \text{average wait-time} = 3.09 \text{ min}$$

Case (ii): random arrivals

For this we use the M/M/1 model, and Eq. (7.8) gives

$$T_W = \text{average wait-time} = 5 \text{ min}$$

Case (iii): hyperexponential

From Table 14.1 we have $\theta = 0.711325$. Using Eq. (14.27) we get

$$T_W - \text{average wait-time} = 6.18 \text{ min}$$

The above results show that the average waiting time can vary a lot according to the pattern of arrivals. A study of the arrival pattern would help the airport management determine whether an additional server should be considered.

We could investigate the waiting time further by looking at percentiles. The formulae for percentiles are a little harder to work with, but using the supplied subroutines we get the results in Table 14.2.

Table 14.2. Waiting-time percentiles for Example 14A

	Waiting time (min)		
	Average	90th percentile	95th percentile
Erlang-2	3.09	10.80	16.40
Random arrivals	5	16.10	23.00
Hyperexponential	6.18	19.10	26.90

15
General arrivals and general service (G/G/1)

Introduction

For the M/M/1 model we made very precise assumptions about the arrival pattern, which was random, and the service-times, which were exponentially distributed. In consequence, we were able to describe in great detail the characteristics of waiting time, queueing time, waiting-line size, number-in-system, and busy period. When we allowed service-times to have a general distribution in M/G/1, the results available were not quite so detailed. Similarly with G/M/1, where arrivals could be non-random, many useful results can be obtained, but not the level of detail available for M/M/1. Now for G/G/1 we are going to allow both arrivals and service-times to depart from the 'pure' case of M/M/1 at the same time. The price we pay for this generality is that almost no precise statements can be made about the behaviour of the system. We have to resort to approximations and bounds. In practice we might decide to use simulation rather than mathematical analysis.

Assumptions and initial calculations for G/G/1

There is a single server with a FIFO queue. The population of potential customers is infinite. Customers arrive at an average rate λ, so that the average inter-arrival time is $T_A = 1/\lambda$. The distribution of inter-arrival time is general, with variance $\sigma_{T_A}^2$, and coeff of variation squared for inter-arrival time is C_A^2. Service-times also have a general distribution, with average value T_S and coeff of variation squared of C_S^2.

Comparisons with other models

G/G/1 allows the coefficients of variation of service-times and inter-arrival times to take on any non-zero value. We have seen in the chapters on M/G/1 and G/M/1 that waiting time increases as either of these coefficients increases. Using this

156

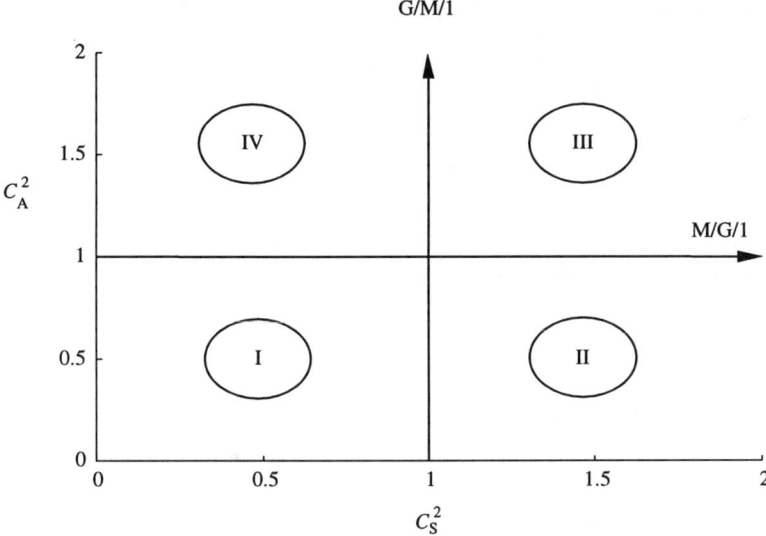

Figure 15.1. Using M/G/1 and G/M/1 to provide bounds on G/G/1.

observation we can make use of the M/G/1 and G/M/1 models to set some bounds on the average waiting time for G/G/1.

In Fig. 15.1 the range of queueing situations covered by G/G/1 is split into four regions, according to whether C_A^2 and C_S^2 are less than or greater than 1. In each of these four regions a different set of comparisons can be made with M/G/1 and G/M/1. In region I waiting time will be less than for M/G/1, since service-times are more regular than exponential. Waiting times will also be less than G/M/1 would predict, since arrivals are more regular than random. We can therefore take M/G/1 and G/M/1 results for waiting time as upper bounds on G/G/1 waiting times in this region. In region II waiting time will again be less than for M/G/1, but now G/M/1 provides a lower bound instead of an upper bound, since service-times are more varied than an exponential distribution. In region III both M/G/1 and G/M/1 provide lower bounds on waiting time. For an upper bound the approximations discussed later in this chapter can be used. In region IV G/M/1 is an upper bound, and now M/G/1 becomes a lower bound. These bounds are summarized in Table 15.1.

Table 15.1. Bounds provided by M/G/1 and G/M/1 on G/G/1 performance

Region	M/G/1	G/M/1
I	Upper bound	Upper bound
II	Upper bound	Lower bound
III	Lower bound	Lower bound
IV	Lower bound	Upper bound

Heavy-traffic approximation for waiting time (HTA)

Under heavy load, waiting time has an approximately exponential distribution, with mean waiting time given by

$$T_{\text{W}} = \text{average waiting time} \approx \frac{\lambda(\sigma_{T_A}^2 + \sigma_{T_S}^2)}{2(1 - \rho)} \tag{15.1}$$

This approximation improves as ρ gets closer and closer to 1. For low utilizations the approximation is extremely poor, as will be seen in the comparisons later.

Marchal's modified upper bound for waiting time (MAR)

Marchal proposed a modification to the heavy-traffic approximation, by scaling it down so that it becomes an exact fit to M/G/1. This gives the approximation:

$$T_{\text{W}} = \text{average waiting time} \approx \frac{(1 + C_S^2)}{(1/\rho^2) + C_S^2} \frac{\lambda(\sigma_{T_A}^2 + \sigma_{T_S}^2)}{2(1 - \rho)} \tag{15.2}$$

Numerical studies have shown that Eq. (15.2) is a good fit to G/M/1. The fit to G/G/1 is fair, getting worse as C_S^2 or C_A^2 increases, but improving as ρ increases.

Lower bound on waiting time (LBD)

The following lower bound on average waiting time, due to Marchal, seems to be the best available that is straightforward to calculate.

$$T_{\text{W}} = \text{average waiting time} \geq \frac{\rho^2 C_S^2 + \rho(\rho - 2)}{2\lambda(1 - \rho)} \tag{15.3}$$

Better lower bounds exist, but they are too complex to calculate to be useful in practice. More detail about lower bounds can be found in Kleinrock (1976).

Allen–Cunneen approximation for waiting time (ACA)

Allen and Cuneen devised the Eq. (15.4) as an approximation for average waiting time. The formula given here is actually a special case of the Allen–Cunneen formula for G/G/m, which is given separately in Chapter 21.

$$T_{\text{W}} = \text{average waiting time} \approx \frac{\rho T_S}{1 - \rho} \left(\frac{C_A^2 + C_S^2}{2} \right) \tag{15.4}$$

Comparison of bounds and estimates for waiting time

To get some idea of how good the various approximations and bounds are, we can compare them with M/G/1 and G/M/1. Starting with M/G/1, with $C_S^2 = 3$, we see

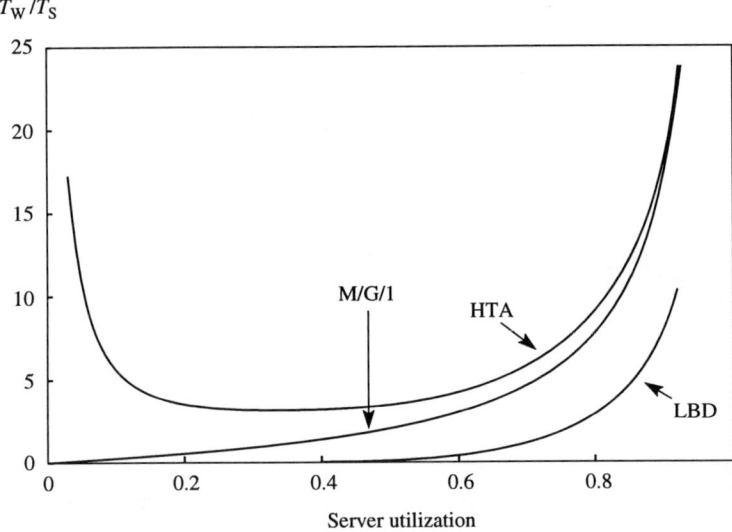

T_W/T_S

Server utilization

Figure 15.2. Comparison of G/G/1 bounds with M/G/1.

in Fig. 15.2 that HTA is very good as ρ gets near 1. LBD is a very weak lower bound. ACA is in fact exact for M/G/1, so it is not shown.

We look now at Fig. 15.3, which shows the comparisons with G/M/1. In this case Ga/M/1 was used, with $C_A^2 = 3$. We can see that ACA is very good. The HTA is still good for high utilizations, although not as good as in the M/G/1 comparison.

Next, in Fig. 15.4, we compare the bounds available for G/G/1 with M/G/1 as we increase the variance of service-time, or C_S^2. The comparison was done with

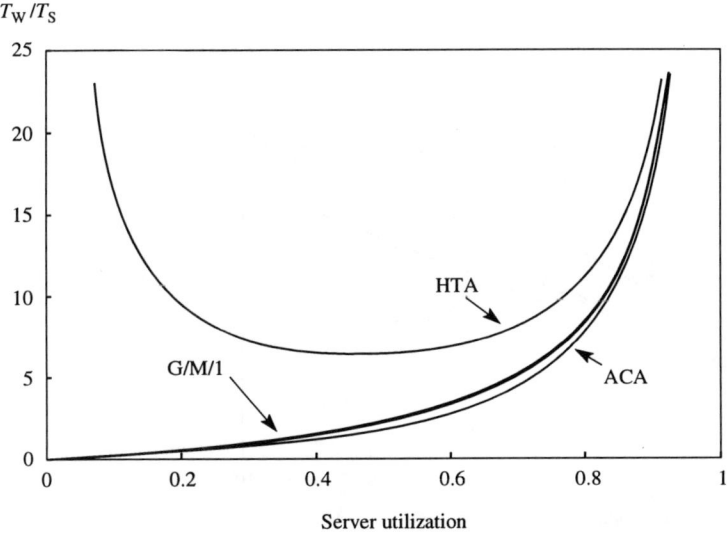

T_W/T_S

Server utilization

Figure 15.3. Comparison of G/G/1 bounds and estimates with G/M/1.

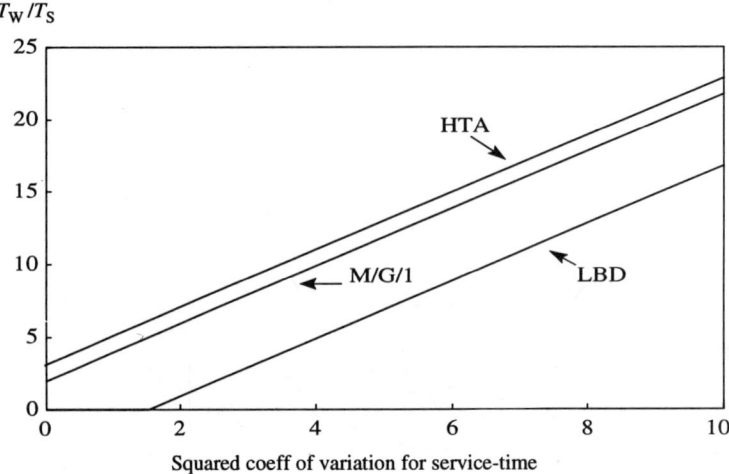

Figure 15.4. G/G/1 bounds compared with M/G/1 as service-time variation increases.

$\lambda = 0.8$, $T_S = 1$. The absolute error in the bounds stays the same while T_W grows. The relative error is therefore getting less as service-time variance gets bigger. Note that Marchal's estimate and the Allen–Cunneen estimate are exact for M/G/1, so they are not shown.

Our final comparison is illustrated by Fig. 15.5, where the G/G/1 bounds are compared with G/M/1 as inter-arrival time variance increases. The comparison was done for $\lambda = 0.8$, $T_S = 1$.

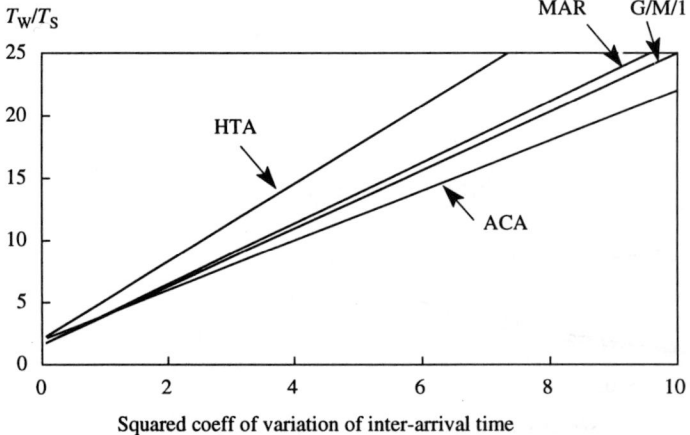

Figure 15.5. G/G/1 bounds compared with G/M/1 as inter-arrival time variance increases.

Other characteristics

The bounds and estimates given so far have all been for T_W, the average waiting time. Other characteristics can be estimated using basic queueing-theory relationships and Little's law. These give us

$$T_Q = \text{average time-in-system} = T_W + T_S \tag{15.5}$$

$$L_W = \text{average number waiting} = \lambda T_W \tag{15.6}$$

$$L_Q = \text{average number-in-system} = \lambda T_Q \tag{15.7}$$

Programs for G/G/1

In order to use the routines for G/G/1, a program must contain the following statements.

```
{$N+,E+ }
{$I \qthprog\QTHTypes.pas }
{$I \qthprog\QTHError.pas }
{$I \qthprog\GG1Calc.pas  }
```

GG1Calc—general calculations for G/G/1

This routine performs the basic calculations for G/G/1. Each of the approximations and bounds for the average wait is provided. The Allen–Cunneen approximation is used as a basis for calculating averages for time-in-system, number waiting, and number-in-system.

```
{--------------------------------------------------------}
{> GG1Calc - Calculations for G/G/1 queueing model }
{ Inputs:                                                }
{      LAMBDA   customer arrival rate                    }
{      CA2      sqd coeff of var inter-arrival times }
{      TS       mean service time                        }
{      COEFF2   sqd coeff of var service times           }
{ Outputs:                                               }
{      RHO      server utilisation                       }
{      TWHTA    average wait, heavy traffic appx         }
{      TWACA    average wait, Allen-Cunneen appx         }
{      TWMAR    average wait, Marchals appx              }
{      TWLBD    average wait, lower bound                }
{      TQ       average time in system                   }
{      LW       average number waiting                   }
{      LQ       average number in system                 }
{      VALID    T=results valid, F=not valid             }
{  Copyright Mike Tanner 1993                            }
{--------------------------------------------------------}
Procedure GG1Calc(LAMBDA,CA2,TS,COEFF2:QTHreal;
                  Var RHO,TWHTA,TWACA:QTHreal;
                  Var TWMAR,TWLBD:QTHreal;
```

```
                  Var TQ,LW,LQ:QTHreal;
                  Var VALID:boolean);
Var VTA,VTS:QTHreal;
begin
   RHO:=LAMBDA*TS;
   VALID:=(RHO<1.0);   If not VALID then Exit;
   {-Variance of inter-arrival and service times---}
   VTA:=CA2/Sqr(LAMBDA);
   VTS:=COEFF2*Sqr(TS);
   {-Heavy-traffic approximation for average wait--}
   TWHTA:=LAMBDA*(VTA+VTS)/(2*(1-RHO));
   {-Marchals approximation for average wait------}
   TWMAR:=(1+COEFF2)/(1/Sqr(RHO)+COEFF2);
   TWMAR:=TWMAR*TWHTA;
   {-Lower bound on waiting time------------------}
   TWLBD:=Sqr(RHO)*COEFF2+RHO*(RHO-2);
   TWLBD:=TWLBD/(2*LAMBDA*(1-RHO));
   If TWLBD<0 then TWLBD:=0;
   {-Allen-Cunneen approximation for average wait--}
   TWACA:=RHO*TS*(CA2+COEFF2)/(2*(1-RHO));
   {-Use the Allen-Cunneen approx for other stats--}
   TQ:=TWACA+TS;
   LW:=LAMBDA*TWACA;
   LQ:=LW+RHO;
end;
```

Example 15A—The Kayos Confectionery Kiosk

Let us revisit the kiosk of Example 12A, where we saw the effect of changing the variability of service-time. Taking Example 12A, calculate the average waiting time when the squared coeff of variation of inter-arrival times has the values 1.7, 1, and 0.33. Do this by using the Allen–Cunneen approximation in Eq. (15.4). Table 15.2 gives the results.

Table 15.2. Average waiting time for Example 15A

C_A^2	Average waiting time		
	$C_S^2 = 1.7$	$C_S^2 = 1$	$C_S^2 = 0.33$
1.70	7.65	6.08	4.57
1.00	6.08	4.50	2.99
0.33	4.57	2.99	1.49

Part Five
Multiple servers

16
Unlimited number of servers
(M/M/infinity)

Introduction

For this queueing system we assume that a server can be made available immediately for every arriving customer, regardless of how many customers are already being served when the new customer arrives. This idea is illustrated in Fig. 16.1. No customers ever have to wait, so the only characteristic we are interested in is the number of customers who are present.

Although it seems unrealistic to think of an unlimited number of servers being available, this simple system is an easy introduction to multiple-server systems, and is useful in practice as a first approximation to systems where the number of servers is large but not unlimited. We shall also use this queueing model to illustrate a 'scaling effect' for the number of servers, rather than the power of single servers.

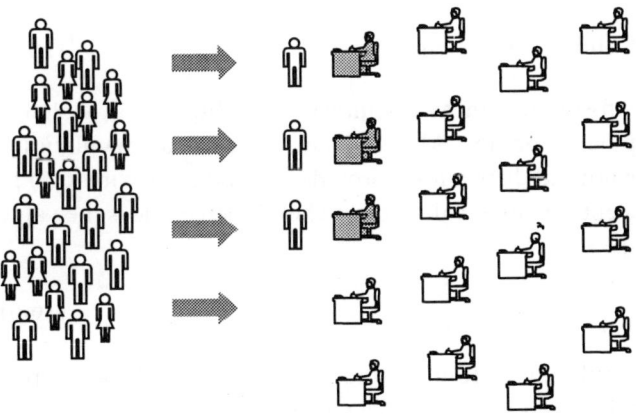

Figure 16.1. Queue with limited number of servers.

Traffic intensity

With multiserver systems we must distinguish between server utilization and traffic intensity. Traffic intensity is defined as

$$u = \text{traffic intensity} = \lambda T_{\text{S}} \qquad (16.1)$$

Readers will note that this is the definition of server utilization for a single-server system. With multiple servers, traffic intensity can exceed 1. If the number of servers is m, then server utilization and traffic intensity are related by

$$\rho = \text{server utilization} = \frac{\text{traffic intensity}}{\text{number of servers}} = \frac{u}{m} = \frac{\lambda T_{\text{S}}}{m} \qquad (16.2)$$

With an unlimited number of servers, m is not constant. Every time a new customer arrives a server is added, and every time a customer finishes service a server is removed. So server utilization is not really meaningful, but in later chapters we shall be dealing with a fixed number of servers and so the above relationship will be important.

Number-in-system

The number of customers in the system is the key characteristic of this queueing model. The number-in-system has a Poisson distribution, so we can fairly easily calculate mean, variance, and percentiles. The probability of a specific number of customers being in the system is given by

$$p_k = \text{probability}(k \text{ customers in system}) = \frac{u^k}{k!}\, e^{-u} \qquad (16.3)$$

The mean and variance are given by

$$L_{\text{Q}} = \text{average number of customers in system} = u \qquad (16.4)$$

$$\sigma^2_{L_{\text{Q}}} = \text{variance of number-in-system} = u \qquad (16.5)$$

Percentiles of the number-in-system can be calculated using Eq. (16.3), which is straightforward in a computer program. Alternatively, for manual calculations, we can use the fact that the normal distribution provides a good approximation to the Poisson distribution except for small values of u. The rth percentile is the value of n that fits the relationship

$$\text{Prob}(k \le n) = \frac{r}{100} \qquad (16.6)$$

The value n of the rth percentile can be found using Eq. (16.7), with the appropriate value of b selected from Table 16.1. So to find a percentile, decide the percentile (value of r) wanted, look up b in Table 16.1, then calculate the percentile using

Table 16.1. Values for estimating percentiles

r	b
80.00	0.80
90.00	1.30
95.00	1.70
99.00	2.40
99.90	3.20

Eq. (16.7). Since we are talking about the number of customers in the system, a percentile should be an integral value, so the result from Eq. (16.7) should be rounded off.

$$r\text{th percentile} = n(r) = u + b\sqrt{u} \tag{16.7}$$

The mean number-in-system, together with the 90th and 95th percentiles, are shown in Figs 16.2 and 16.3. Separate graphs are given for low and high traffic intensities because of the range of values involved. Readers should note that for high traffic intensities the percentiles are closer to the average than for low traffic intensities. This is another scaling effect, and is more clearly seen in Fig. 16.4, where the ratio of percentile to mean is plotted against increasing traffic intensity. The ratio can be seen to decrease as traffic intensity gets larger. If we were planning a service facility, we might use the M/M/infinity model and provide the number of servers indicated by, say, the 95th percentile. If so, the higher the traffic intensity being handled, the fewer servers per unit of traffic needed, and the more efficient our facility would be.

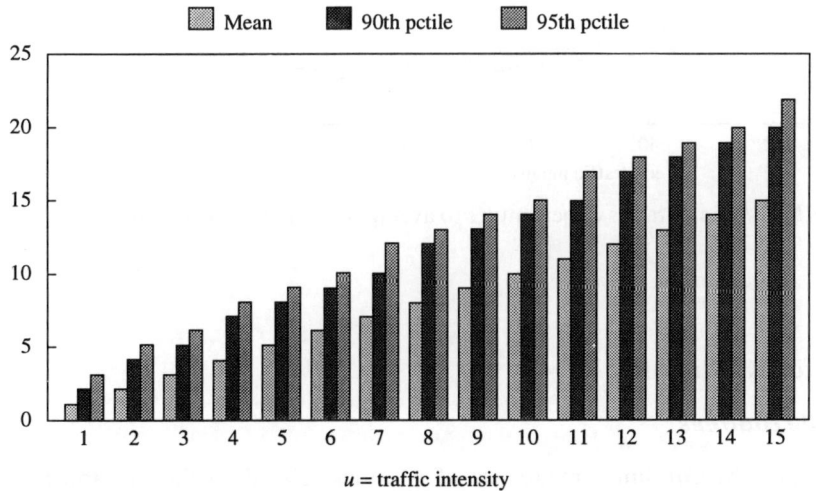

Figure 16.2. M/G/infinity mean and percentiles of number-in-system for low traffic intensities.

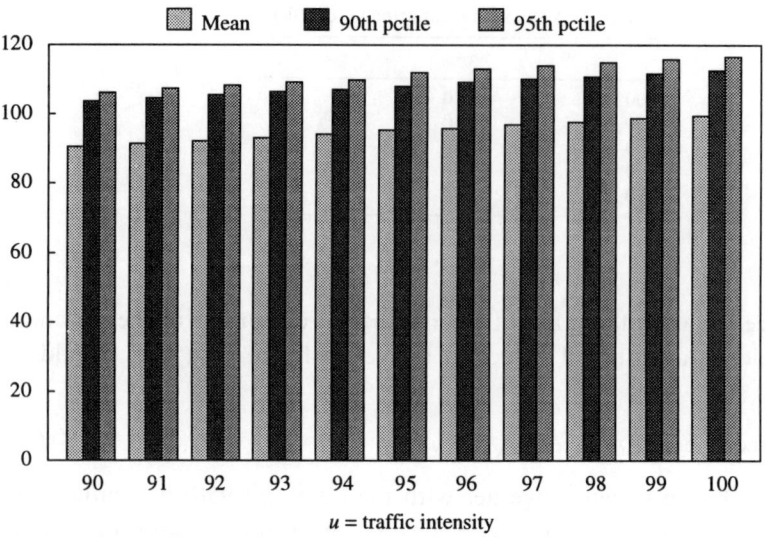

Figure 16.3. M/G/infinity mean and percentiles of number-in-system for high traffic intensities.

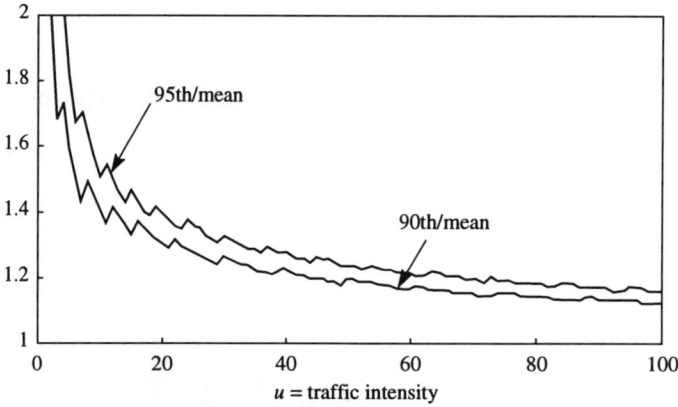

Figure 16.4. M/G/infinity ratios of percentiles to averages of number-in-system.

Programs for M/G/infinity

Prerequisite routines

Calculations for M/G/infinity make use of routines related to the Poisson and normal distributions, so the following set of 'include file' statements would be used in Turbo-Pascal.

```
{ $N+,E+ }
{ $I \qthprog\qthtypes.pas  }
{ $I \qthprog\qthfile.pas   }
{ $I \qthprog\qtherror.pas  }
{ $I \qthprog\normal.pas    }
{ $I \qthprog\poisson.pas   }
{ $I \qthprog\mg0calc.pas   }
```

MGOCalc—general calculations for M/G/infinity

This is a very simple routine, and the calculations could easily be incorporated directly into the calling program. However, a distinct routine is provided for consistency with other queueing models.

```
{-----------------------------------------------------}
{> MGOCalc - Calculations for M/G/infinity            }
{ Inputs:                                             }
{     LAMBDA      customer arrival rate               }
{     TS          mean service time                   }
{ Outputs:                                            }
{     U           traffic intensity                   }
{     LQ          average number in system            }
{     SDVLQ       std deviation of number in system   }
{   Copyright Mike Tanner 1993                        }
{-----------------------------------------------------}
Procedure MGOCalc(LAMBDA,TS:QTHreal;
                  Var U,LQ,SDVLQ:QTHreal);
begin
   U:=LAMBDA*TS;
   LQ:=U;
   SDVLQ:=Sqrt(U);
end;
```

Using Poisson distribution routines for M/G/infinity

Noting that the number of customers in the system has a Poisson distribution, we can use the following routines.

Prob(k customers in system) = PoissonPdf(k, u)

Prob($\leqslant k$ customers in system) = PoissonCdf(k, u)

rth percentile of number-in-system = PoissonPct(r, u)

Example 16A—The telephone exchange

The telephone exchange for a large office handles calls that arrive randomly at an average rate of 80 per hour. (This is the sum of both outgoing and incoming calls.) Calls last on average 4 min, and a large number of telephone lines has been provided. Estimate the average number of lines in use, and the 90th and 95th percentiles.

First we have to convert the arrival rate and call duration to the same basis, i.e.

$$\text{arrival rate} = \lambda = 80 \text{ calls/hour} = \frac{80}{3600} \text{ calls/sec} = 0.0222 \text{ calls/sec}$$

$$\text{average call duration} = T_S = 4 \text{ min} = 240 \text{ secs}$$

The next step is to calculate the traffic intensity, which is also the average number of calls in progress, i.e. the average number of lines in use:

$$\text{traffic intensity} = u = \lambda T_S = (0.0222 \text{ calls/sec}) \times (240 \text{ secs}) = 5.333 \text{ calls}$$

In order to get the percentiles of the number of lines in use we shall use the Normal approximation, so that

$$\text{90th percentile} \approx u + 1.3\sqrt{u} = 5.333 + 1.3\sqrt{5.333} = 8.34$$

$$\text{95th percentile} \approx u + 1.7\sqrt{u} = 5.333 + 1.7\sqrt{5.333} = 9.26$$

The percentiles must be integral values. We get fractional values because we are using the normal approximation, which means we are using a continuous distribution to approximate a discrete distribution. The 90th and 95th percentile values therefore must be rounded to get 8 and 9 respectively. Using the subroutine provided, which for small traffic intensities calculates the percentiles directly, we get the results 8 and 9 for the 90th and 95th percentiles.

Example 16B—The construction site

In a busy city centre there is a construction site. People walk past the site at an average of 100 per minute. One in ten people decide to stop and observe the work on the site. Each person who decides to stop spends on average 3 min watching. What is the average number of people watching? For what proportion of the time will there be 35 or more people watching? For what proportion of time will there be 40 or more people watching?

As usual, we shall start by stating the parameters of our system clearly, making sure we are using the same time units throughout. A 'customer' is a passer-by who stops to watch the site, and the 'service-time' is the length of time he or she stay watching the site.

$$\text{average arrival time} = \lambda = \frac{100}{10} = 10 \text{ customers/min}$$

$$\text{average service-time} = T_S = 3 \text{ min}$$

Next we calculate traffic intensity, which is also the average number watching.

$$\text{traffic intensity} = \lambda T_S = (10 \text{ customers/min}) \times (3 \text{ min}) = 30 \text{ customers}$$

So the answer to the first part of the problem is that there are on average 30 people watching the work on the construction site. To find out what proportion of time there are more than a specified number of people watching, we can use the Poisson Cdf function.

$$\text{Prob}(\geq 35 \text{ people watching}) = 1 - \text{Prob}(\leq 34 \text{ people watching})$$
$$= \text{Poisson Cdf}(34, u) = 0.20$$

$$\text{Prob}(\geq 40 \text{ people watching}) = 1 - \text{Prob}(\leq 39 \text{ people watching})$$
$$= \text{Poisson Cdf}(39, u) = 0.046$$

17
Multiserver with queueing (M/M/m)

Introduction

The M/M/m model has wide practical use. The situation of several servers dealing with customers who form a single queue is one we all encounter many times in banks, post offices, and similar places. When we ring up to make a travel reservation, or to book theatre or sports tickets, we know there are a number of agents taking calls, and a recording often informs us that our call is in a queue that will be served in arrival order (Fig. 17.1). Multiple-server situations are also common in computer systems and telecommunciation systems.

Mathematically, the assumptions of random arrivals and exponential service make it possible to derive useful results. These results are not as easy to use as the M/M/1 results, and either tables or a simple program are needed in practice.

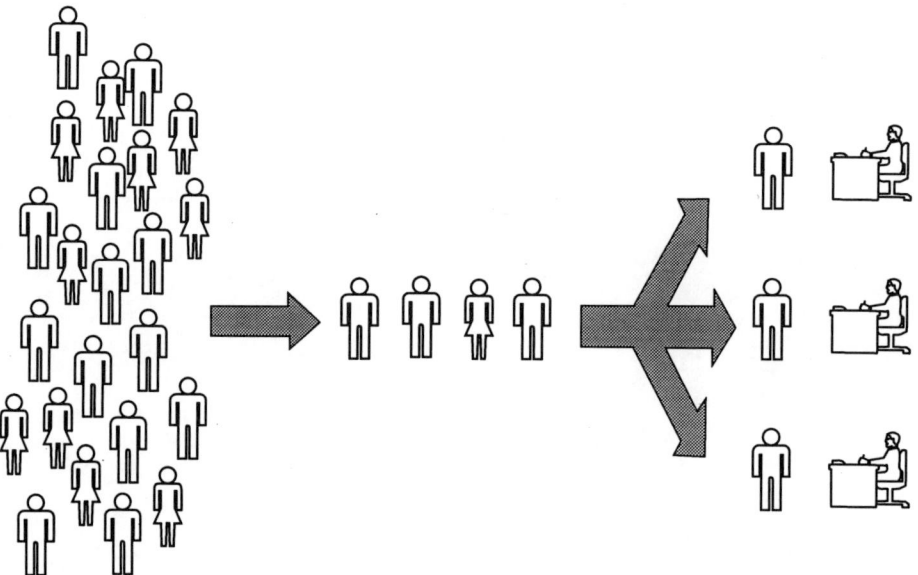

Figure 17.1. Multiple servers with a single queue.

Assumptions for the M/M/m model

The M/M/m model is a generalization of the M/M/1 model from a single server to multiple servers. The same assumptions about random arrivals, exponential service times, and an infinite population of potential customers apply. Customers form a single queue and are served in first-come-first-served order.

Parameters and initial calculations for M/M/m

The parameters that need to specified for M/M/m are

λ, the average customer arrival rate, e.g. customers per minute
T_S, the average service time for a customer, e.g. 2 min
m, the number of servers

The first things to calculate are the traffic intensity and server utilization. These are easily obtained from

$$\text{traffic intensity} = u = \lambda T_S \tag{17.1}$$

$$\text{server utilization} = \rho = \frac{u}{m} \tag{17.2}$$

For the system to be stable, the servers must be able to deal with the average arrival rate of customers. This means that the traffic intensity must be less than the number of servers, or equivalently the server utilization must be less than 1, i.e.

$$u < m \text{ and } \rho < 1 \tag{17.3}$$

Many of the useful results for M/M/m involve the Erlang-C function $E_C(m,u)$, which is defined as

$$E_C(m, u) = \frac{1 - R(m, u)}{1 - \rho R(m, u)} \tag{17.4}$$

where $R(m, u)$ is the Poisson ratio function, and is described in Appendix 2. This is in fact the probability that all the servers are busy, and therefore also the probability that an arriving customer will have to wait. The Erlang-C formula plays an important role in the performance of telephone systems. Most texts define the Erlang-C formula directly without using the Poisson ratio function, i.e.

$$E_C(m, u) = \frac{\frac{u^m}{m!}}{\frac{u^m}{m!} + (1 - \rho) \sum_{k=0}^{m-1} \frac{u^k}{k!}} \tag{17.5}$$

In Chapter 29 we shall encounter the Erlang-B function, which is also of great importance in telephone systems, and is defined as $E_B(m,u) = 1 - R(m, u)$. The Erlang-B function is mentioned here so that the relationship between the Erlang-B and Erlang-C functions can be given, i.e.

$$E_C(m, u) = \frac{E_B(m, u)}{1 - \rho[1 - E_B(m, u)]} \qquad (17.6)$$

Waiting time

A suitable characteristic to look at first is the average waiting time. Once $E_C(m,u)$ has been calculated, average waiting time is easy to calculate as

$$T_W = \text{average wait-time} = \frac{E_C(m, u)\,T_S}{m(1 - \rho)} \qquad (17.7)$$

The shape of the curve for average wait is illustrated in Fig. 17.2 for a range of values of m, the number of servers. Again we see the rapid deterioration in service as the system approaches 100 per cent load that is characteristic of many queueing systems. The graph uses server utilization as the horizontal axis, so that a particular value of server utilization implies a different arrival rate for each value of m. This allows the different shapes of the curves for different values of m to be seen. As the number of servers increases, deterioration in waiting time happens at higher utilizations, but the speed of deterioration is greater. As the arrow indicates, increasing the number of servers pushes the curve further into the bottom right-hand corner of the chart.

The actual distribution of waiting time is a modified exponential, with the cdf

$$\text{Probability(waiting time} < t) = W(t) = 1 - E_C(m, u)\mathrm{e}^{-(m-u)\frac{t}{T_S}} \qquad (17.8)$$

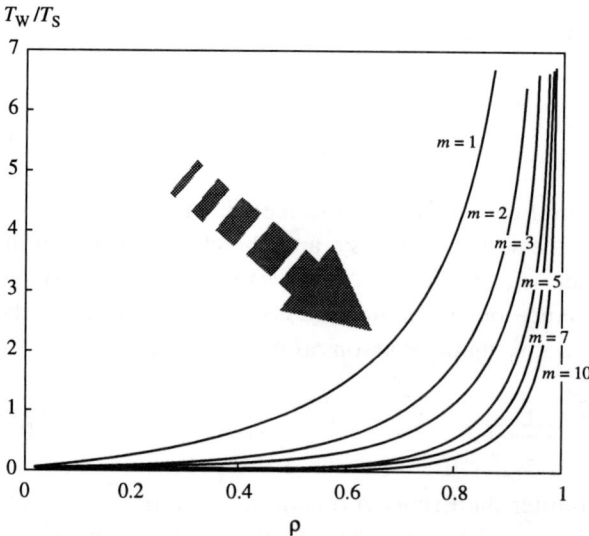

Figure 17.2. Wait-time versus server utilization for M/M/m.

and from this we can see that the probability of a customer not having to wait is $W(0) = 1 - E_C(m, u)$. Conversely, the probability that a customers does have to wait is, therefore, $E_C(m, u)$. In a multi-server system, there is often a significant chance that a customer will not have to wait, so let us look at how the probability of zero wait is related to traffic intensity and the number of servers. The relationship is illustrated in Fig. 17.3. For $m = 1$ we have the M/M/1 system, where the probability of zero wait is just $1 - \rho$. Notice how the graph is constructed. The x-axis is ρ, the server utilization, rather than u, the traffic intensity. This is a more useful way of comparing the behaviour with different numbers of servers, as we shall see. The probability of zero wait increases as the number of servers increases, for the same server utilization. For example, a single server 80 per cent loaded gives a probability of 0.2 of no waiting. Ten servers 80 per cent loaded gives a probability of about 0.6 of no waiting. These two situations, one server and ten servers, are dealing with the same workload relative to the total server capacity, so the difference in the chance of not waiting is remarkable.

Other things we may want to know about waiting time are its variance and percentiles. The variance can be calculated from

$$\sigma_{T_w}^2 = \text{variance of waiting time} = \frac{[2 - E_C(m, u)]E_C(m, u)T_S^2}{m^2(1 - \rho)^2} \tag{17.9}$$

Prob(wait-time = 0)

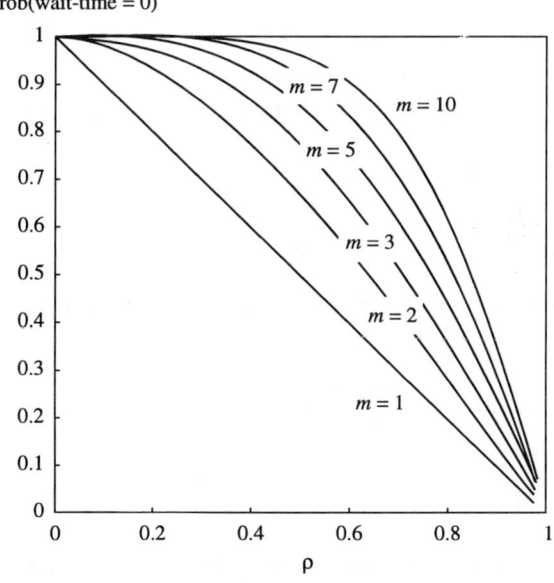

Figure 17.3. Probability of zero wait.

Percentiles can be easily obtained from the definition above of $W(t)$. Straight-forward rearrangement gives

$$r\text{th percentile of waiting time} = \frac{T_S}{m-u} \ln \left(\frac{E_C(m, u)}{\left(1 - \frac{r}{100}\right)} \right) \tag{17.10}$$

A little care is needed when calculating percentiles, since Eq. (17.10) may give negative values, in which case the correct value is zero.

Time-in-system (or queueing time)

The mean and variance of time-in-system or queueing time are best calculated using the basic relationships

$$T_Q = T_W + T_S \tag{17.11}$$

$$\sigma^2_{T_Q} = \sigma^2_{T_W} + \sigma^2_{T_S} \tag{17.12}$$

where expressions for T_W and $\sigma^2_{T_W}$ have been given above. The actual distribution of time-in-system is moderately complicated. The probability distribution function is defined in one of two ways, depending on whether $u = m - 1$ or not, i.e.

$$\text{Prob(queueing time} < t) = 1 - \left(1 + \frac{t}{T_S} E_C(m, u) \right) e^{-\frac{t}{T_S}} \text{ for u} = m - 1$$
$$\tag{17.13}$$

or

$$= 1 + \left(\frac{B + E_C(m, u)}{B} \right) e^{-\frac{t}{T_S}} + \frac{E_C(m, u)}{B} e^{-(m-u)\frac{t}{T_S}} \text{ for } u \neq m - 1 \tag{17.14}$$

where $B = m - 1 - u$. Two forms of the pdf are needed, since the second form is not defined when $u = m - 1$.

Queue size (number-in-system)

The probability that there will be exactly k customers in the system at a particular instant is ρ_k, where

$$\text{Prob(}k\text{ customers in system)} = p_k = \frac{u^k}{k!} p_0 \text{ for } k \leq m \tag{17.15}$$

$$p_k = \frac{u^k}{m! m^{k-m}} p_0 \text{ for } k \geq m \tag{17.16}$$

and p_0 is the probability that there are no customers in the system, i.e. the system is idle. The value of p_0 is obtained by using the definitions of p_k in Eqs (17.15) and

(17.16) and the fact that $\Sigma p_k = 1$. The result is

$$\text{Prob(zero customers in system)} = p_0 = \left(\frac{u^m}{(1-\rho)m!} + \sum_{k=0}^{m-1} \frac{u^k}{k!} \right)^{-1}$$

(17.17)

An interesting difference between a single-server and a multiserver system can be seen when we look at the probability distribution function of queue size. Figure 17.4 below shows the pdf for a seven-server system with a utilization of 0.7 (and so a traffic intensity of 4.9). The pdf has a peak, so that we are more likely to find, say, four customers in the system than no customers. When we looked at the queue size distribution for the M/M/1 model, we saw that queue size has a geometric distribution, which means that for M/M/1 a larger queue size (meaning total in system) is always less likely than a smaller queue size.

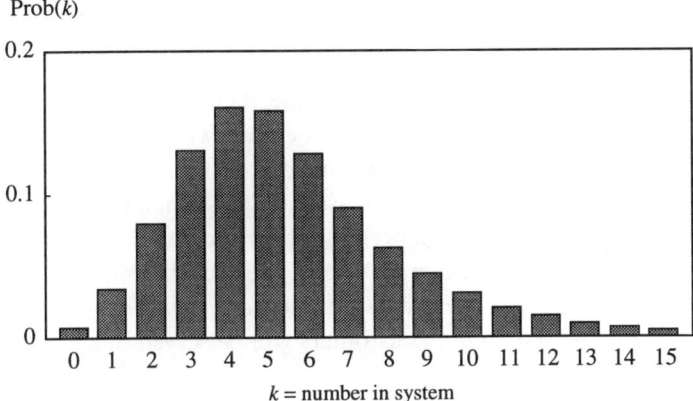

Figure 17.4. Pdf for queue size $m = 7$ and $\rho = 0.7$.

The average number of customers in the system is L_Q, where

$$L_Q = \text{average number in system} = u + \frac{\rho E_C(m, u)}{1 - \rho}$$

(17.18)

Waiting-line size

The probability distribution of the number of customers waiting is defined using Eqs (17.15) and (17.16). We just have to note that there will be zero customers waiting as long as there are m or fewer customers in the system, and that if there are x customers waiting there must be $x + m$ customers in the system. So we have

$$\text{Prob(no customers waiting)} = \sum_{k=0}^{m} p_k$$

(17.19)

$$\text{Prob(}x \text{ customers waiting)} = p_{x+m} \text{ for } x > 0$$

(17.20)

The mean and variance of the number waiting are L_W and $\sigma^2_{L_W}$, and are given by

$$L_W = \text{average number waiting} = \frac{\rho E_C(m, u)}{1 - \rho} \tag{17.21}$$

$$\sigma^2_{L_W} = \text{variance of number waiting} = \frac{\rho E_C(m, u)[1 + \rho - \rho E_C(m, u)]}{(1 - \rho)^2} \tag{17.22}$$

Separate queues versus common queue

One of the 'classic' comparisons made in queueing theory is between multiple servers with a common queue for all servers, and multiple servers each with its own separate queue. In this chapter we will show the straightforward comparison, but the reader should also study Chapter 18 on variations of queue discipline for multiple servers.

The single common queue is generally held to be the more efficient and fairer of the two methods. To see why, consider the case illustrated in Fig. 17.5, where there is one customer who needs an unusually long service-time. This customer is represented by a man carrying a large pile of documents to be dealt with. We shall refer to him as 'Mr Big' for convenience. When Mr Big reaches a server, we could think of that server as having been removed—where we assume the service-time needed by Mr Big is many times the average. So for a period of time the number of servers is reduced from m to $(m - 1)$, or from three to two in the illustration. This will slow down the rate at which customers are processed, but no one will have to wait behind Mr Big before getting served, and all customers will be served in the order in which they arrived.

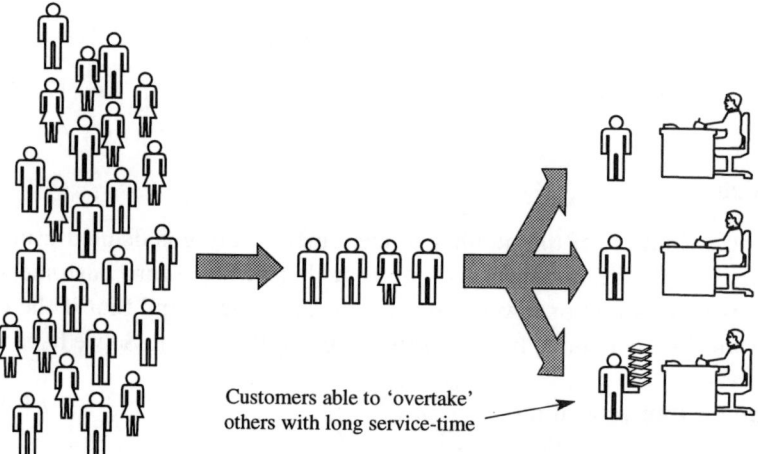

Customers able to 'overtake'
others with long service-time

Figure 17.5. Benefit of a common queue.

Now look at what happens with a separate queue for each server. Let us first look closely at how the queues operate. When a customer arrives, how does that customer decide which queue to join? In real systems, of course, the customer would choose the shortest queue, or if more than one server were idle choose one at random. The theory behind the formal comparison assumes that an arriving customer chooses a queue/server purely randomly. This is usually a wildly unrealistic assumption for a genuine group of servers, but perfectly reasonable where the servers are dispersed and a customer has to choose the queue/server without being able to observe the waiting-line for each server.

Customers unlucky enough to have joined the queue behind Mr Big will have to wait for him to be dealt with before they can get served (Fig. 17.6). In the meantime, customers who arrived later and joined other queues will get served first. Worse still, customers may be waiting behind Mr Big while other servers are actually idle.

A quantitative comparison of the two queueing arrangements is shown in Fig. 17.7, where average queueing time (or time-in-system) is plotted against arrival rate. For this illustration there are three servers, and $T_S = 1$. With separate queues we have, in effect, three separate single-server queues, each with one-third of the total arrival rate. Each of these separate queues is an M/M/1 system. For the common queue we use the M/M/m formulae given in this chapter. The comparison is clearly in favour of the common queue, although it would be more accurate to regard this as a comparison between dispersed individual servers and servers grouped together.

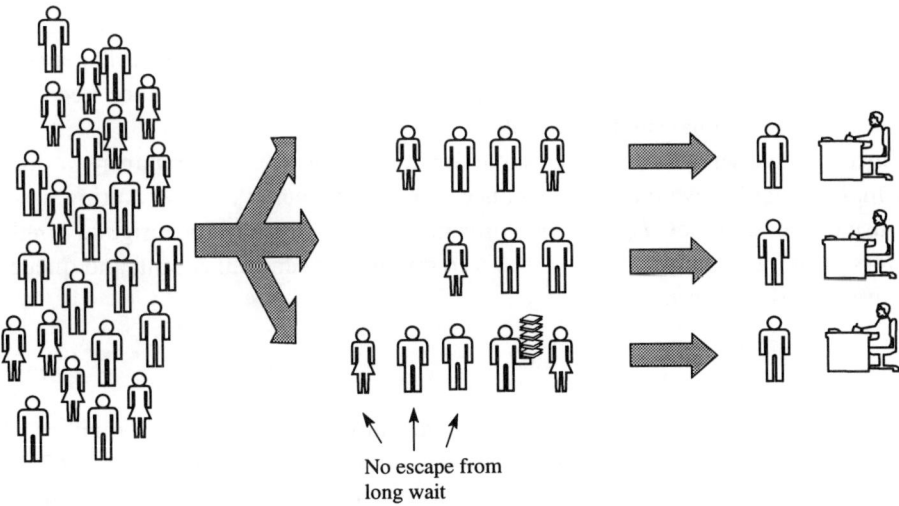

No escape from
long wait

Figure 17.6. Disadvantage of separate queues.

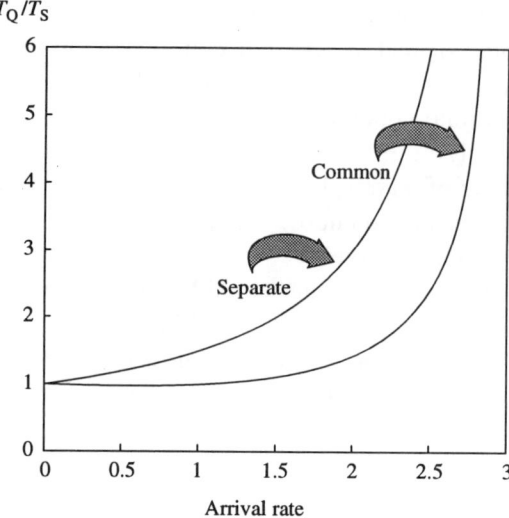

Figure 17.7. Separate versus common queues.

Loading different numbers of servers for same performance

Suppose we have a single server (M/M/1) system which is giving acceptable performance at an arrival rate that means the server is 40 per cent utilized. Now there happen to be three of these systems operating, and it is practical to combine them into a group of servers. If we do that then the arrival rate of the group of servers will be treble the single-server arrival rate, and the utilization of the servers will be the same as before. From the previous topic, readers will realize that grouping the servers will reduce the average waiting-time for customers. Suppose the average arrival rate is slowly increasing, e.g. from week to week, but average service-time is not changing. We know that customers are satisfied with a single server 40 per cent loaded. How much can the arrival rate increase before performance becomes worse than the original single servers?

Since average service-time is the same for the systems we are comparing, it does not matter whether we take performance to mean queueing time or waiting time. As usual, we shall set $T_S = 1$ for convenience. This means the original single servers must have an arrival rate $\lambda = 0.4$ to give 40 per cent utilization, and queueing time will therefore be

$$T_Q = \frac{T_S}{1 - \rho} = \frac{1}{1 - 0.4} = 1.67 \qquad (17.23)$$

Turning now to the multiserver system, we know that $T_S = 1$, the number of servers is $m = 3$, and we have to find the arrival rate λ that results in $T_Q = 1.67$. In principle this is easy, since all we have to do is rearrange the formula for average time-in-system to calculate the traffic intensity, and so the arrival rate. This is not

Table 17.1. Loading multiple servers for same performance

Single-server utilization	Utilization for M servers				
	M = 2	M = 3	M = 5	M = 7	M = 10
0.10	0.32	0.45	0.59	0.67	0.74
0.20	0.45	0.57	0.70	0.76	0.82
0.30	0.55	0.66	0.77	0.82	0.87
0.40	0.63	0.73	0.82	0.86	0.90
0.50	0.71	0.79	0.86	0.90	0.93
0.60	0.77	0.84	0.90	0.93	0.96
0.70	0.84	0.89	0.93	0.95	0.96
0.80	0.89	0.93	0.96	0.97	0.98
0.90	0.95	0.97	0.98	0.98	0.99

possible to do algebraically, but it is simple enough to program a numerical method to find the answer.

Table 17.1 shows how much a multiserver system can be loaded compared with a single-server system. Using this table for the example discussed above, we look down the 'single-server utilization' column to find the row for a utilization of 0.40 or 40 per cent. Looking along this row to the column headed 'M = 3', we see that a three-server system can be loaded to 0.73 or 73 per cent to give the same average waiting time. Looking at the column headed 'M = 10', we see that a ten-server system could be loaded 0.90 or 90 per cent.

The data in Table 17.1 is shown graphically in Fig. 17.8. Several interesting things can be seen from this graph. If only very low server utilization is tolerable

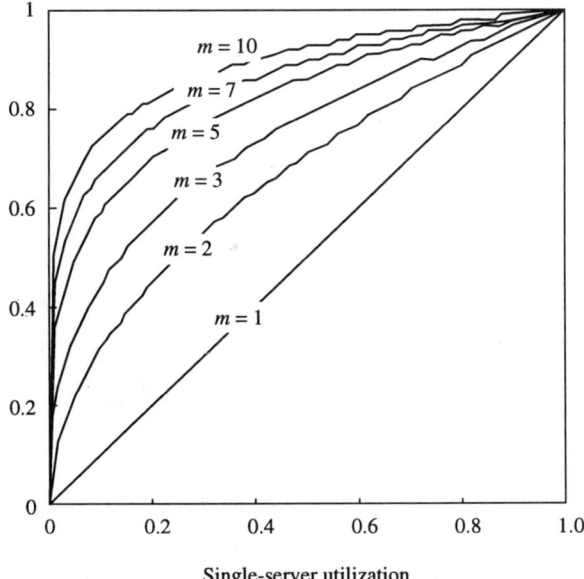

Figure 17.8. Loading multiple servers for same performance.

with a small number of servers, then as the number of servers is increased the tolerable server loading rises dramatically. On the other hand, if with few servers it is found that high utilization gives satisfactory performance, then more servers will not allow a large increase in utilization. If the utilization per server is already high anyway, it cannot be increased much further. The final observation from this graph is that each additional server has a smaller impact on tolerable utilization than the previous additional server.

Different numbers of servers with fixed total capacity

Another interesting comparison that can be made between single servers and multiple servers is with fixed total capacity. Consider a single server with a capacity of C customers per unit time, as shown in Fig. 17.9. This means that the average service time must be $T_S = 1/C$. (Readers should not confuse the use of C to represent server capacity with its more usual use as coefficient of variation.)

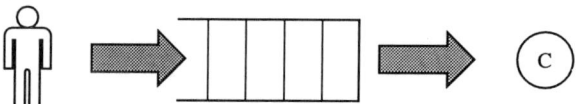

Figure 17.9. Single server with capacity C.

Now suppose we can choose to have the capacity C as a single server, or as a number of servers with the capacity equally divided between them, as in Fig. 17.10. Which is the better arrangement?

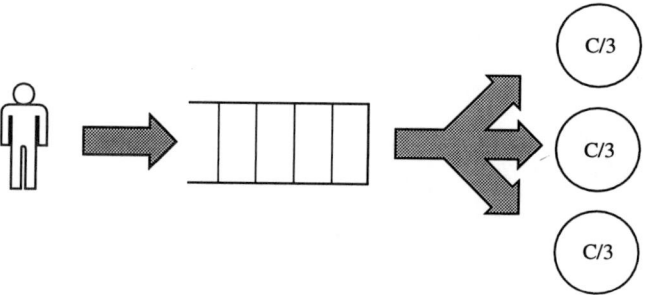

Figure 17.10. Multiple servers with total capacity C.

If we look first at T_Q, the average time-in-system, we get the results shown in Fig. 17.11. Obviously, as the capacity of the servers is further and further subdivided, the average service-time increases. At low utilizations there is very little waiting, and since a single customer can make use of only one server, several of the

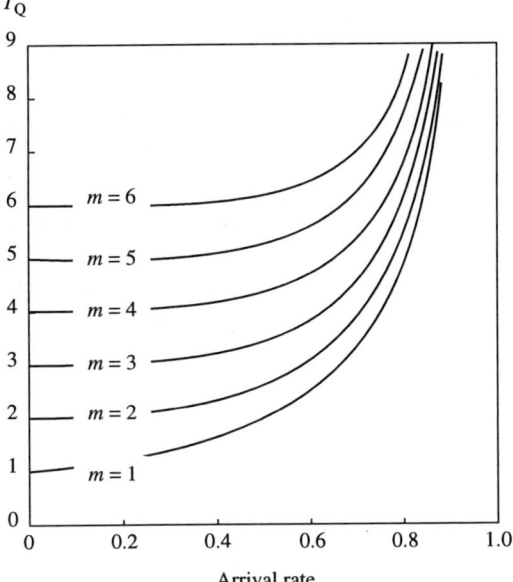

Figure 17.11. Time-in-system for multiple servers with fixed total capacity.

servers are idle. On this basis there would seem little to be gained by subdividing a fixed total capacity between more than one server. On the other hand, let us now investigate waiting time instead of time-in-system. This is shown in Fig. 17.12. Note that the greater the number of servers, the lower is the average waiting time.

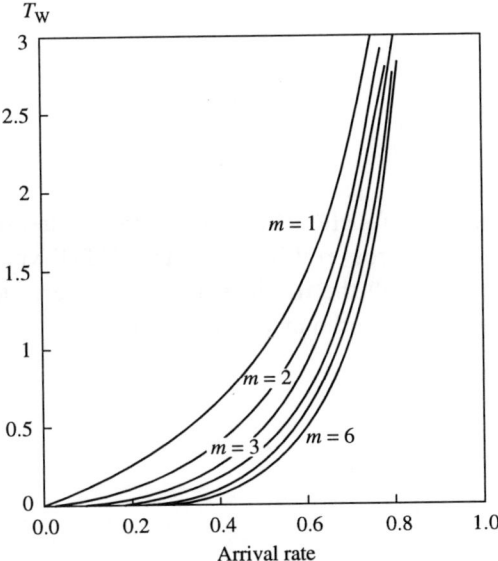

Figure 17.12. Waiting time for multiple servers with fixed total capacity.

If we carried the subdivision of capacity to extremes, we could have so many servers that no customer would ever have to wait. On the other hand, the capacity of each server would be so low that service-times would be very long.

So, whether we should subdivide server capacity depends on whether waiting time or time-in-system is more important. We can reduce average waiting time at the expense of average time-in-system. In general we are more interested in time-in-system, but there may be occasions when it is better for the customer to commence service as soon as possible, even if that means the actual service will take longer. Figure 17.13 illustrates the difference, with a single server compared to the same capacity spread across three servers at a particular customer arrival rate.

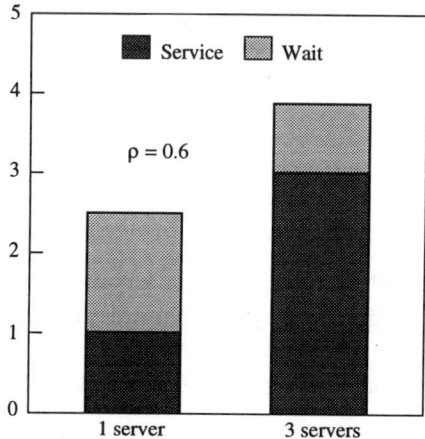

Figure 17.13. Multiple servers with fixed total capacity.

Programs for M/M/m

Prerequisite routines

Most of the routines for M/M/m require the function PRatio, which is listed in Appendix 2, to be available. The QTHError routine and the QTHTypes definitions are also required. In order to use all the routines in this chapter, the following set of 'include file' statements would be used in Turbo-Pascal.

```
{$E+,N+ }
{$I \qthprog\qthtypes.pas      }
{$I \qthprog\qtherror.pas      }
{$I \qthprog\qthfile.pas       }
{$I \qthprog\pratio.pas        }
{$I \qthprog\mmmcalc.pas       }
{$I \qthprog\mmmauxy.pas       }
{$I \qthprog\mmmqsze.pas       }
{$I \qthprog\mmmfind.pas       }
```

MMmCalc—general calculations for M/M/m

The routine MMmCalc calculates most of the statistics of interest for M/M/m.
The variable VALID is set to TRUE if the results are meaningful, and to FALSE
if server utilization is more than 100 per cent.

```
{-----------------------------------------------------}
{> MMmCalc - Calculations for M/M/m queueing model }
{ Inputs:   LAMBDA     customer arrival rate        }
{           TS         mean service time            }
{           M          number of servers            }
{ Outputs:  U          traffic intensity            }
{           RHO        server utilisation           }
{           TQ/SDVTQ   mean/std dev time in system  }
{           TW/SDVTW   mean/std dev  waiting time   }
{           PZW        probability of zero wait     }
{           LQ         average number in system     }
{           LW/SDVLW   mean/std dev number waiting  }
{           VALID      T=results valid, F=invalid   }
{ Copyright Mike Tanner 1994                         }
{-----------------------------------------------------}
Procedure MMmCalc(LAMBDA,TS:QTHreal;M:integer;
                  Var U,RHO,TQ,SDVTQ,TW:QTHreal;
                  Var SDVTW,PZW,LQ,LW,SDVLW:QTHreal;
                  Var VALID:boolean);
Var RM,RMU,PMU:QTHreal;
begin
   RM:=M; U:=LAMBDA*TS; RHO:=U/RM;
   VALID:=(RHO<1.0); If not VALID then Exit;
   {-Calculate Poisson ratio function and prob of--}
   {-all servers busy----------------------------}
   RMU:=Pratio(M,U);
   PMU:=(1-RMU)/(1-RHO*RMU))
   {-Waiting time-------------------------------}
   TW :=PMU*TS/(RM*(1-RHO));
   SDVTW:=Sqrt((2-PMU)*PMU)*TS/(RM*(1-RHO));
   PZW:=1-PMU;
   {-Time in system-----------------------------}
   TQ:=TW+TS;
   SDVTQ:=Sqrt(Sqr(SDVTW)+Sqr(TS));
   {-Number waiting-----------------------------}
   LW:=RHO*PMU/(1-RHO);
   SDVLW:=Sqrt(RHO*PMU*(1+RHO-RHO*PMU))/(1-RHO);
   {-Number in system---------------------------}
   LQ:=LW+U;
end;
```

MMmWaitCdf—cdf of waiting time

The routine MMmWaitCdf calculates the cumulative distribution function for
waiting time. This is done by direct calculation of the relevant formula.

```
{--------------------------------------------------}
{> MMmWaitCdf  -- Cdf of waiting time for M/M/m    }
{ Inputs: T          time for which cdf required   }
{         LAMBDA     customer arrival rate         }
{         TS         mean service time             }
{         M          number of servers            }
{ Returns: Prob(t<T/TS)                            }
{ Copyright Mike Tanner 1994                       }
{--------------------------------------------------}
Function MMmWaitCdf(T,LAMBDA,TS:QTHreal;
                    M:integer):QTHreal;
Var RM,U,RHO,RMU,PMU:QTHreal;
begin
   RM:=M;   U:=LAMBDA*TS; RHO:=U/RM;
   If RHO>=1.0 then QTHError('MMmWaitCdf error');
   RMU:=Pratio(M,U);
   PMU:=(1-RMU)/(1-RHO*RMU);
   MMmWaitCdf:=1-PMU*Exp(-(RM-U)*T/TS);
end;
```

MMmQtmeCdf—cdf of queueing time

The MMmQtmeCdf calculates the cumulative distribution function for the queueing time, or time-in-system. This is done by direct calculation, bearing in mind that there are two forms of the cdf according to the value of u, the traffic intensity.

```
{--------------------------------------------------}
{> MMmQtmeCdf  -- Cdf of time in system for M/M/m  }
{ Inputs: T          time for which cdf required   }
{         LAMBDA     customer arrival rate         }
{         TS         mean service time             }
{         M          number of servers            }
{ Returns:  Prob(t<T/TS)                           }
{ Copyright Mike Tanner 1994                       }
{--------------------------------------------------}
Function MMmQtmeCdf(T,LAMBDA,TS:QTHreal;
                    M:integer):QTHreal;
Var RM,U,RHO,RMU,PMU,G,C1,C2:QTHreal;
begin
   RM:=M; U:=LAMBDA*TS; RHO:=U/RM;
   If RHO>=1.0 then QTHError('MMmQtmeCdf error');
   RMU:=Pratio(M,U);
   PMU:=(1-RMU)/(1-RHO*RMU);
   G:=T/TS;
   If U=(RM-1)
   then MMmQtmeCdf:=1-(1+PMU*G)*Exp(-G)
   else begin
           C1:=(U-RM+1-PMU)/(RM-1-U);
           C2:=PMU/(RM-1-U);
           MMmQtmeCdf:=1+C1*Exp(-G)
                        +C2*Exp(-(RM-U)*G);
        end;
end;
```

MMmWaitPct—percentiles of waiting time

The routine MMmWaitPct calculates any required percentile of waiting time. This can be done by direct calculation. Since the formula will produce negative values of percentiles for low utilizations, the routine substitutes zero for a negative result.

```
{---------------------------------------------------}
{> MMmWaitPct  -- Pctile waiting time for M/M/m    }
{ Inputs:  PCT        percentile required          }
{          LAMBDA     customer arrival rate         }
{          TS         mean service time             }
{          M          number of servers             }
{ Returns: T where Prob(wait<T)=PCT/100            }
{ Copyright Mike Tanner 1994                        }
{---------------------------------------------------}
Function MMmWaitPct(PCT,LAMBDA,TS:QTHreal;
                    M:integer):QTHreal;
Var RM,U,RHO,RMU,PMU,PCTILE:QTHreal;
begin
    {-Preliminary calculations and checks----------}
    RM:=M;  U:=LAMBDA*TS;  RHO:=U/RM;
    If RHO>=1.0 then QTHError('MMmWaitPct error');
    RMU:=Pratio(M,U);
    PMU:=(1-RMU)/(1-RHO*RMU);
    {-Calculate percentile directly----------------}
    PCTILE:=TS*Ln(PMU/(1-PCT/100))/(RM-U);
    If PCTILE<0 then MMmWaitPct:=0
                else MMmWaitPct:=PCTILE;
end;
```

MMmQtmePct—percentiles of queueing time

The routine MMmQtmePct calculates any required percentile of queueing time, or time-in-system. Direct calculation of queueing-time percentiles is not practical, so a binary search of the possible range of values is used, looking for the queueing time at which MMmQtmeCdf is equal to the percentage required. The lower bound of this possible range is taken to be the corresponding percentile of waiting time, and the higher bound is found by simply starting at the lower bound plus average service-time and incrementing the upper bound by average service-time until the cdf exceeds the required percentage value.

```
{---------------------------------------------------}
{> MMmQtmePct  -- Pctile queuing time for M/M/m    }
{ Inputs:  PCT       percentile required           }
{          LAMBDA    customer arrival rate          }
{          TS        mean service time              }
{          M         number of servers              }
{ Returns: T where Prob(qtme<T)=PCT/100            }
{ Copyright Mike Tanner 1994                        }
{---------------------------------------------------}
```

```
Function MMmQtmePct(PCT,LAMBDA,TS:QTHreal;
                         M:integer):QTHreal;
Const ACC=0.00001;      { accuracy for percentile     }
Var RM,U,RHO,RMU:QTHreal;
Var P,LOW,HGH,CDFLOW,CDFHGH,X,CDF,ACCACH:QTHreal;
Label L1;
begin
   {-Preliminary calculations and checks----------}
   RM:=M;  U:=LAMBDA*TS;  RHO:=U/RM;
   If RHO>=1.0 then QTHError('MMmQtmePct error');
   P:=PCT/100;
   {-Establish range------------------------------}
   LOW:=MMmWaitPct(PCT,LAMBDA,TS,M);
   CDFLOW:=MMmQtmeCdf(LOW,LAMBDA,TS,M);
   HGH:=LOW;
   Repeat
       HGH:=HGH+TS;
       CDFHGH:=MMmQtmeCdf(HGH,LAMBDA,TS,M);
   until CDFHGH>=P;
   {-Binary search--------------------------------}
L1:X:=(LOW+HGH)/2;
   CDF:=MMmQtmeCdf(X,LAMBDA,TS,M);
   If Abs(CDF-P)<ACC
   then begin MMmQtmePct:=X; Exit; end;
   If CDF<P then begin LOW:=X;  CDFLOW:=CDF; end
            else begin HGH:=X;  CDFHGH:=CDF; end;
   Goto L1;
end;
```

MMmQszePO—probability of zero customers in the system

The routine MMmQszePO calculates the probability that the queue size is zero, i.e. that there are no customers in the system. This characteristic may be of direct interest, and is required by some of the following routines.

```
{----------------------------------------------------}
{> MMmQszePO    -- prob zero queue for M/M/m         .}
{ Inputs:   LAMBDA    customer arrival rate          }
{           TS        mean service time              }
{           M         number of servers              }
{ Returns: Prob(queue size = 0)                      }
{ Copyright Mike Tanner 1994                         }
{----------------------------------------------------}
Function MMmQszePO(LAMBDA,TS:QTHreal;
                       M:integer):QTHreal;
Var RM,RI,U,RHO,PO,Z:QTHreal;I:integer;
begin
   RM:=M;  U:=LAMBDA*TS;  RHO:=U/RM;
   If RHO>=1.0 then QTHError('MMmQszePO·error');
   Z:=1;
   For I:=1 to M do begin
      RI:=I;
      Z:=Z*U/RI;
   end;
```

```
   Z:=Z/(1-RHO); P0:=Z+1;  Z:=1;
   For I:=1 to (M-1) do begin
      RI:=I;
      Z:=Z*U/RI;
      P0:=P0+Z;
   end;
   MMmQszeP0:=1/P0;
end;
```

MMmQszePdf/MMmQszeCdf—pdf/cdf of queue size (number-in-system)

```
{-----------------------------------------------------}
{> MMmQszePdf  -- pdf of queue size for M/M/m         }
{> MMmQszeCdf  -- cdf of queue size for M/M/m         }
{ Inputs: K         queue size for which pdf wanted   }
{         LAMBDA    customer arrival rate             }
{         TS        mean service time                 }
{         M         number of servers                 }
{ Copyright Mike Tanner 1994                          }
{-----------------------------------------------------}
Function MMmQszePdf(K:integer;LAMBDA,TS:QTHreal;
                    M:integer):QTHreal;
Var RM,RI,U,RHO,PK,Z:QTHreal; I:integer;
begin
   RM:=M;  U:=LAMBDA*TS;  RHO:=U/RM;
   If RHO>=1.0 then QTHError('MMmQszePdf error');
   PK:=MMmQszeP0(LAMBDA,TS,M);
   For I:=1 to K do begin
      RI:=I;
      PK:=PK*U;
      If I<M then PK:=PK/RI
             else PK:=PK/RM;
   end;
   MMmQszePdf:=PK;
end;
{-----------------------------------------------------}
Function MMmQszeCdf(K:integer;LAMBDA,TS:QTHreal;
                    M:integer):QTHreal;
Var RM,RI,U,RHO,PK,Z,CDF:QTHreal; I:integer;
begin
   RM:=M;  U:=LAMBDA*TS;  RHO:=U/RM;
   If RHO>=1.0 then QTHError('MMmQszePdf error');
   PK:=MMmQszeP0(LAMBDA,TS,M);
   CDF:=PK;
   For I:=1 to K do begin
      RI:=I;
      PK:=PK*U;
      If I<M then PK:=PK/RI
             else PK:=PK/RM;
      CDF:=CDF+PK;
   end;
   MMmQszeCdf:=CDF;
end;
```

MMmWlnePdf/MMmWlneCdf—pdf/cdf of waiting-line size

```
{---------------------------------------------------}
{> MMmWlnePdf  -- pdf of waitline size for M/M/m   }
{> MMmWlneCdf  -- cdf of waitline size for M/M/m   }
{ Inputs: K        number for which pdf/cdf wanted  }
{         LAMBDA customer arrival rate              }
{         TS       mean service time                }
{         M        number of servers                }
{ Copyright Mike Tanner 1994                        }
{---------------------------------------------------}
Function MMmWlnePdf(K:integer;LAMBDA,TS:QTHreal;
                    M:integer):QTHreal;
Var RM,RHO:QTHreal;
begin
   RM:=M;   RHO:=LAMBDA*TS/RM;
   If RHO>=1.0 then QTHError('MMmWlnePdf error');
   If K=0
   then MMmWlnePdf:=MMmQszeCdf(M,LAMBDA,TS,M)
   else MMmWlnePdf:=MMmQszePdf(M+K,LAMBDA,TS,M);
end;
{---------------------------------------------------}
Function MMmWlneCdf(K:integer;LAMBDA,TS:QTHreal;
                    M:integer):QTHreal;
Var RM,RHO:QTHreal;
begin
   RM:=M; RHO:=LAMBDA*TS/RM;
   If RHO>=1.0 then QTHError('MMmWlneCdf error');
   MMmWlneCdf:=MMmQszeCdf(M+K,LAMBDA,TS,M);
end;
```

MMmTq—short version of MMmCalc

The MMmTq function is a short version of MMmCalc which calculates only the average time in the system, or queueing time. This may be more convenient than MMmCalc for some purposes, and indeed is included for use by the MMmFindRho function described next.

```
{---------------------------------------------------}
{> MMmTQ -- short form of MMmCalc for MMmFindRho   }
{---------------------------------------------------}
Function MMmTQ(LAMBDA,TS:QTHreal;M:integer):QTHreal;
Var RHO,U,RM,RMU,PMU:QTHreal;
begin
   RM:=M;  U:=LAMBDA*TS;  RHO:=U/RM;
   RMU:=Pratio(M,U);
   PMU:=(1-RMU)/(1-RHO*RMU);
   MMmTQ:=TS+PMU*TS/(RM*(1-RHO));
end;
```

MMmFindRho—calculate server utilization for specified T_Q

This function finds the server utilization ρ at which the average time in system T_Q will have a specified value. There is no direct way of calculating this, so a straight-forward binary search procedure is used. The accuracy of the result is specified by the constant ACC, which is internal to the routine and set to 0.00001 in the listing. For a different accuracy change ACC or make it a parameter.

```
{------------------------------------------------------}
{> MMmFindRho - Find server utilisation that gives }
{               a specified time-in-system            }
{ Inputs: TARG  target time in system                 }
{         TS    mean service time                     }
{         M     number of servers                     }
{ Copyright Mike Tanner 1994                          }
{------------------------------------------------------}
Function MMmFindRho(TARG,TS:QTHreal;M:integer):QTHreal;
Const ACC=0.00001;     { accuracy for utilisation    }
Var RM,P,LOW,HGH,TQLOW,TQHGH,X,TQ:QTHreal;
Label L1;
begin
   RM:=M;
   LOW:=0.0;           TQLOW:=MMmTQ(LOW,TS,M);
   HGH:=RM*0.99/TS;    TQHGH:=MMmTQ(HGH,TS,M);
L1:X:=(LOW+HGH)/2;     TQ:=MMmTQ(X,TS,M);
   If Abs(TQ-TARG)<ACC
   then begin MMmFindRho:=X*TS/RM; Exit; end;
   If TQ<TARG then begin LOW:=X;   TQLOW:=TQ; end
              else begin HGH:=X;   TQHGH:=TQ; end;
   Goto L1;
end;
```

Example 17A—The doughnut stalls

At a large agricultural show the Stodgee Donut company has three stalls. Experience suggests that customers are discouraged if the waiting time exceeds 40 secs. It takes on average 20 seconds to serve a customer. If the three stalls are sited remotely from each other, so that each operates independently, what is the maximum rate at which customers could arrive? If the three stalls are gouped together, what then is the maximum customer arrival rate that might occur?

With separated stalls, we have $T_S = 20\,s$, $m = 1$, and we need to find λ such that $T_W \leqslant 40$ seconds. Using MMmCalc and simple trial and error, we find that $\lambda = 2$ customers per minute is the answer. Note this is the arrival rate to each stall, so the total arrival rate is six customers per minute.

With grouped stalls, we move $m = 3$, the service-time is the same, and we want to find λ such that $T_W \leqslant 40$ seconds. Again using trial and error, we find that $\lambda = 7.84$ customers per minute is the maximum arrival rate. By grouping the doughnut stalls together the potential customer arrival rate has been increased by $100 \times 7.84/6 = 31\%$.

Example 17B—The bank branch

In a branch of the Hordit Bank there are four cashier positions open. Customers arrive at the rate of five per minute, and take on average 45 seconds to be served. A common queue is used for all the cashiers. What is the average waiting time? If an additional cashier position were opened, what then would be the waiting time?

This is a very straightforward question. We have $m = 4$, $T_S = 0.75$ min, and $\lambda = 5$ customers per minute. Applying Eq. (17.7) (or using MMmCalc or the Q-Calc package) we find that $T_W = 2.59$ min. If we now set $m = 5$, then waiting time is reduced to $T_W = 0.28$ min.

18
Multiserver queueing rules

Introduction

In this chapter we take a look at the different ways in which customers may queue for a group of servers. In the chapter on M/M/m the classic queueing-theory comparison was given between a single queue for all servers and a separate queue for each server. The weakness of this comparison is the assumption that an arriving customer chooses between separate queues at random. There may be situations where this is a realistic model, but generally it is not.

The comparisons presented in this chapter are based on simulations of an M/M/3 system. Although analytic results can be used to obtain many of the results used, simulation offers a convenient way of experimenting with queueing rules, and of measuring some characteristics related to fairness of a system.

Ways of selecting a queue

There are a range of queue selection rules that may occur in real situations. Some of them are appropriate for non-people situations such as messages queueing for transmission links in a telecommunication system. Other selection rules are more appropriate when the customers are people rather than things. The rules we shall investigate are as follows.

(a) A single queue for all servers.
(b) A separate queue for each server. Each arriving customer chooses a queue/server at random.
(c) A separate queue for each server. Arriving customers are allocated to a queue/server cyclically, i.e. first customer assigned to server 1, second customer to server 2, mth customer to server m, $(m + 1)$th customer to server 1, and so on.
(d) A separate queue for each server. An arriving customer chooses the shortest queue. This is quite a common real-life situation.

(e) A separate queue for each server. An arriving customer is able to see how much time a customer being served will need to complete service, and the service-times required by each customer waiting. The arriving customer can therefore select the server/queue that will provide the minimum waiting time. The most common example of this is the supermarket. A customer's service-time will be quite closely related to the number of items to be handled at the check-out, and it is possible to see how much each customer has in his or her basket or trolley, and how far the customer in service has progressed.

The obvious characteristic to look at first is the average waiting time for each rule. This is shown in Fig. 18.1, where the different rules are compared at each of a range of server utilizations. The single queue appears to be the best, and that is what most readers will have expected, but later we shall see why this is not always the case. In fact, separate queues with customers able to estimate waiting times accurately for each queue are logically identical to a single queue. Choosing a queue at random is the worst approach—obviously, because a customer may choose a long queue even when other servers are idle. Assigning customers to queues cyclically is a definite improvement on random choice, but still quite bad. Selecting the shortest queue is, as common sense and experience would suggest, a good method which is only slightly worse than a single queue.

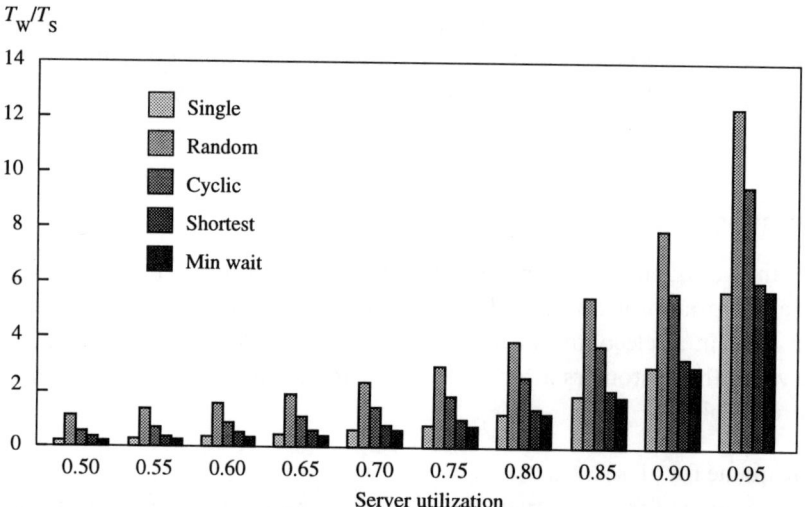

Figure 18.1. Average waiting time without queue-hopping.

Queue-hopping

It is important to realize that different rules produce different average waiting times because once a customer has selected a queue, he or she remains in that queue even if another server becomes free. We now look at the effect of 'hopping',

$T_{\mathrm{w}}/T_{\mathrm{s}}$

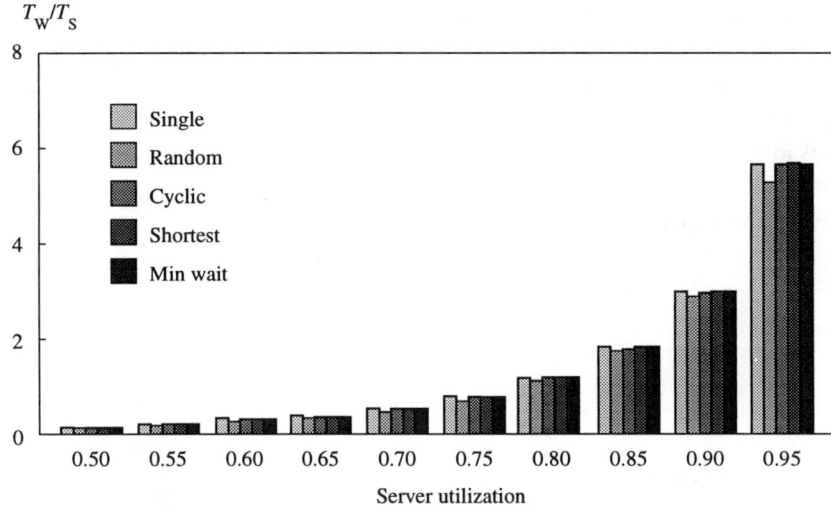

Figure 18.2. Average waiting time with queue-hopping.

i.e. customers changing queue after they have selected a queue and waited for a time. The reasoning that people use in deciding whether to hop queues is likely to be very subtle and complex, involving observation of the rate at which each server works as well as simple queue lengths and estimated service-times for other people in the queues. For our analysis here, a straightforward queue-hopping mechanism is assumed. Whenever a customer completes service, the person who is last in the longest queue will hop to the shortest queue if that improves his or her queueing position. (There is no point in hopping if the queue lengths are, for example, 2 and 1.)

Figure 18.2 shows the average waiting time for each of the queueing rules when customers hop queues. Slight differences between the results for each rule are due to inevitable variation in simulation results: the conclusion is that average waiting time is the same regardless of the way a queue is initially selected. If we were to look at, say, the 90th percentile of waiting time, that would also be identical between the possible rules. Different queueing rules are different ways of selecting the next customer to be served. As long as this choice is made without reference to the individual service-times of the customers waiting, the average waiting time will be the same. It is possible to reduce the average waiting time by serving customers in the order of shortest-time first, and this is investigated when we look at priority queues.

Single queue with overhead

Nowadays it is common practice in banks, building societies and post offices to have a single queue for several counter clerks. Most of us prefer this to separate queues, because it seems fairer. The single queue, however, can be less efficient than separate queues. When a server becomes free, the next customer has to walk from the head of the queue to the counter position. The time taken for this walk is effectively added to the customer's service-time, although it is unproductive in that the server is idle. To illustrate this with the simulations, we assume that the 'walk'-time is 10 per cent of the average service-time. The comparison of a single queue with and without this overhead is shown in Fig. 18.3.

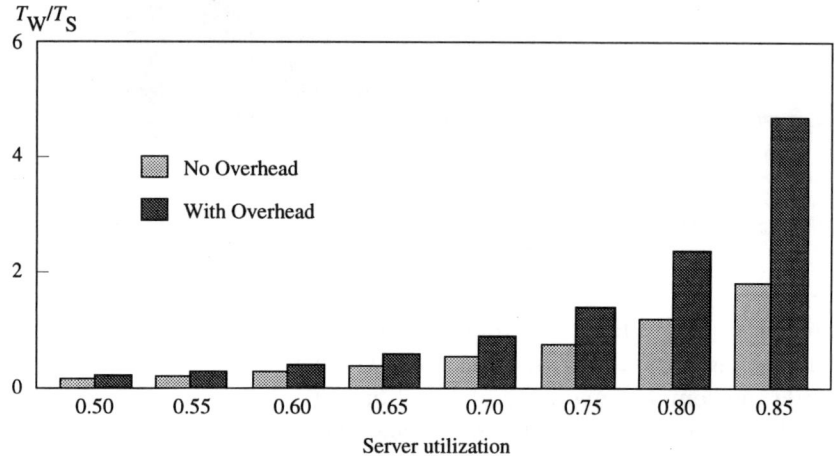

Figure 18.3. Single queue with and without overhead.

At high utilizations the effect of the walk-time overhead is significant. We have seen that if customers select between separate queues in a sensible way, and particularly if customers hop between queues, then separate queues, produce the same waiting times as a single queue without walk-time overhead. It follows that separate queues are better (in the sense of average waiting time) than a single queue if the walk-time is more than a very short time. The typical supermarket choice of separate queues is, therefore, perfectly rational.

Fairness in queueing

Inanimate things such as messages in communication systems and tasks within a computer do not have feelings. People, on the other hand, attach importance to characteristics such as fairness, predictability, and consistency in the time it takes to get served. In multiserver systems with separate queues, people do not like someone who arrived after they did getting served first. A simple measure of this is the percentage of customers that get 'overtaken' by one or more other customers. By overtaking, we mean that customer X is overtaken by customer Y if X arrived before Y, but Y starts service before X does. More sophisticated measures of fair-

% overtaken

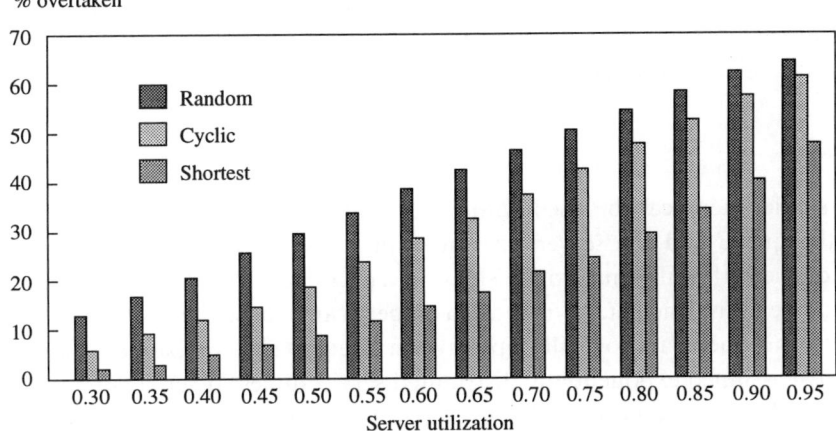

Figure 18.4. Overtaking level without queue-hopping.

ness might be useful, since, for example, a customer may be overtaken by others but at the same time do some overtaking. The simple measurement will, however, be sufficient for the point being made.

If customers do not hop between queues, the overtaking levels are shown in Fig. 18.4. Obviously, for a single queue the overtaking level is zero. Given that selecting a queue on the basis of accurate predictions of waiting time is equivalent to a single queue, overtaking will be zero in this case also. Figure 18.4 shows that the remaining systems all have a high level of overtaking, and selecting the shortest queue is not markedly better than random selection of a queue.

When we looked at average waiting times we saw that queue-hopping had a big effect in reducing waiting times and making them the same regardless of how the initial choice between separate queues was made. Looking at Fig. 18.5 we can see that queue-hopping does equalize the overtaking level between random, cyclic and

% overtaken

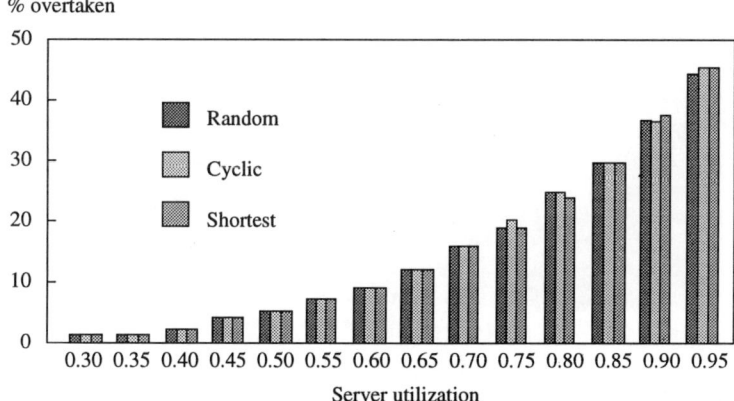

Figure 18.5. Overtaking level with queue-hopping.

shortest-queue selection. However, the overtaking level for these rules is still high, even with queue-hopping. There is a marked difference between these queue selection methods and a single queue, which by definition has zero overtaking.

Conclusions

Where a single queue can be used without a significant walk-time between the head of the queue and the servers, a single queue is undoubtedly superior to separate queues. If walk-time for a single queue is non-trivial, then separate queues may be more efficient, as long as customers can choose the shortest queue or shortest-wait queue and/or will hop between queues. The comparisons in this chapter have assumed exponentially distributed service-times. If service-times are less variable than this, especially if they are nearly constant, than selecting the shortest queue is closer to selecting the shortest-wait queue, with its zero over-taking level.

19
Multiserver loss (M/M/m/m)

Introduction

With this queueing system there are multiple servers, but customers are not allowed to wait if no server is available. Customers arriving to find all the servers busy are lost. Such lost customers depart immediately, never to return (Fig. 19.1). The main practical application of this queueing model is in the design of telephone networks. For example, a large office will have a number of external telephone lines, and if somebody in the office wishes to make an outside call, then it is possible that all the external lines will already be in use. Of course in a real situation the

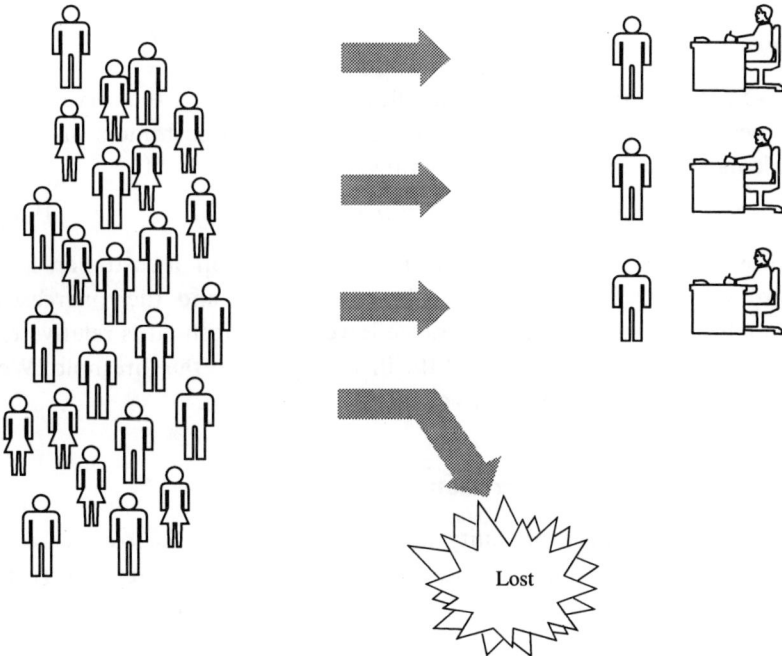

Figure 19.1. Multiserver loss system.

caller would try again either immediately or after a short time. In practice the assumption that customers (calls) are lost forever is a reasonable design assumption, and call retries are an unnecessary complication.

Assumptions and initial calculations for M/M/m/m

There are m servers. Customers arrive at random at an average rate λ. Service-times are exponentially distributed with an average T_S. If, when a customer arrives, all the servers are occupied, then that customer immediately departs and does not return.

First we calculate the traffic intensity

$$u = \text{traffic intensity} = \lambda T_S \tag{19.1}$$

Next we must calculate the Erlang-B function, which occurs in many of the formulae for M/M/m/m, and is defined by Eq. (19.2). An alternative definition is given in Eq. (19.3) in terms of the Poisson ratio function. We have already encountered the Erlang-C function, in Chapter 17. The relationship between the Erlang-B and Erlang-C functions is defined by Eq. (17.6).

$$E_B(m, u) = \text{Erlang-}B = \frac{\frac{u^m}{m!}}{\sum_{k=0}^{m} \frac{u^k}{k!}} \tag{19.2}$$

$$E_B(m, u) = \text{Erlang-}B = 1 - R(m, u) \tag{19.3}$$

Lost customers

Since customers are lost, rather than being allowed to wait if all the servers are busy, the key performance characteristic is the proportion of customers that are lost. This is calculated by Eq. (19.4) and illustrated in Fig. 19.2.

$$\text{Prob(arriving customer is lost)} = E_B(m, u) \tag{19.4}$$

The shape of this probability curve is important. The proportion of customers lost stays near zero until traffic intensity reaches a threshold, and then increases rapidly. The smaller the number of servers, the lower this threshold is relative to the number of servers, and the more rapid the increase beyond the threshold. We shall explore this scaling effect later in this chapter.

Actual arrival rate and server utilization

Once we have calculated the probability an arriving customer is lost, we can very simply calculate the actual arrival rate λ_A, as in Eq. (19.5). By actual arrival rate we mean the rate seen by the servers, rather than the arrival rate of potential customers.

$$\lambda_A = \text{arrival rate excluding lost customers} = \lambda[1 - E_B(m, u)] \tag{19.5}$$

Erlang-*B*

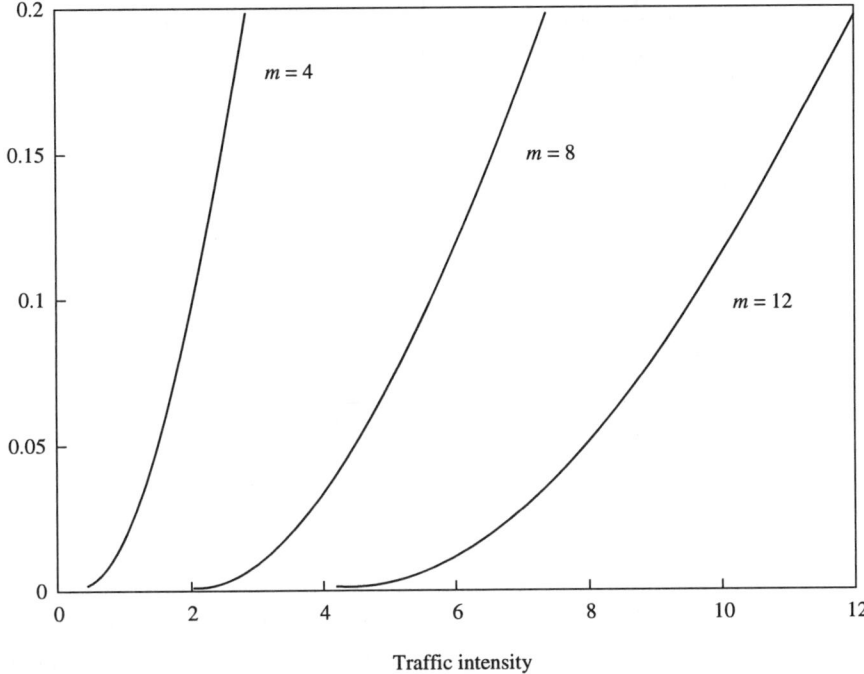

Figure 19.2. Probability that a customer is lost for M/M/m/m.

The server utilization depends, of course, on the actual arrival rate rather than the potential or offered rate λ. Server utilization is calculated by either Eq. (19.6) or Eq. (19.7).

$$\rho = \text{server utilization} = \frac{\lambda_A T_S}{m} \tag{19.6}$$

$$\rho = \text{server utilization} = \frac{u}{m}[1 - E_B(m, u)] \tag{19.7}$$

The average number of servers busy is the same as the average number of customers in the system. We can apply Little's law to calculate this, as in Eq. (19.8).

$$L_S = \text{average number of servers busy} = \lambda_A T_S = u[1 - E_B(m, u)] \tag{19.8}$$

Now let us investigate how server utilization is related to the customer arrival rate. We shall use the term 'offered traffic' for λ. Figure 19.3 is a plot of server utilization against offered traffic relative to the capacity of the servers, which is $\lambda T_S/m$. A little reflection will show the reader that this is the server utilization that would be achieved if no customers were lost and they were allowed to wait. The line labelled 'maximum possible' shows what this utilization with no lost customers would be. It goes up in a straight line until the servers are 100 per cent busy, and

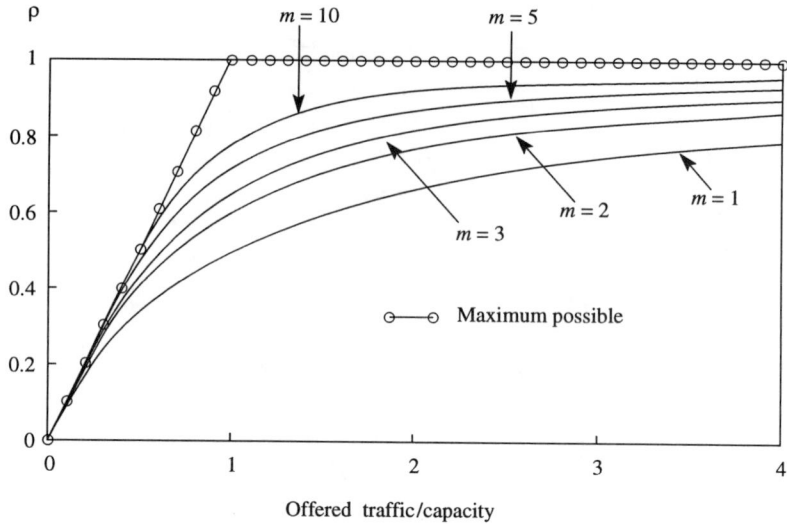

Figure 19.3. Server utilization against offered traffic/capacity for M/M/m/m.

then, of course, stays at 100 per cent. When the 'relative traffic intensity' is 1, there are enough customers arriving to keep the servers fully utilized. In fact this does not happen. Even if the average rate of arrivals is less than the servers could on average deal with, intervals of high arrival rates will still occur and some customers will still be lost. Looking at the line labelled $m = 1$, we see that server utilization is only 50 per cent when the offered rate would saturate the server if waiting were allowed. The offered rate would have to be many times the servers' capacity for the utilization to get anywhere near 100 per cent. With more servers, the situation is the same, but utilization increases more quickly as offered traffic increases. The other way of looking at this is that the proportion of lost customers increases less rapidly for more servers, as we saw in Fig. 19.2.

Number-in-system

The probability of there being a specific number of customers in the system can be calculated from Eq. (19.9), together with the definition in Eq. (19.10) which simply states mathematically that no waiting is allowed.

$$p_k = \text{Prob}(k \text{ customers in system}) = \frac{u^k}{k!} p_0 \text{ for } k \leq m \qquad (19.9)$$

$$p_k = 0 \text{ for } k > m \qquad (19.10)$$

Equation (19.9) involves the quantity p_0, which is the probability that there will be no customers present in the system. Using the fact that the sum of the p_k must be 1, this probability can be calculated by

$$p_0 = \text{Prob(system is empty)} = \left(\sum_{k=0}^{m} \frac{u^k}{k!} \right)^{-1} \qquad (19.11)$$

The average number of customers in the system is identical to the average number of servers busy, and a formula for calculating this value has already been given as Eq. (19.8). Another way is

$$L_Q = \text{average number in system} = \rho m \qquad (19.12)$$

Since we know how to calculate the individual p_k, we can calculate the variance of the number of customers in the system from first principles, as in

$$\sigma_{L_Q}^2 = \text{variance of number in system} = \sum_{k=0}^{m} (k - L_Q)^2 p_k \qquad (19.13)$$

Calculating number of servers needed given arrival and loss rates

If we know the arrival rate and the average service-time, we will often need to calculate how many servers we need to achieve a particular loss rate. Mathematically we want to find

$$\text{minimum } m \text{ such that } E_B(m, \lambda T_S) \leq q, \text{ given } \lambda \text{ and } q \qquad (19.14)$$

A computer-oriented solution to this is trivial, since we just start with $m = 1$ and calculate the loss rate. If the loss rate is more than q we just increase m by 1 and recalculate the loss rate. This is repeated until m is just big enough to satisfy our loss rate criterion. An improved method might be to devise a better starting value for m than 1, so that fewer iterations are needed. Alternatively, a form of integer binary search could be used. However, since the simple approach works satisfactorily we may as well use it. An implementation of this method is in subroutine MMmLsSvr.

Calculating maximum arrival rate given numbers of servers and loss rate

Another common requirement is to find the maximum allowable arrival rate, given the average service-time, the number of servers, and the maximum proportion of customers to be lost. Mathematically we want to find

$$\text{maximum } \lambda \text{ such that } E_B(m, \lambda T_S) \leq q, \text{ given } m \text{ and } q \qquad (19.15)$$

Since arrival rate is a non-integral quantity, we shall need to use a binary search method. First we have to establish lower and upper bounds for the arrival rate. The lower bound is simple, we can just use zero. The upper bound is not so obvious. A simple way is to take $\lambda = T_S/m$ and keep doubling λ until the loss rate is exceeded. A binary search can then proceed in the usual way. An implementation of this method is in subroutine MMmLsTrf.

Scaling effect

Suppose for a given number of servers we do not want the proportion of customers lost to be more than, say 5 per cent. We want to know the maximum permitted utilization such that this proportion is not exceeded. We can use the method described above for solving Eq. (19.15) to find the arrival rate, and hence the utilization. Doing this for a range of numbers of servers, and for several loss rates, we can plot the graph shown in Fig. 19.4. From this we can see, for example, that with 10 servers if we want a loss rate not greater than 5 per cent, the server utilization must be less than about 0.6. The message from Fig. 19.4 is that as the number of servers increases we can load up each server more, while still keeping the same loss rate. Larger groups of servers are therefore more efficient than smaller groups.

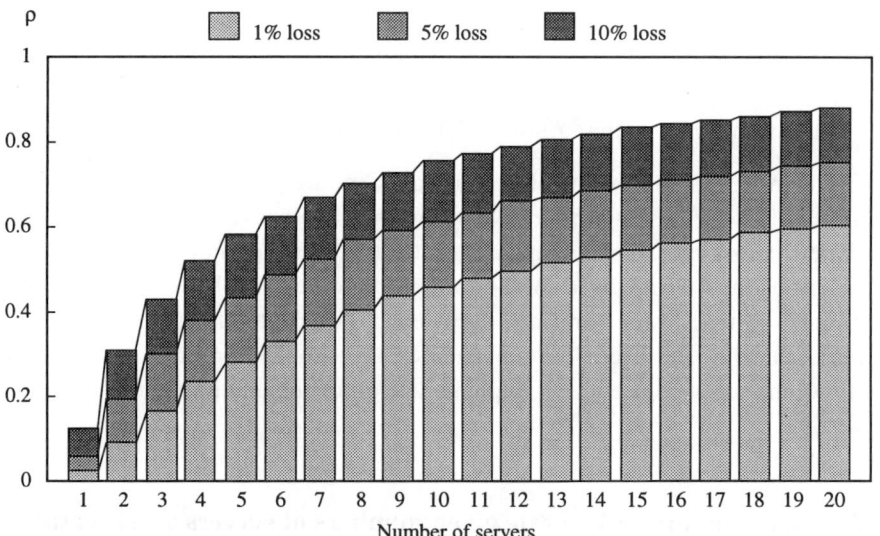

Figure 19.4. Maximum permitted server utilization for given rate of lost customers for M/M/m/m.

Programs for M/M/m/m

Prerequisite routines

The following set of 'include' statements are required to make use of the routines from this chapter.

```
{$N+,E+ }
{$I \qthprog\qthtypes.pas     }
{$I \qthprog\pratio.pas       }
{$I \qthprog\MMmLaClc.pas     }
{$I \qthprog\MMmLsAux.pas     }
```

MMmLsClc—general calculations for M/M/m/m

The subroutine MMmLsClc performs all the general calculations for an M/M/m/m system.

```
{-----------------------------------------------------}
{> MMmLsClc - Calculations for M/M/m loss model      }
{ Inputs:                                             }
{     LAMBDA    customer arrival rate                 }
{     TS        mean service time                     }
{     M         number of servers                     }
{ Outputs:                                            }
{     U         traffic intensity                     }
{     RHO       server utilisation                    }
{     PLOSS     probability arriving customer lost    }
{     ARVRTE    arrival rate excluding lost           }
{               customers                             }
{   Copyright Mike Tanner 1993                        }
{-----------------------------------------------------}
Procedure MMmLsClc(LAMBDA,TS:QTHreal;M:integer;
                   Var U,RHO,PLOSS,ARVRTE,P0,
                   LQ,SDVLQ:QTHreal);
Var RM,RMU,ELGB,P,TOTP,RK:QTHreal;K:integer;
begin
   RM:=M;  U:=LAMBDA*TS;
   RMU:=Pratio(M,U);  ELGB:=1-RMU;
   PLOSS:=ELGB;
   ARVRTE:=LAMBDA*(1-ELGB);
   RHO:=(U/RM)*(1-ELGB);
   LQ:=RHO*RM;
   P:=1;   TOTP:=1;
   SDVLQ:=Sqr(LQ);
   For K:=1 to M do begin
      RK:=K;
      P:=P*U/RK;
      TOTP:=TOTP+P;
      SDVLQ:=SDVLQ+P*Sqr(RK-LQ);
   end;
   P0:=1/TOTP;
   SDVLQ:=Sqrt(SDVLQ*P0);
end;
```

MMmLsSvr—calculate servers needed for M/M/m/m

The subroutine MMmLsSvr calculates the number of servers needed for a given arrival rate, average service-time, and proportion of customers lost. It solves Eq. (19.14).

```
{-----------------------------------------------------}
{> MMmLsSvr - Calculate the number of servers         }
{>            needed for M/M/m given arrival rate     }
{>            and prob. of lost customer              }
{ Inputs:                                             }
{      LAMBDA      customer arrival rate              }
{      TS          mean service time                  }
{      PTARG       target prob(customer lost)         }
{ Outputs:                                            }
{      M           number of servers needed           }
{      U           traffic intensity                  }
{      RHO         server utilisation                 }
{      PLOSS       actual prob(customer lost)          }
{  Copyright Mike Tanner 1993                         }
{-----------------------------------------------------}
Procedure MMmLsSvr(LAMBDA,TS,PTARG:QTHreal;
             Var M:integer;Var U,RHO,PLOSS:QTHreal);
Var RM:QTHreal;
begin
   U:=LAMBDA*TS;
   M:=0;
   Repeat
      M:=M+1;
      PLOSS:=1-Pratio(M,U);
   until PLOSS<=PTARG;
   RM:=M;   RHO:=U/RM;
end;
```

MMmLsTrf—calculate maximum traffic for M/M/m/m

The subroutine MMmLsTrf calculates the maximum arrival rate that a specified number of servers can handle, given the average service-time and maximum proportion of customers that may be lost. It solves Eq. (19.15)

```
{-----------------------------------------------------}
{> MMmLsTrf - Calculate the maximum arrival rate      }
{>            for M/M/m given number of servers       }
{>            and prob. of lost customer              }
{ Inputs:                                             }
{      TS          mean service time                  }
{      PTARG       target prob(customer lost)         }
{      M           number of servers                  }
{ Outputs:                                            }
{      LAMBDA      customer arrival rate              }
{      U           traffic intensity                  }
{      RHO         server utilisation                 }
{  Copyright Mike Tanner 1993                         }
{-----------------------------------------------------}
Procedure MMmLsTrf(TS,PTARG:QTHreal;M:integer;
                Var LAMBDA,U,RHO:QTHreal);
Const ACC=0.0001;                  { accuracy required }
Var LAMBDALOW,LAMBDAHGH,PLOSS,PLOW,PHGH,RM:QTHreal;
Label L1;
```

```
begin
   RM:=M;      { type conversion   }
   {-Establish range------------------------------}
   LAMBDALOW:=0; PLOW:=0;
   LAMBDAHGH:=RM/TS;
   U:=LAMBDAHGH*TS;   PHGH:=1-Pratio(M,U);
   While PHGH<PTARG do begin
       LAMBDAHGH:=2*LAMBDAHGH;
       U:=LAMBDAHGH*TS;   PHGH:=1-Pratio(M,U);
   end;
   {-Binary search-------------------------------}
L1:LAMBDA:=(LAMBDALOW+LAMBDAHGH)/2;
   U:=LAMBDA*TS;   PLOSS:=1-Pratio(M,U);
   If Abs(PTARG-PLOSS)<ACC
   then begin RHO:=U/RM; Exit; end;
   If PLOSS<PTARG
   then begin LAMBDALOW:=LAMBDA;   PLOW:=PLOSS;  end
   else begin LAMBDAHGH:=LAMBDA;   PHGH:=PLOSS;  end;
   Goto L1;
end;
```

Example 19A—The Busyline Ticket Agency

The Busyline theatre ticket agency has 25 agents taking calls. Calls arrive at the rate of seven per minute, and it takes on average four minutes to deal with a call. No call queueing system is used, so if all the agents are busy further calls get a line-busy tone and are lost. How busy are the servers, and what percentage of calls will be lost? If the manager wants to reduce lost calls to 5 per cent of the total, how many agents will be needed?

The parameters are $\lambda = 7$ calls/min, $T_S = 4$ min, and $m = 25$. The traffic intensity is $u = \lambda T_S = 28$. We must next calculate $E_B(m,u) = 0.206$, using Eq. (19.3) and the subroutine for the Poisson ratio function given in Appendix 2. From Eq. (19.4) we know that this is the probability of losing a customer, so 20.6 per cent of customers will be lost. Equation (19.7) then gives the server utilization, which is 89 per cent.

The number of servers needed to reduce the loss rate to 5 per cent or less is the solution of Eq. (19.14), with $q = 0.05$. Using the MMmLsSvr routine, or Q-Calc, or trial and error with increasing numbers of servers, we can easily find out that 34 agents will be needed. The server utilization will then be 82.4 per cent, and the actual loss rate will be 4.2 per cent. (With 33 agents the loss rate would be 5.3 per cent.)

20
Multiserver with limited waiting-line (M/M/m/K)

Introduction

In Chapter 19 we looked at multiserver systems where no waiting was allowed. In this chapter we look at situations where customers can wait if all the servers are busy, but there is a limit on how many customers can be waiting. In reality this may be because there is a finite amount of space in, say, a shop or a bank. Figure 20.1 illustrates this type of system. The multiserver loss system is a special case of M/M/m/K, where K, the maximum number allowed in the system, is the same as the number of servers m, so no waiting-line is allowed. On the other hand, if K were very large compared to m, so that long waiting-lines could be accommodated, we would expect behaviour to be similar to M/M/m where there is no limit on the waiting-line.

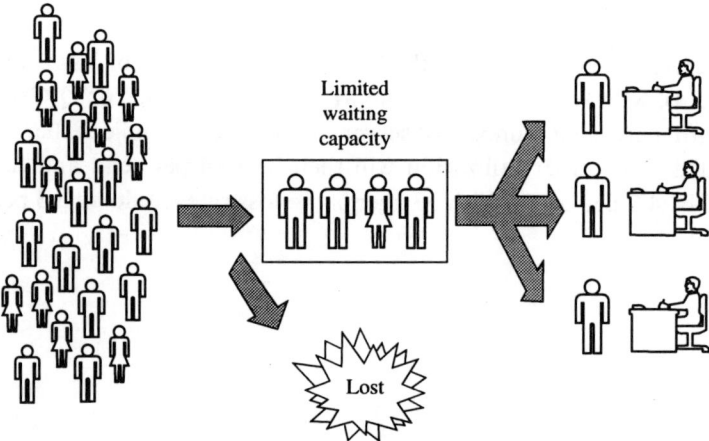

Figure 20.1. Queue with limited waiting capacity.

Assumptions, parameters, and initial calculations for M/M/m/K

There are m servers. Customers arrive at random, forming a single waiting-line if all servers are busy. Any customer arriving to find K customers already present, so that m customers are being served and $K - m$ are waiting, immediately departs and is lost to the system. Service-times are exponentially distributed. The following parameters therefore need to be specified to define the system.

> m, the number of servers
> λ, the average arrival rate
> T_s, the average service-time
> K, the maximum number of customers in the system

We first calculate the traffic intensity, i.e.

$$u = \text{traffic intensity} = \lambda T_S \tag{20.1}$$

Stability

The system is by definition stable since, if more customers arrive than the servers can deal with, a proportion of the customers are 'lost'. Even if the traffic intensity is less than the number of servers, some customers will still be lost due to short-term peaks, even though on average the servers could have dealt with all the arriving customers.

Number-in-system

In order to calculate the usual range of performance characteristics we have to evaluate the probability distribution of the number of customers in the system. This is done with the following formulae.

$$p_k = \text{Prob}(k \text{ customers in system}) = p_0 \frac{u^k}{k!} \text{ for } k = 1, 2, \ldots, m \tag{20.2}$$

$$p_k = p_0 \frac{u^m}{m!} \left(\frac{u}{m}\right)^{k-m} \text{ for } k = m+1, \ldots, K \tag{20.3}$$

$$p_k = 0 \text{ for } k > K \tag{20.4}$$

Equations (20.2)–(20.4) assume we know p_0, the probability of the system being empty of customers. To calculate p_0 we use

$$p_0 = \left[\sum_{k=0}^{m} \frac{u^k}{k!} + \frac{u^m}{m!} \sum_{k=m+1}^{K} \left(\frac{u}{m}\right)^{k-m}\right]^{-1} \tag{20.5}$$

Once we know the p_k we can calculate the mean and variances of the number of customers in the system. We do this directly from first principles.

$$L_Q = \text{average number-in-system} = \sum_{k=0}^{K} kp_k \tag{20.6}$$

$$\sigma_{L_Q}^2 = \text{variance of number-in-system} = \sum_{k=0}^{K} (k - L_Q)^2 p_k \tag{20.7}$$

Lost customers versus maximum waiting-line

For any system in which some customers may be lost, a key characteristic is the probability of losing an arriving customer, or equivalently the proportion of arriving customers that are lost. This probability is simply

$$\text{Prob(arriving customer is lost)} = p_K \tag{20.8}$$

We must now define λ_A as the actual arrival rate. This is the arrival rate as seen by the servers, or the throughput rate. Calculation of the actual arrival rate is very simple, from

$$\lambda_A = \text{actual arrival rate} = \lambda(1 - p_K) \tag{20.9}$$

Let us look at how the percentage of arriving customers that are lost is related to the maximum waiting-line size. This is shown in Fig. 20.2 for a single-server

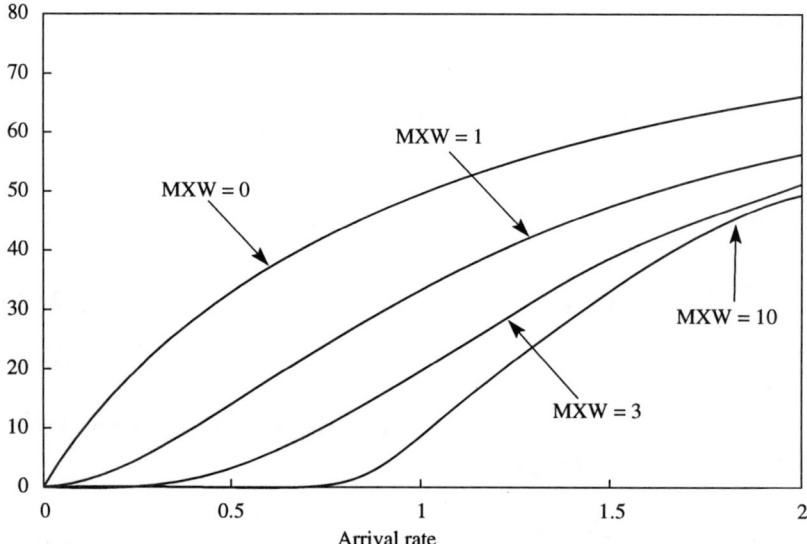

Figure 20.2. Percentage of customers lost versus arrival rate and maximum waiting-line.

system. MXW is the maximum waiting line allowed, i.e. $K - m$. The line labelled MXW = 0 corresponds to no waiting allowed, and a high proportion of customers is lost even when the arrival rate is well below the throughput capacity of the server. As the limit of the number of customers waiting is increased, the proportion lost goes down. Note that since $T_S = 1$ if the arrival rate is, say, two customers per unit time, the server will be able to handle only one per unit time on average, and 50 per cent of the customers will be lost.

Another way of looking at the loss rate in relation to the maximum waiting-line is shown in Fig. 20.3. Three cases are illustrated. In each case the traffic intensity

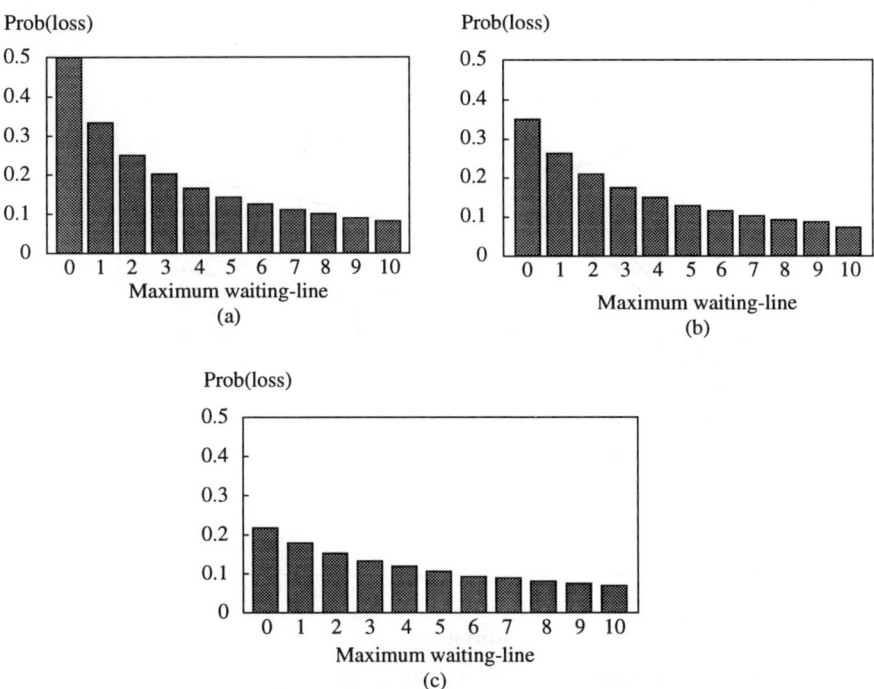

Figure 20.3. Loss rate against maximum waiting-line: (a) $\lambda = m = 1$, (b) $\lambda = m = 3$, (c) $\lambda = m = 10$.

(or arrival rate since T_S was chosen to be 1) is the same as the number of servers. In other words the customer arrival rate is the same as the throughput capacity of the servers. This is not essential to the point being made, but is just a convenient assumption for our illustration. In each case the histogram shows how the loss rate declines as the maximum waiting line increases. In case (a) with a single server, the effect is more marked than in case (b) with three servers, which in turn is more marked than in case (c) with 10 servers.

Server utilization versus maximum waiting-line

Server utilization depends on the actual arrival rate rather than the offered rate, and the calculation of utilization is

$$\rho = \text{server utilization} = \frac{\lambda_A T_S}{m} \tag{20.10}$$

Figure 20.4 shows the relationship between maximum waiting-line, offered arrival rate and server utilization, for the single-server case. For a given arrival rate a longer maximum waiting-line allows a higher server utilization to be achieved (since a lower proportion of customers will be lost).

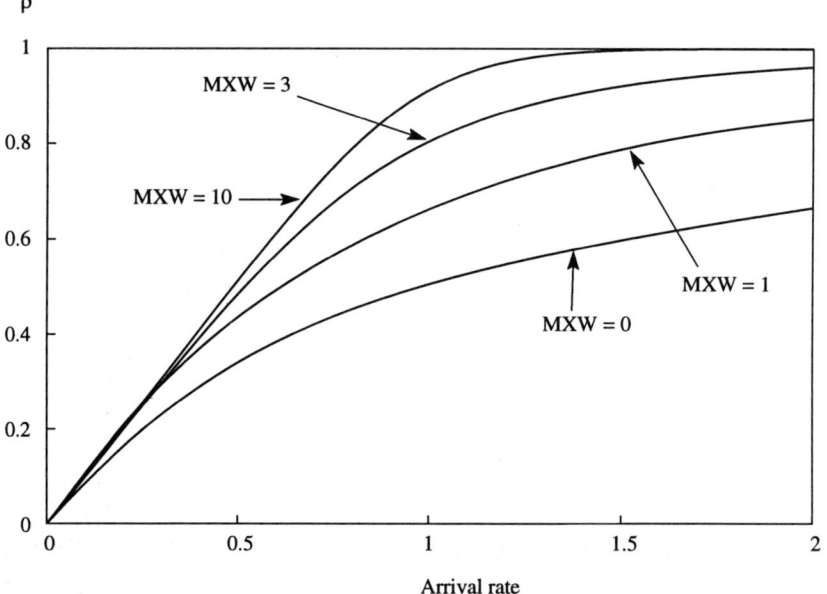

Figure 20.4. Utilization against arrival rate for a single server.

The average number of customers being served can be arrived at in two ways. From first principles Eq. (20.11) gives us L_S. Alternatively, we can apply Little's law to the servers (excluding the waiting-line). Remembering that λ_A is the relevant arrival rate, we get Eq. (20.12).

$$L_S = \text{average number being served} = \sum_{k=0}^{m-1} k p_k + m\left(1 - \sum_{k=0}^{m-1} p_k\right) \tag{20.11}$$

$$L_S = \lambda_A T_S \tag{20.12}$$

Number waiting

The mean and variance of the number of customers waiting can be calculated directly once the p_k values have been found. The appropriate formulae are

$$L_W = \text{average number waiting} = \sum_{k=m+1}^{K} (k-m)p_k \qquad (20.13)$$

$$\sigma_{L_W}^2 = \text{variance of number waiting} = \sum_{k=m+1}^{K} (k-m-L_W)^2 p_k \qquad (20.14)$$

An alternative way of calculating the average number waiting is Eq. (20.15). A drawback of this formula is that it is undefined for $u = m$.

$$L_W = \frac{u^m p_0 \theta}{m!(1-\theta)^2}[1 - \theta^{K-m+1} - (K-m+1)(1-\theta)\theta^{K-m}], \qquad (20.15)$$

where

$$\theta = \frac{u}{m}$$

Waiting time

Average waiting time is most easily obtained using Little's law applied to the servers and waiting-line, i.e. not including the part of the system where arriving customers are forced to depart if there is no further room for customers to wait. The result is

$$T_W = \text{average waiting time} = \frac{L_W}{\lambda_A} \qquad (20.16)$$

We have already calculated p_k, the probability that an arriving customer is lost. There are some other probabilities that we may wish to know. First of all, the probability that an arriving customer is served immediately and has zero wait time is given by Eq. (20.17). This is the probability that at least one server is free.

$$p_{ZW} = \text{Prob(customer served immediately)} = \sum_{k=0}^{m-1} p_k \qquad (20.17)$$

Another probability of interest is that of an arriving customer having to wait. This is the probability that all the servers are busy, but less than the permitted maximum number of customers in the system is present, and is given by

$$p_{DLY} = \text{Prob(arriving customer must wait)} = \sum_{k=m}^{K-1} p_k \qquad (20.18)$$

We may also want to know the average waiting time for the customers that do actually have to wait. The average waiting time T_W is the average across all the customers that are served, excluding those that are lost. Equation (20.19) gives the average across the customers that are served, but after a non-zero wait.

$$T_{WD} = \text{average wait for delayed customers} = T_W \frac{p_{ZW} + p_{DLY}}{p_{DLY}} \quad (20.19)$$

Time-in-system

In Eq. (20.6) we calculated the average number of customers in the system from first principles. Using Little's law we can use this to arrive at the average time-in-system, as in

$$T_Q = \text{average time-in-system} = \frac{L_Q}{\lambda_A} \quad (20.20)$$

Programs for M/M/m/K

Prerequisite routines

Apart from the type definitions, only the specific subroutines for M/M/m/K are needed, so that the set of 'include' statements required to make use of the routines from this chapter is as follows.

```
{$N+,E+ }
{$I \qthprog\qthtypes.pas      }
{$I \qthprog\mmmkcalc.pas      }
```

The MMmKCalc routine

This routine calculates most of the statistics likely to be of interest for an M/M/m/K queueing system. The subroutine generates the probabilities p_k and uses these to calculate directly the average and standard deviation of waiting-line size and number of customers in the system. The limit on how many customers may be in the system is expressed as MXW, the maximum waiting-line size, rather than the maximum either waiting or being served. This is a more natural way of expressing the limit. Readers who prefer to specify maximum in system can easily modify the subroutine by making MXW a local variable, replacing it with MXIS in the procedure header, and adding the statement

```
MXW:=MXIS-M;
```

at the beginning of the executable statements.

   The calculations are done in two passes, i.e. the averages are obtained first, and variances on a second pass. It would be possible to calculate variances in the first pass, by summing squared values of, for example, waiting-line size. This, however, can result in floating-point overflow, so a second pass is used and squared deviations from the mean are summed, keeping the numerical values to be dealt with to much lower values. This avoids numerical problems for parameter values likely to be used.

```
{------------------------------------------------}
{> MMmKCalc - Calculations for M/M/m/K loss model }
{ Inputs:                                         }
{      LAMBDA    customer arrival rate            }
{      TS        mean service time                }
{      M         number of servers                }
{      MXW       maximum number waiting           }
{ Outputs:                                        }
{      U         traffic intensity                }
{      RHO       server utilisation               }
{      PLOSS     probability arriving customer lost }
{      ARVRTE    arrival rate excluding lost      }
{                customers                         }
{      P0        probability all servers idle      }
{      PZW       prob. customer served immediately }
{      PDLY      prob. customer served after delay }
{      LW/SDVLW  mean/std dev no. customers waiting }
{      LQ/SDVLQ  mean/std dev no. in system        }
{      TW        average wait time for customers   }
{                served i.e. not lost              }
{      TWD       average wait time for customers   }
{                served but who have to wait       }
{      TQ        average time in system for        }
{                customers served i.e. not lost    }
{  Copyright Mike Tanner 1993                      }
{------------------------------------------------}
Procedure MMmKCalc(LAMBDA,TS:QTHreal;M,MXW:integer;
             Var U,RHO,PLOSS,ARVRTE,P0,PZW,PDLY,LW,
                SDVLW,LQ,SDVLQ,TW,TWD,TQ:QTHreal);
Var RM,P,RR,Z,TOTP:QTHreal;R:integer;
begin
   RM:=M;    U:=LAMBDA*TS;
   {-Probability calculations and average number---}
   {-in system and waiting------------------------}
   P:=1;  TOTP:=1;   PZW:=1;   PDLY:=0;
   LW:=0;   LQ:=0;
   For R:=1 to (M+MXW) do
   begin
      RR:=R;    { type conversion }
      {-Calculate Prob(R in system) relatively-----}
      If R<=M then P:=P*U/RR
              else P:=P*U/RM;
      TOTP:=TOTP+P;
      If R<M then PZW:=PZW+P;
      If (R>=M) and (R<(M+MXW)) then PDLY:=PDLY+P;
      LQ:=LQ+P*RR;
      If R>M then LW:=LW+P*(RR-RM);
   end;
   {-Adjust for value of TOTP--------------------}
   P0:=1/TOTP;
```

```
PLOSS:=P*P0;   PZW:=PZW*P0;   PDLY:=PDLY*P0;
LW:=LW*P0;     LQ:=LQ*P0;
{-Generate Prob(R customers in system)----------}
SDVLW:=P0*Sqr(LW);   SDVLQ:=P0*Sqr(LQ);
P:=P0;
For R:=1 to (M+MXW) do begin
   RR:=R;     { type conversion }
   If R<=M then P:=P*U/RR
           else P:=P*U/RM;
   If R>M then SDVLW:=SDVLW+P*Sqr(RR-RM-LW);
   SDVLQ:=SDVLQ+P*Sqr(RR-LQ);
end;
SDVLW:=Sqrt(SDVLW);   SDVLQ:=Sqrt(SDVLQ);
{-Actual arrival rate, and server utilisation---}
ARVRTE:=LAMBDA*(1-PLOSS);
RHO:=U*(1-PLOSS)/RM;
{-Average waiting time and time in system------}
If ARVRTE>0 then begin
                    TW:=LW/ARVRTE;
                    TQ:=LQ/ARVRTE;
                end
            else begin
                    TW:=0;
                    TQ:=0;
                end;
{-Waiting time for delayed customers-----------}
If PDLY>0 then TWD:=TW*(PDLY+PZW)/PDLY
          else TWD:=0;
end;
```

Example 20A—The Busyline Ticket Agency with call-queueing

Recall Example 19A, the ticket agency that wanted to reduce the percentage of lost customers. Suppose the manager has decided, instead of simply increasing the number of agents, to install extra telephone equipment so that calls could be queued until an agent became available. The new equipment allows a maximum of 50 calls to be either waiting or being dealt with. What percentage of calls would be lost with the same number of agents? How many agents would be needed to reduce losses to less than 5 per cent? If, in addition, customer waiting time must be less than 30 s, how many agents would be needed?

From Example 19A we have $\lambda = 7$ calls/min, $T_S = 4$ min, and $m = 25$ agents. Using Eq. (20.2), (20.3), (20.5) and (20.8) (in practice the subroutines given, or QCalc), we can calculate that 11 per cent of customers would be lost. The server utilization in this case would be 99.7 per cent. Experimenting with different numbers of agents shows that 28 agents would be needed, giving a loss rate of 3.4 per cent. Using Eqs (20.13) and (20.16) to calculate average wait, we can experiment with numbers of agents to discover that 32 agents are needed to reduce the average wait to less than 30 seconds. In fact the average wait would be 20.4 seconds.

21
Multiserver with general arrivals and general service-times (G/G/m)

Introduction

In Chapter 15 we dealt with the G/G/1 queueing model, and in Chapter 17 we covered the M/M/m model. In this chapter we are concerned with a generalization of both those models. With G/G/m the arrival pattern is arbitrary, there is a general distribution of service-time, and there are multiple servers. For G/G/1 we could only get bounds and approximations for the queueing statistics. So obviously for G/G/m we shall again have to be content with bounds and approximations.

Assumptions and initial calculations

There are m servers with a single FIFO queue. The population of potential customers is infinite. Customers arrive at an average rate λ, so that the average inter-arrival time is $T_A = 1/\lambda$. The distribution of inter-arrival time is general, with variance $\sigma_{T_A}^2$ and coefficient of variation squared C_A^2. Service-times have a general distribution, with mean T_S, variance $\sigma_{T_S}^2$, and coefficient of variation squared C_S^2.

Heavy-traffic approximation for average wait (HTA)

This approximation was derived by Kollerstrom. Under heavy load, the distribution of waiting-time has an approximately exponential distribution, with mean waiting time given by Eq. (21.1). For low utilization this approximation is poor, but at high utilizations (say over 90 per cent) the approximation is reasonable.

$$T_W = \text{average waiting time} \approx \frac{\lambda\left(\sigma_{T_A}^2 + \frac{1}{m^2}\sigma_{T_S}^2\right)}{2(1-\rho)} \quad \text{as } \rho \to 1 \qquad (21.1)$$

Kingman's upper bound for average wait (KUB)

Kingman derived the following upper bound for the average wait. Comparisons later in this chapter demonstrate that this upper bound is generally pessimistic. However, it is useful because it is a true upper bound rather than an approximation.

$$T_W = \text{average waiting time} \le \frac{\left(\sigma_{T_A}^2 + \frac{1}{m^2}\sigma_{T_S}^2 + \frac{(m-1)}{m^2}T_S^2\right)}{2T_A(1-\rho)} \tag{21.2}$$

Allen–Cunneen approximation for average wait (ACA)

Allen and Cunneen devised the following approximation for average waiting time.

$$T_W = \text{average waiting time} \approx \frac{E_C(m,u)}{m(1-\rho)}\left(\frac{C_A^2 + C_S^2}{2}\right)T_S \tag{21.3}$$

where $E_C(m, u)$ is the Erlang-C function, defined by Eqs (17.4) and (17.5). This approximation is exact for M/M/m, and comparisons later in this chapter will demonstrate that the Allen–Cunneen approximation is very good.

Published tables of results for G/G/m and G/G/m/K

Although exact solutions for G/G/m are not available, results have been published in Seelen *et al.* (1985) for particular forms of inter-arrival and service-time distributions. For these tables the distributions can be deterministic, mixed-Erlangian or hyperexponential-2. Deterministic just means constant. We have seen the Erlang distribution in Chapter 11. An Erlang-k distribution has a squared coefficient of variation of $1/k$, where k is an integer. This allows a limited set of coefficient values to choose from. A mixed-Erlangian distribution is a suitably weighted average of two Erlang distributions, one with parameter k and the other with parameter $k - 1$. This gives any coefficient of variation we want between 0 and 1. The hyperexponential-2 distribution has also been encountered already, in Chapter 13, where we saw that the distribution could have any squared coefficient of variation greater than 1. This range of distributions for inter-arrival and service-times is not completely general, but is extremely useful none the less.

As explained in Chapters 11 and 13, Erlang and hyperexponential distributions are combinations of exponential distributions. Exponential distributions are very tractable mathematically, because of their 'memoryless' property (see Chapter 6). The mathematical analysis of queueing systems based on exponential distributions gives rise to a set of 'equilibrium equations'. For queueing models such as M/M/1, M/M/m, these equations can be solved algebraically to produce usable formulae for calculating the queueing statistics we want. As the distributions become more complicated combinations of exponential distributions, it becomes increasingly difficult and less useful to solve the equilibrium equations algebraically. We then resort to numerical solutions of the equations, and even this approach involves non-trivial numerical analysis problems. Luckily for us, however, Seelen *et al.* developed a set of computer programs to solve the equilibrium equations for the inter-arrival and service-time distributions described. These programs were then used to produce the tables in Seelen *et al.* (1985). The tables cover $G/G/m$ and $G/G/m/K$.

Comparison A

In this comparison we choose $m = 6$ servers, $T_S = 1$, and $C_A^2 = C_S^2 = 0.5$. For a range of server utilizations exact results were obtained from the tables in Seelen *et al.* (1985), and the bounds and approximations given above were calculated. The results are given in Table 21.1 and shown in Fig. 21.1. The Allen–Cunneen approximation is very close to the exact result, the heavy traffic approximation is reasonable at high utilization, but the Kingman upper bound is rather pessimistic.

Table 21.1. Results for comparison A

Server utilization	TWHTA (heavy traffic approximation)	TWACA (Allen–Cunneen approximation)	TWKUB (Kingman's upper bound)	EXACT (from Seelen *et al.* (1985))
0.40	0.201	0.006	0.479	0.002
0.50	0.208	0.017	0.625	0.008
0.60	0.236	0.041	0.861	0.026
0.65	0.261	0.062	1.034	0.044
0.70	0.296	0.093	1.268	0.071
0.75	0.347	0.141	1.597	0.115
0.80	0.427	0.216	2.094	0.186
0.85	0.563	0.347	2.924	0.313
0.90	0.838	0.617	4.588	0.555
0.95	1.669	1.443	9.586	1.401
0.98	4.168	3.938	24.584	3.895

Comparison B

The parameters chosen for this comparison were the same as for comparison A, except that $C_A^2 = C_S^2 = 4$. The results are given in Table 21.2 and shown in Fig. 21.2. This comparison leads to the same conclusions as comparison A. Allen–Cunneen is very good, the heavy traffic approximation is reasonable at high utilization, and the Kingman upper bound is pessimistic.

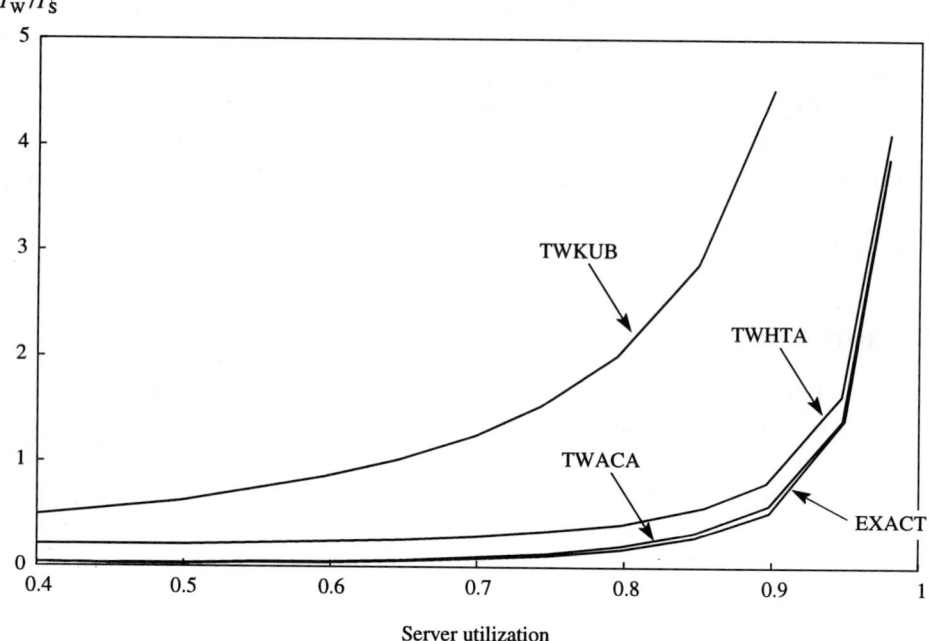

Figure 21.1. Graph of results of comparison A.

Table 21.2. Results for comparison B

Server utilization	TWHTA (heavy-traffic approximation	TWACA (Allen–Cunneen approximation)	TWKUB (Kingman's upper bound) .	EXACT (from Seelen *et al.* (1985))
0.40	1.611	0.044	1.889	0.063
0.50	1.667	0.132	2.083	0.170
0.60	1.889	0.328	2.514	0.393
0.65	2.084	0.497	2.858	0.579
0.70	2.365	0.747	3.337	0.846
0.75	2.778	1.124	4.028	1.242
0.80	3.417	1.726	5.083	1.862
0.85	4.503	2.774	6.864	2.930
0.90	6.704	4.934	10.454	5.111
0.95	13.351	11.541	21.268	11.734
0.98	33.340	31.506	53.757	31.720

Programs for G/G/m

In order to use the routines for G/G/m, a program must contain the following statements

```
{$N+,E+ }
{$I \qthprog\QTHTypes.pas }
{$I \qthprog\QTHError.pas }
{$I \qthprog\PRatio.pas   }
{$I \qthprog\GGmCalc.pas  }
```

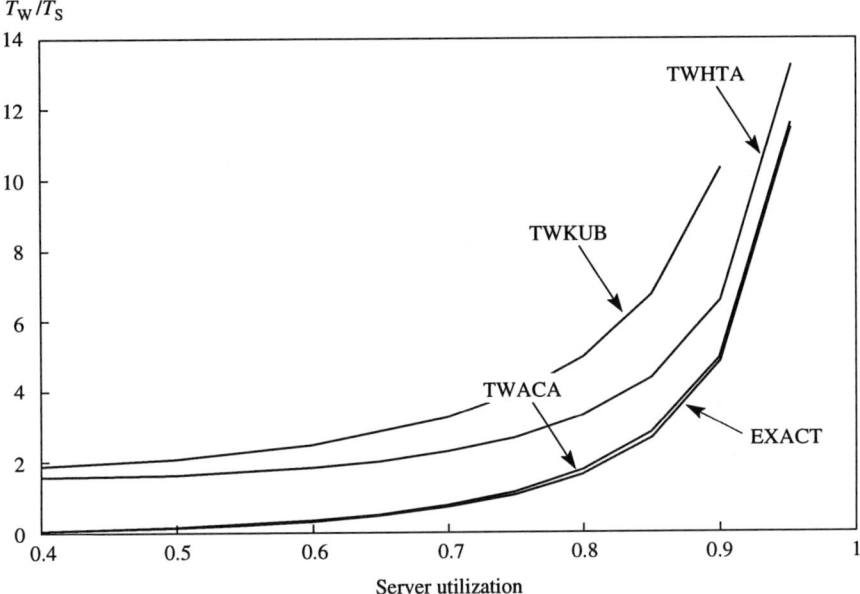

Figure 21.2. Graph of results for comparison B.

GGmCalc—general calculations for G/G/m

The routine performs calculations for G/G/m. Each of the approximations and bounds for average wait are provided. The Allen–Cunneen result is used as a basis for calculating averages for time-in-system, number waiting, and number-in-system.

```
{-------------------------------------------------}
{> GGmCalc - Calculations for G/G/m queueing model }
{ Inputs:                                          }
{      LAMBDA   customer arrival rate              }
{      CA2      sqd coeff of var inter-arrival times }
{      TS       mean service time                  }
{      COEFF2   sqd coeff of var service times     }
{      M        number of servers                  }
{ Outputs:                                         }
{      RHO      server utilisation                 }
{      TWHTA    average wait, heavy traffic appx   }
{      TWACA    average wait, Allen-Cunneen appx   }
{      TWKUB    average wait, Kingmans upper bound  }
{      TQ       average time in system             }
{      LW       average number waiting             }
{      LQ       average number in system           }
{      VALID    T=results valid, F=not valid       }
{   Copyright Mike Tanner 1993                     }
{-------------------------------------------------}
```

```
Procedure GGmCalc(LAMBDA,CA2,TS,COEFF2:QTHreal;
                  M:integer;
                  Var RHO,TWHTA,TWACA,TWKUB:QTHreal;
                  Var TQ,LW,LQ:QTHreal;
                  Var VALID:boolean);
Var VTA,VTS,U,RM,RMU,ELGC,Z:QTHreal;
begin
    RM:=M;                              { type conversion    }
    {-Calculate utilisation and check stability-----}
    U:=LAMBDA*TS;   RHO:=U/RM;
    VALID:=(RHO<1.0) and (LAMBDA>0);
    If not VALID then Exit;
    {-Variance of inter-arrival and service times---}
    VTA:=CA2/Sqr(LAMBDA);
    VTS:=COEFF2*Sqr(TS);
    Z:=VTA+VTS/Sqr(RM);
    {-Heavy-traffic approximation for average wait--}
    TWHTA:=LAMBDA*Z/(2*(1-RHO));
    {-Kingmans upper bound------------------------}
    TWKUB:=LAMBDA*(Z+(RM-1)*Sqr(TS/RM))/(2*(1-RHO));
    {-Allen-Cunneen approximation for average wait--}
     RMU:=Pratio(M,U);
    ELGC:=(1-RMU)/(1-RHO*RMU);
    TWACA:=ELGC*TS*(CA2+COEFF2)/(2*RM*(1-RHO));
    { Use Allen-Cunneen as base for other stats-----}
    TQ:=TWACA+TS;
    LW:=LAMBDA*TWACA;
    LQ:=LAMBDA*TQ;
end;
```

Example 21A—The Hordit Bank

The Hordit Bank plans the number of cashiers to have on duty so that the average number of customers waiting (not including those being served) will be the same as the number of cashiers. For example, with four cashiers, the average number waiting should be four. The average service-time is 30 secs with $C_S^2 = 0.2$. Customers arrive rather more bunched together than with 'random' arrivals, and we assume $C_A^2 = 2$. Use the Allen–Cunneen approximation (Eq. (21.3)) and Little's law (Eq. (5.6)) to find the arrival rate that can be dealt with by two, three, four and five cashiers.

The results are calculated by a little trial and error using Q-Calc, and are given in Table 21.3. Alternatively a small program could be constructed using the

Table 21.3. Results for Example 21A

Cashiers	Target average waiting-line	Maximum customers per min	Average wait in min (s)	Server utilization (%)
2	2	2.97	0.68 (41)	74.30
3	3	4.84	0.62 (37)	80.60
4	4	6.75	0.59 (35)	84.30
5	5	8.48	0.47 (28)	84.80

GGMCalc routine. The idea of planning for size of waiting-line to equal the number of cashiers seems a reasonable one, since average waiting time and server utilization are maintained at very roughly the same levels. Intuitively we would expect the average wait to be similar to the average service-time.

(This rule is not the same as the method used by a number of banks to respond to changing arrival rates, where cashiers are added if the waiting-line exceeds the number of cashiers. This example is about a steady-state situation, rather than dynamic responses to changing arrival rates of transient peaks in arrival rates.)

Part Seven
Limited number of customers

22
Priority queues

Introduction

Priority systems are very common. In computer systems users or jobs are assigned a priority based on their importance. Tasks within an operating system get a high priority if they need to handle a 'real-time' event. On communication links short messages are often given priority over long messages, and 'status' or 'control' messages are given priority over data messages. Airlines and car-rental companies allow you to buy priority in check-in procedures. At the supermarket 'one basket' customers may have special check-outs denied to customers with a trolley-full of goods.

There are two types of priority scheme. With head-of-line systems, also known as non-preemptive systems, once a customer has commenced service he or she will not be interrupted by a higher priority customer who arrives after service has started. The higher priority customer will join the waiting-line ahead of any lower priority customers present, and within the same priority class customers are served first-come-first-served. A head-of-line system is illustrated in Fig. 22.1.

The other type of priority scheme is called pre-emptive, where a higher priority customer will pre-empt the server if the server is dealing with a lower priority customer. Once the higher priority customer has been served, service of the lower priority customer will resume where it left off. A pre-emptive system is shown in Fig. 22.4. Complex sequences of the server switching between customers of several different priorities may occur, where an arriving customer pre-empts the server, only to be pre-empted in turn by a later-arriving yet-higher priority customer. Other types of pre-emptive system may be found, such as pre-emptive-restart where a customer whose service is interrupted must start the service activity from the beginning again when the server next becomes available. We shall deal only with pre-emptive-resume, since that is the most widespread system. Pre-emptive-resume is used for most cpu scheduling algorithms.

It is important to note that both these priority schemes require priority to be assigned to classes of customers independent of individual service-times. It may be that classes of customers do in fact have different service-times, either constant per class or different distribution per class. If customers are assigned a priority based on their individual service-time requirements, then a different analysis is needed. One method of doing such an analysis is explained in this chapter.

Unfortunately, easy-to-use results are available only for single-server systems with priorities. Multiple-server systems with priorities are complex to analyse, and simulation will often be used to tackle such queues.

Perhaps it would be worth while to point out the slightly confusing terminology of priority systems. We use the phrases 'higher priority' and 'lower priority' in the sense that customers with a higher priority will be given preference over those with a lower priority. When it comes to the mathematics, we usually label the classes of customer 1, 2, 3, etc. meaning that class 1 has the highest priority, class 2 has the next highest priority, and so on.

Assumptions and preliminary calculations

There are n classes of customer. Class 1 has priority over class 2, which in turn has priority over class 3, and so on. For each class of customer we specify the parameters

$$\lambda_{(j)} = \text{arrival rate, } j = 1 \ldots n \tag{22.1}$$

$$T_{S(j)} = \text{average service-time, } j = 1 \ldots n \tag{22.2}$$

$$C_{S(j)}^2 = \text{coeff of variation squared for service-time, } j = 1 \ldots n \tag{22.3}$$

We shall want the total arrival rate, which is

$$\lambda = \text{total arrival rate } = \lambda_{(1)} + \lambda_{(2)} + \ldots + \lambda_{(n)} = \sum_{j=1}^{n} \lambda_{(j)} \tag{22.4}$$

A key role in the calculation is played by what we might term the 'partial traffic intensities'. The jth partial traffic intensity is the traffic intensity up to and including the jth customer class, i.e.

$$u_{(j)} = \text{traffic intensity up to } j\text{th class} = \sum_{i=1}^{j} \lambda_{(i)} T_{S(i)} \text{ for } j = 1 \ldots n \tag{22.5}$$

For mathematical and programming convenience we define $u_{(0)} = 0$, and the total server utilization is clearly $\rho = u_n$. Another key item we need is the second moment of service-time for each customer class, and overall. These values are calculated with the following formulae.

$$s_{2(j)} = \text{second moment of service-time for class } j$$
$$= T_{S(j)}^2 (1 + C_{S(j)}^2) \tag{22.6}$$

$$s_2 = \text{second moment of overall service-time} = \sum_{i=1}^{n} \frac{\lambda_{(i)}}{\lambda} s_{2(i)} \tag{22.7}$$

$$s_2 = \frac{\lambda_{(1)}}{\lambda} s_{2(1)} + \frac{\lambda_{(2)}}{\lambda} s_{2(2)} + \ldots + \frac{\lambda_{(n)}}{\lambda} s_{2(n)} \tag{22.8}$$

Non-preemptive (head-of-line) priority system

With a head-of-line priority scheme, once a customer has commenced service that customer will not be interrupted by a customer of greater priority who arrives later (Fig. 22.1).

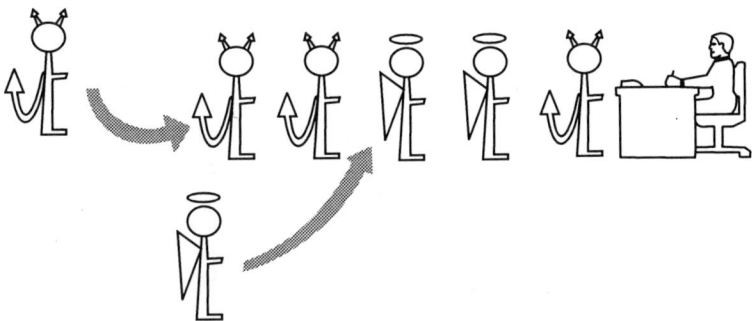

Figure 22.1. Head-of-line priority system.

For the system to be stable we need, as usual, $\rho < 1$. The average waiting time for each class is

$$T_{W(j)} = \text{average wait for } j\text{th class}$$

$$= \frac{\lambda s_2}{2(1 - u_{(j-1)})(1 - u_{(j)})} \quad \text{for } j = 1 \ldots n \tag{22.9}$$

The average waiting time over all classes is then just the average of these values weighted by the arrival rate for each class.

$$T_W = \text{average wait overall} = \sum_{j=1}^{n} \frac{\lambda_{(j)}}{\lambda} T_{W(j)} \tag{22.10}$$

Average time-in-system is obtained using the basic relationship that time-in-system is service-time plus waiting time. This is applied to each class, and then the weighted average is taken to get the overall average time-in-system.

$$T_{Q(j)} = \text{average time-in-system for } j\text{th class} = T_{W(j)} + T_{S(j)} \tag{22.11}$$

$$T_Q = \text{average time-in-system overall} = \sum_{j=1}^{n} \frac{\lambda_{(j)}}{\lambda} T_{Q(j)} \tag{22.12}$$

The effect of a head-of-line priority scheme is shown in Fig. 22.2. Ten per cent of arriving customers are 'angels', with a constant service-time of 0.5 min. Angels are given head-of-line priority over 'devils', who form 90 per cent of the customers and have a constant service-time of 5 min. Figure 22.2 shows that angels continue to have a low waiting time even at high utilizations. The line marked 'All' shows

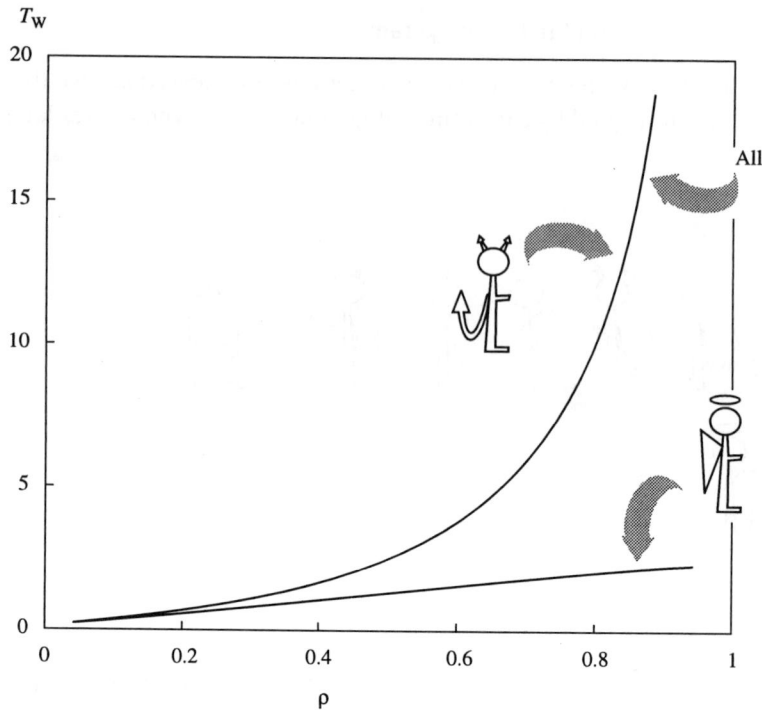

Figure 22.2. Performance of HOL system with small class of high priority customers.

the average waiting time for all customers for first-come-first-served scheduling. This curve is almost identical to the devils' average wait-time with the priority scheme. The angels' low waiting times have been achieved with negligible effect on overall waiting times. Of course this is because the angels form only a small proportion of the customers, and have a low average service-time.

If we now change the mix of customers, so that 50 per cent are angels and 50 per cent are devils, and assume that all customers, both angels and devils, have a constant service-time of 2.5 min, then the resulting average waiting times are shown in Fig. 22.3. Low waiting times for the angels are still achieved, but at the cost of significant worsening of the devils' waiting time compared to what would happen without any priority scheme.

Pre-emptive priority system

With pre-emptive priority (Fig. 22.4), customers in a higher priority class are never affected by lower priority customers. Stability of the system then becomes an interesting question. The capacity of the server may be inadequate to handle the total arrival rate across all priority classes, while being able to cope with at least the highest priority class. For the top priority customers such a system will be stable,

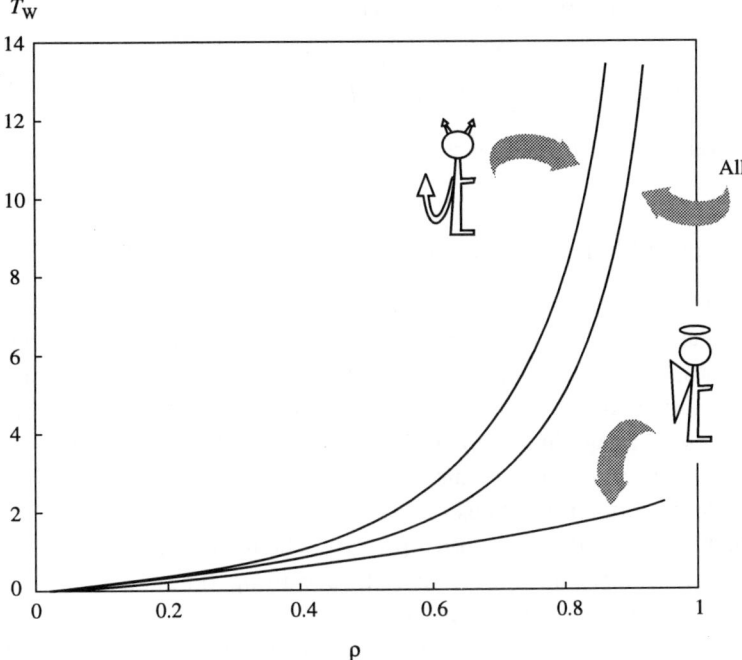

Figure 22.3. Performance of HOL system with large class of high priority customers.

but there will be lower priority customers who will never get served, and the overall waiting-line will grow without limit.

With the head-of-line scheme we worked out the average waiting time for each class, and then derived the average time-in-system by adding average service-times. For pre-emptive priority the formulae are more convenient if we calculate average time-in-system first, and then subtract average service-time to get the average wait.

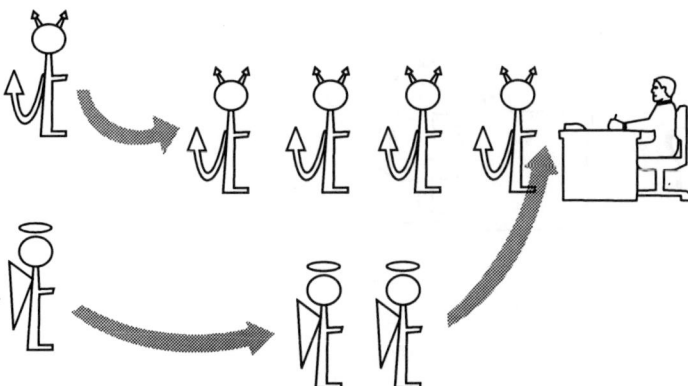

Figure 22.4. Pre-emptive priority system.

$T_{Q(j)}$ = average time-in-system for jth class

$$= \frac{1}{(1 - u_{(j-1)})} \left(T_{S(j)} + \frac{\sum_{i=1}^{j} \lambda_{(i)} S_{2(i)}}{2(1 - u_{(j)})} \right) \tag{22.13}$$

$T_{W(j)}$ = average wait for jth class = $T_{Q(j)} - T_{S(j)}$ $\hspace{2cm}$ (22.14)

Priority based on individual customer service-times

The priority schemes we have looked at so far require priority to be assigned on the basis of a pre-defined set of customer classes. The classes may have different average service-times, and we may decide to assign higher priority to classes that have lower average service-times. However, we do not need to do this, and in practice we are quite likely to assign higher priority to categories of customer with particular importance, authority, need, status, etc. without reference to the average service-times of different categories of customer.

On the other hand, we may decide to use a priority scheme to reduce the overall average waiting time. We would expect to reduce waiting times if we always selected the next customer to be served on the basis of shortest service-time. As a trivial example, Fig. 22.5 shows a server with two customers waiting to be served. The devil will take up 5 min of the server's time, while the angel will take only 30 seconds. It does not matter which of them arrived first, since whatever waiting time has so far been incurred, the question is how to minimize additional waiting time. If the devil is served first then the angel will have to wait for 5 min, and on average each will wait an additional 2.5 min. If the angel is served first then the devil will wait for 30 seconds, and on average they will each wait a further 15 seconds.

5 min

30 s

Figure 22.5. Choosing next customer to serve on the basis of individual service-times.

We can analyse a more realistic situation as follows. As usual, we shall take the M/M/1 system as our basis for comparison. For the priority scheme, we shall approximate the exponential service-time distribution with a large number of customer classes. Each class has a constant service-time, and the proportion of customers in each class is chosen so that the overall service-time distribution closely follows the exponential distribution.

$$n = \text{number of customer classes} \qquad (22.15)$$

$$\delta = \text{range of service-time represented by a single class} = \frac{10T_S}{n} \qquad (22.16)$$

$$
\begin{aligned}
p_j &= \text{proportion of customers in class } j \\
&= F(j\delta) - F[(j-1)\delta] \text{ for } j = 2 \ldots n
\end{aligned}
\qquad (22.17)
$$

where $F(x)$ is the cumulative distribution function of the exponential distribution with mean T_S.

$$T_{S(j)} = \text{service-time for } j\text{th class} = \left(j - \frac{1}{2}\right)\delta \qquad (22.18)$$

The calculations required are done in the program STmePrty, a listing of which is given later in this chapter. Table 22.1 shows the results produced by that program. Column RHO is the server utilization or, since $T_S = 1$ was chosen, the arrival rate. Column TWMM1 is the average waiting time for an M/M/1 system, calculated using Eq. (7.08). Column TWFCFS is the average waiting time calculated using our approximation to the exponential with many classes of customers, but with first-come-first-served priority. The purpose of this column is to demonstrate the validity of our approximation by obtaining almost identical results to TWMM1. The final column, TWPRTY, is the interesting one, and shows the average waiting time achieved by using a shortest-service-time-first head-of-line priority scheme. Figure 22.6 shows the same results in graphical form. Observe that at high utilizations we can more than halve the average waiting time.

Table 22.1 Results from program STmePrty

RHO	TWMM1	TWFCFS	TWPRTY
0.100	0.111	0.111	0.105
0.200	0.250	0.249	0.223
0.300	0.428	0.427	0.358
0.400	0.666	0.665	0.517
0.500	0.999	0.997	0.711
0.600	1.499	1.495	0.960
0.700	2.331	2.326	1.309
0.800	3.994	3.985	1.878
0.900	8.974	8.953	3.189

T_W/T_S

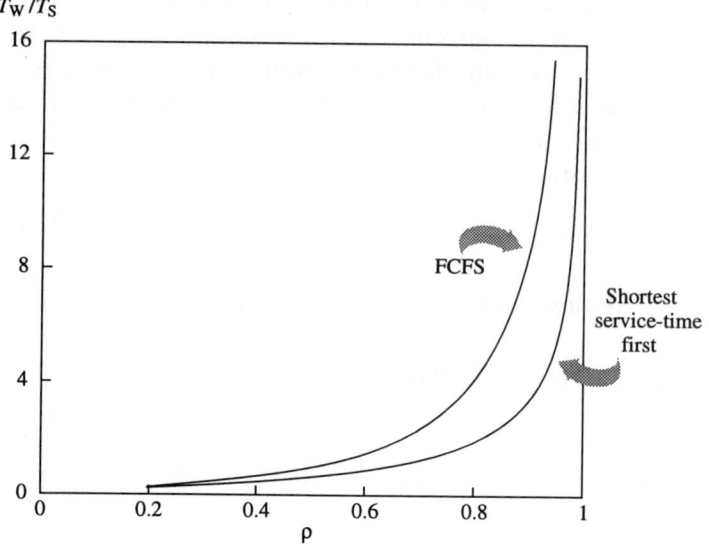

Figure 22.6. Comparison of M/M/1 systems with FCFS scheduling and shortest-service-time-first.

Programs for priority queues

In order to use the routines for priority queues, a program must contain the following statements.

```
{$N+,E+,R+ }
{$I \qthprog\qthtypes.pas   }
{$I \qthprog\qtherror.pas   }
{$I \qthprog\qthfile.pas    }
{$I \qthprog\holcalc.pas    }
{$I \qthprog\premtclc.pas   }
{$I \qthprog\prty0clc.pas   }
{$I \qthprog\pratio.pas     }
```

HOLCalc—calculations for head-of-line priority

This routine calculates the performance characteristics for head-of-line, or non-preemptive, priority.

```
{----------------------------------------------------}
{> HolCalc - Calculations for M/G/1 with             }
{>           non-preemptive priority(head-of-line)   }
{ Inputs:                                            }
{     N            number of customer classes        }
{     LAMBDAC[.] arrival rate for each class         }
{     TSC[.]      mean service time for each class    }
{     CS2C[.]     sqd. coeff of variation of service }
{                 time for each class                }
```

```
{ Outputs:                                                    }
{      LAMBDA      total arrival rate                         }
{      RHO         server utilisation                         }
{      TIC[.]      traffic intensity per class                }
{      CTIC[.]     cumulative traffic intensity               }
{                  per class                                  }
{      TW          mean waiting time all classes              }
{      TWC[.]      mean waiting per class                     }
{      TQ          mean time in system all classes            }
{      TQC[.]      mean time in system per class              }
{      VALID       T=results valid, F=not valid               }
{   Copyright Mike Tanner 1993                                }
{-------------------------------------------------------------}
Procedure HolCalc(N:integer;
           Var LAMBDAC,TSC,CS2C:QTHRealArray;
           Var LAMBDA,RHO:QTHreal;
           Var TIC,CTIC:QTHRealArray;Var TW:QTHreal;
           Var TWC:QTHRealArray;Var TQ:QTHreal;
           Var TQC:QTHRealArray;Var VALID:boolean);
Var ES2C:QTHRealArray;TS,ES2:QTHReal;J:integer;
begin
    LAMBDA:=0; CTIC[0]:=0; TS:=0; ES2:=0;
    For J:=1 to N do begin
       LAMBDA:=LAMBDA+LAMBDAC[J];
       TIC[J]:=LAMBDAC[J]*TSC[J];
       CTIC[J]:=CTIC[J-1]+TIC[J];
       TS:=TS+LAMBDAC[J]*TSC[J];
       ES2C[J]:=Sqr(TSC[J])*(1+CS2C[J]);
       ES2:=ES2+LAMBDAC[J]*ES2C[J];
    end;
    TS:=TS/LAMBDA; ES2:=ES2/LAMBDA;
    RHO:=CTIC[N];
    If RHO>=1.0 then begin VALID:=false; Exit; end
                else VALID:=true;
    TW:=0;
    For J:=1 to N do begin
       TWC[J]:=LAMBDA*ES2
               /(2*(1-CTIC[J-1])*(1-CTIC[J]));
       TW:=TW+LAMBDAC[J]*TWC[J];
       TQC[J]:=TWC[J]+TSC[J];
    end;
    TW:=TW/LAMBDA; TQ:=TW+TS;
end;
```

PreemptCalc—calculations for pre-emptive (resume) priority

This subroutine calculates all the useful performance characteristics for a pre-emptive priority scheme.

```
{-------------------------------------------------------------}
{> PreemptCalc - Calculations for M/G/1 with                  }
{                preemptive-resume priority                   }
{ Inputs:                                                     }
{      N           number of customer classes                 }
```

```
{      LAMBDAC[.]  arrival rate for each class          }
{      TSC[.]      mean service time for each class      }
{      CS2C[.]     sqd. coeff of variation of service   }
{                  time for each class                   }
{ Outputs:                                               }
{      LAMBDA      total arrival rate                     }
{      RHO         server utilisation                     }
{      TIC[.]      traffic intensity per class            }
{      CTIC[.]     cumulative traffic intensity           }
{                  per class                              }
{      TW          mean waiting time all classes          }
{      TWC[.]      mean waiting per class                 }
{      TQ          mean time in system all classes        }
{      TQC[.]      mean time in system per class          }
{      VALID       T=results valid, F=not valid           }
{   Copyright Mike Tanner 1993                            }
{-------------------------------------------------------}
Procedure PreemptCalc(N:integer;
          Var LAMBDAC,TSC,CS2C:QTHRealArray;
          Var LAMBDA,RHO:QTHreal;
          Var TIC,CTIC:QTHRealArray;Var TW:QTHreal;
          Var TWC:QTHRealArray;Var TQ:QTHreal;
          Var TQC:QTHRealArray;Var VALID:boolean);
Var ES2C,CES2C:QTHRealArray;TS,ES2:QTHReal;J:integer;
begin
   LAMBDA:=0; CTIC[0]:=0; TS:=0; ES2:=0; CES2C[0]:=0;
   For J:=1 to N do begin
      LAMBDA:=LAMBDA+LAMBDAC[J];
      TIC[J]:=LAMBDAC[J]*TSC[J];
      CTIC[J]:=CTIC[J-1]+TIC[J];
      TS:=TS+LAMBDAC[J]*TSC[J];
      ES2C[J]:=Sqr(TSC[J])*(1+CS2C[J]);
      CES2C[J]:=CES2C[J-1]+LAMBDAC[J]*ES2C[J];
      ES2:=ES2+LAMBDAC[J]*ES2C[J];
   end;
   TS:=TS/LAMBDA;
   ES2:=ES2/LAMBDA;
   RHO:=CTIC[N];
   If RHO>=1.0 then begin VALID:=false; Exit; end
              else VALID:=true;
   {-Average wait time and time in system----------}
   TQ:=0;
   For J:=1 to N do begin
      TQC[J]:=(TSC[J]+CES2C[J]/(2*(1-CTIC[J])))
             /(1-CTIC[J-1]);
      TQ:=TQ+LAMBDAC[J]*TQC[J];
      TWC[J]:=TQC[J]-TSC[J];
   end;
   TQ:=TQ/LAMBDA; TW:=TQ-TS;
end;
```

Prty0Calc—calculations ignoring priority

When analysing a priority system, we often want to compare the performance with priorities against the same queueing system without a priority scheme. We could use MG1Calc for this, but it is convenient to have a routine that accepts the same set of parameters as our priority calculation routines. This is what Prty0Calc does.

```
{------------------------------------------------}
{> Prty0Calc - Calculations for M/G/1 with       }
{             priorities ignored.                }
{ Inputs:                                        }
{     N           number of customer classes     }
{     LAMBDAC[.]  arrival rate for each class     }
{     TSC[.]      mean service time for each class }
{     CS2C[.]     sqd. coeff of variation of service }
{                 time for each class            }
{ Outputs:                                       }
{     LAMBDA      total arrival rate             }
{     RHO         server utilisation             }
{     TW          mean waiting time all classes  }
{     TWC[.]      mean waiting per class          }
{     TQ          mean time in system all classes }
{     TQC[.]      mean time in system per class   }
{     VALID       T=results valid, F=not valid    }
{  Copyright Mike Tanner 1993                    }
{------------------------------------------------}
Procedure Prty0Calc(N:integer;
          Var LAMBDAC,TSC,CS2C:QTHRealArray;
          Var LAMBDA,RHO,TW:QTHreal;
           Var TWC:QTHRealArray;Var TQ:QTHreal;
          Var TQC:QTHRealArray;Var VALID:boolean);
Var ES2C:QTHRealArray;TS,CS2,ES2:QTHReal;J:integer;
begin
   LAMBDA:=0; TS:=0;  ES2:=0;
   For J:=1 to N do begin
      LAMBDA:=LAMBDA+LAMBDAC[J];
      TS:=TS+LAMBDAC[J]*TSC[J];
      ES2C[J]:=Sqr(TSC[J])*(1+CS2C[J]);
      ES2:=ES2+LAMBDAC[J]*ES2C[J];
   end;
   TS:=TS/LAMBDA;
   ES2:=ES2/LAMBDA;
   CS2:=(ES2-Sqr(TS))/Sqr(TS);
   RHO:=LAMBDA*TS;
   If RHO>=1.0 then begin VALID:=false; Exit; end
               else VALID:=true;
   {-Average wait time and time in system M/G/1----}
   TW:=TS*(RHO/(1-RHO))*(1+CS2)/2;;
   For J:=1 to N do begin
      TWC[J]:=TW;
      TQC[J]:=TWC[J]+TSC[J];
   end;
   TQ:=TW+TS;
end;
```

STmePrty—calculations for service in service-time order

Unlike most of the program routines in this book, STmePrty is a complete
program rather than a subroutine. STmePrty calculates the average waiting time
for an M/M/1 system where customers are selected for service based on their individual waiting times, shortest first. The exponential distribution is approximated
by using many classes of customer, each class having a constant service-time. The
proportion of customers in each class is derived from the exponential distribution.
This program was used to produce Table 22.1, and also, with modifications to
output more data points, Fig. 22.6.

Readers may wish to modify this program to handle other service-time distributions. All that is needed is a subroutine to evaluate the cdf of service-time, and a
judgement of the maximum service-time to be considered and the number of customer classes to use. Note that the file QTHTYPES.PAS is not included, but the
relevant statements from that file are contained in the program. This allows the
constant MaxASize to be set to a big enough value for the number of customer
classes to be used.

```
{----------------------------------------------------}
{> STmePrty -- Calculations to show effect of        }
{>            service in shortest service time        }
{>            first order.                            }
{  Copyright Mike Tanner 1993                         }
{----------------------------------------------------}
Program STmePrty;
{$N+,E+,R+ }
Uses Dos,Crt;
{ This section is instead of QTHTYPES.PAS----------}
Const MaxASize=200;  { larger than default value   }
Type QTHreal=double;
Type QTHRealArray=array[0..MaxAsize] of QTHreal;
{-Include other QTH subroutines required----------}
{$I \qthprog\qtherror.pas   }
{$I \qthprog\qthfile.pas    }
{$I \qthprog\holcalc.pas    }
{$I \qthprog\premtclc.pas   }
{$I \qthprog\prty0clc.pas   }
{$I \qthprog\expondst.pas   }
Var N:integer;RN:QTHreal; { number of classes       }
Var LAMBDA:QTHreal;      { total arrival rate       }
Var LAMBDAC:QTHRealArray; { arrival rate per class  }
Var P:QTHRealArray;      { proportion cust per class }
Var RHO   :QTHreal;      { server utilisation        }
Var TIC:QTHRealArray;    { traffic int per class     }
Var CTIC:QTHRealArray;   { cum traf int per class    }
Var TS:QTHreal;          { overall avg service time  }
Var TSC:QTHRealArray;    { avg svce time per class   }
Var CS2C:QTHRealArray;   { svce coeff var per class  }
Var TW:QTHreal;          { overall average wait      }
Var TWO:QTHreal;         { average wait no priority  }
Var TWMM1:QTHreal;       { average wait for M/M/1    }
```

```
Var TWC:QTHRealArray;    { average wait per class    }
Var TQ:QTHreal;          { overall avg time in sys   }
Var TQC:QTHRealArray;    { avg tis per class         }
Var J:integer;
Var ARVRATE,RJ,DELTA:QTHReal;
Var VALID:boolean;
Const TAB=chr(09);
begin
    QTHOpenFile('\qthprog\stmeprty.prn');
    Writeln(OPF,'    RHO ',TAB,'    TWMM1',
            TAB,'   TWFCFS',TAB,'    TWPRTY');
    {-Specify the overall mean service time and the-}
    {-number of customer classes to be used--------}
    TS:=1.0;                 { overall mean service time  }
    N:=200; RN:=N;           { number of classes          }
    DELTA:=10*TS/RN;         { see text                   }
    For J:=1 to N do begin
        RJ:=J;               { type conversion            }
        {-Calculate service time for class j, and----}
        {-set coeff. of variation to zero-----------}
        TSC[J]:=(RJ-0.5)*DELTA;
        CS2C[J]:=0;
        {-Calc proportion of customers in class j----}
        P[J]:=ExponCdf(RJ*DELTA,TS)
              -ExponCdf((RJ-1)*DELTA,TS);
    end;
    {-Calc performance for a range of arrival rates-}
    LAMBDA:=0.1;
    While LAMBDA<0.99 do begin
        {-Arrival rate for each class---------------}
        For J:=1 to N do LAMBDAC[J]:=P[J]*LAMBDA;
        {-Performance with no priority scheme--------}
        Prty0Calc(N,LAMBDAC,TSC,CS2C,
                  ARVRATE,RHO,TW,TWC,TQ,TQC,VALID);
        TW0:=TW;
        {-Performance with priority-----------------}
        HolCalc(N,LAMBDAC,TSC,CS2C,
                ARVRATE,RHO,TIC,CTIC,TW,TWC,TQ,TQC,VALID);
        {-M/M/1 wait-time for comparison------------}
        TWMM1:=RHO*TS/(1-RHO);
        Writeln(OPF,RHO:8:3,TAB,TWMM1:8:3,
                    TAB,TW0:8:3,TAB,TW:8:3);
        LAMBDA:=LAMBDA+0.1;
    end;
    QTHWaitKey;
    QTHCloseFile;
end.
```

Example 22A—The Helpful Hardware Store

The Helpful Hardware Store is a small place, with just one person serving. There are two types of customers. 'Purchasers' actually buy something, with a service-time exponentially distributed averaging 150 seconds. 'Enquirers' have a simple query, with a service-time exponentially distributed averaging 12 seconds. On average one purchaser and two enquirers arrive every 5 min. The store owner believes he can improve service by having a separate queue for enquiries, and by giving enquirers priority. Calculate the waiting times for both categories of customers with first-come-first-served, head-of-line priority, and pre-emptive priority.

The easiest way to do this is to construct a simple program using the Prty0Calc, HolCalc and PreemptCalc routines. The results are given in Table 22.2. Alternatively, the hand calculations for just two classes are not too laborious. Use Eqs (22.1)–(22.8) to establish the basic terms needed Equation (22.9) gives the head-of-line result, and Eqs (22.13) and (22.14) give the pre-emptive priority result. For no priority all customers form one category, with a service-time that is a mixture of two exponential distributions. Referring to Chapter 13, readers will see that this is an M/H2/1 model. Alternatively, Q-Calc can be used.

Table 22.2. Results for Example 22A

Queueing rule	Average waiting time (s)		
	All customers	Enquirers	Purchasers
First-come-first-served	180.86		
Head-of-line priority to enquirers	120.57	82.57	196.58
Pre-emptive priority to enquirers	70.57	1.04	209.63

Part Six
Priority queues

23
Single server with finite number of customers (M/M/1/K/K), i.e. machine repairman model

Introduction

This queueing system has one big difference from the queueing systems looked at so far. In previous models there has been an infinitely large population of potential customers. Here we have a fixed number of customers each going through a cycle of a period of time when service is not required, requiring service and joining the waiting-line, being served, and returning to a period of not requiring service. This system is stable, because if all customers are waiting or being served then there are no further customers to join the waiting-line (Fig. 23.1).

This system is often called the 'machine repairman' model, because it could represent a set of machines (the customers) that run for some time and then

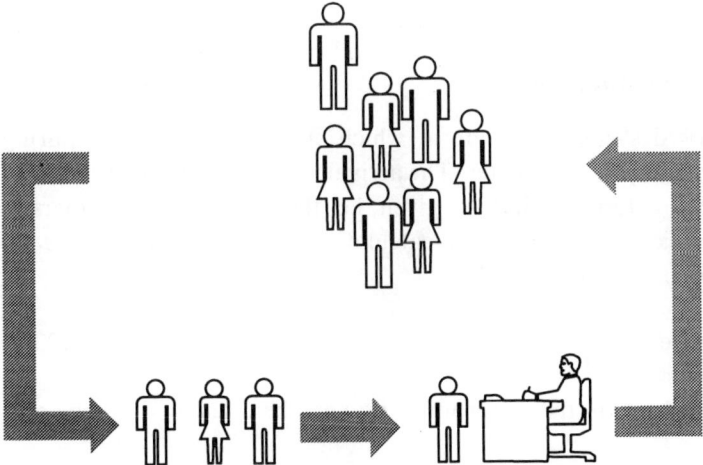

Figure 23.1. Single server with finite number of customers.

break down (requiring service), and must wait for the repairman (the server) to fix them.

The machine repairman model is a closed system, since no customers ever enter or leave the system. It can also be regarded as the simplest possible network of queues, and forms a link between the theory of isolated queues and the theory of queueing networks.

Assumptions, parameters and initial calculations for M/M/1/K/K

There is a fixed number K of customers in the system. Customers neither enter nor leave the system. It is this factor that makes the behaviour of the system self-limiting. If service-times are very long, then customers might spend a large proportion of their time either waiting to be served or being served, but the number waiting can never exceed $K - 1$. Service-times are exponentially distributed with an average T_S. Customers are served in FCFS order. When customers complete service, they then spend a period of time, referred to as the 'think' time or 'operating' time depending on the context, when they do not require service. These periods are assumed to be exponentially distributed with an average of T_A.

The calculations for this model involve the ratio of operating time to service-time, referred to as the service ratio

$$z = \text{service ratio} = \frac{T_A}{T_S} \tag{23.1}$$

Some authors define the service ratio as T_S/T_A, which in some ways is more natural. The next calculation required is the Poisson ratio function, which is defined as

$$R(K, z) = \text{Poisson ratio function} = 1 - \frac{\frac{z^K}{K!}}{\sum_{j=0}^{K} \frac{z^j}{j!}} \tag{23.2}$$

Arrival rate and server utilization

Because this is a closed system, the rate at which customers 'arrive' requiring service is not a given, or input, parameter. The arrival rate is something we have to work out from the characteristics that are given, i.e. the number of customers in the system and the service ratio. The actual calculation is simple enough, once the Poisson ratio function has been calculated.

$$\lambda = \text{arrival rate} = \frac{R(K, z)}{T_S} \tag{23.3}$$

Once the arrival rate is known, the server utilization can be calculated in the usual way, i.e.

$$\rho = \text{server utilization} = \lambda T_S = R(K, z) \tag{23.4}$$

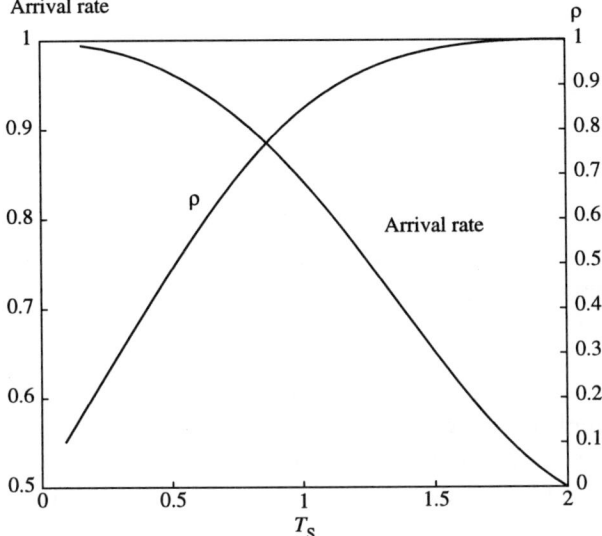

Figure 23.2. Arrival rate and server utilization versus service time for $T_A = 20$ and $K = 20$.

Figure 23.2 shows an example of how service-time, operating time, arrival rate, utilization, and population size are interrelated. There is a population of 20 customers, each with an average operating time of 20 time units. The graph shows how arrival rate and server utilization change as average service-time varies from nearly zero to 2 time units. As service-time increases, utilization goes up more and more slowly, and approaches 100 per cent only for very large average service-time. This is because the arrival rate is decreasing as service-time goes up.

Waiting time

The average waiting time is given by Eq. 23.5. This is a very simple formula, since the 'hard' calculations are in working out λ.

$$T_W = \text{average wait-time} = \frac{K}{\lambda} - (T_A + T_S) \tag{23.5}$$

The probability that a customer will have to wait is just ρ, the server utilization. The average waiting time, given that a customer does have to wait, is

$$T_{WD} = \text{average wait for delayed customers} = \frac{T_W}{\rho} \tag{23.6}$$

Time-in-system (queueing time)

The average time-in-system is given by Eq. 23.7. Again this is a simple formula once λ is known.

$$T_Q = \text{average time-in-system} = \frac{K}{\lambda} - T_A \qquad (23.7)$$

An obvious way to depict the behaviour of a queueing system is to plot T_Q, the average time spent in the system, against customer arrival rate or, equivalently, server utilization. If we do this we get the graph in Fig. 23.3, which also shows the M/M/1 characteristic for comparison. In a moment we shall see how this approach can be quite misleading.

For M/M/1, as server utilization gets close to 1, queueing time increases markedly and is 'infinite' for $\rho = 1$. Now in our closed system we have a fixed number of customers. Suppose the average operating or think time T_A is zero. This means that as soon as a customer completes service, that customer immediately joins the queue again for another service. In this situation the utilization of the server will be 1, but the queueing time will not be infinite, since it will consist of $K - 1$ service-times for all the other customers. When we plot T_Q against ρ for M/M/1, we are plotting a measure of the performance of the system against something that we can, at least in principle, control. We can often decide how much workload the server will be subjected to, i.e. we can decide the average customer arrival rate we want to handle, which translates directly into server utilization since $\rho = \lambda T_S$. However, for the closed system M/M/1/K/K, arrival rate and server utilization depend in a complex way on K, T_A and T_S. A more instructive way of depicting the performance of the closed system is to plot queueing time against the number of customers in the system. This is shown in Fig. 23.4, which also shows server utilization versus number of customers.

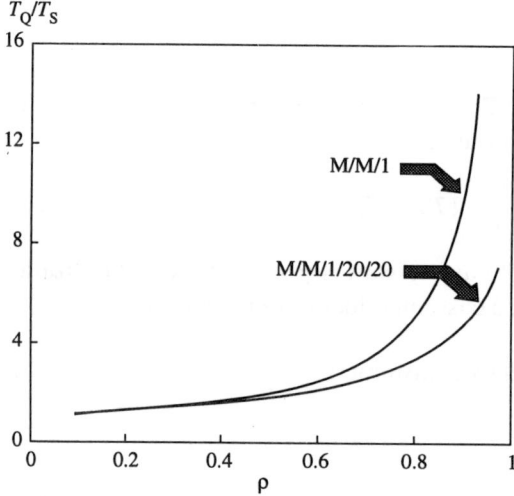

Figure 23.3. Queueing time versus server utilization for M/M/1 and M/M/1/20/20.

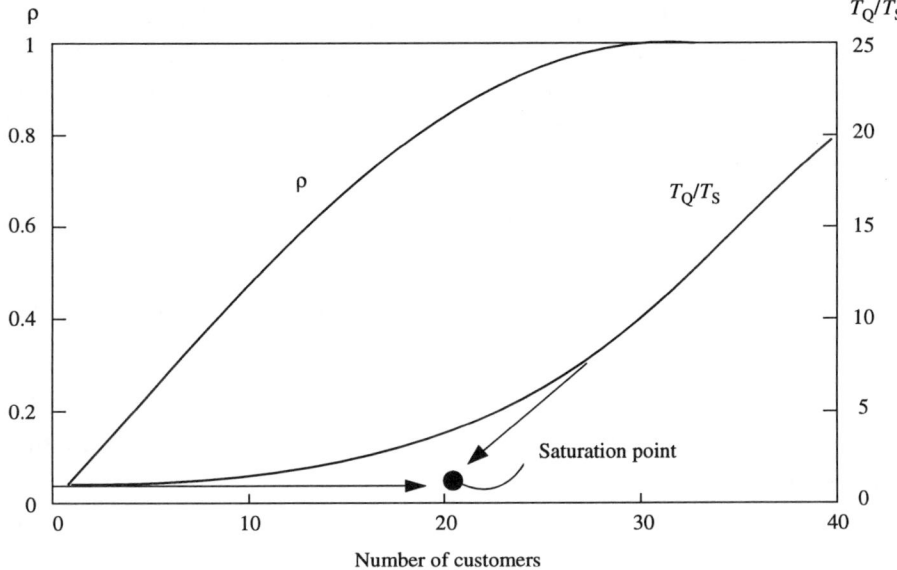

Figure 23.4. Server utilization and queueing time versus number of customers.

Let us look first at the T_Q/T_S curve. With few customers there is very little waiting, and the average queueing time is very nearly just the average service-time. As more customers are added, say in the range 10–20 in the particular system illustrated, more waiting occurs and average queueing time becomes significantly greater than average service-time. As even more customers are added, say from 30 customers upwards in Fig. 23.4, the curve becomes a straight line. Customers are spending a large proportion of their time in the waiting line, and the operating or think time becomes relatively insignificant. Each additional customer added to the system makes the queueing time just a little bit worse for all the customers. Contrast this with an open system such as M/M/1, where a slight increase in customer arrival rate at very high utilization would mean an explosive increase in queueing time.

Now we turn to the utilization curve in Fig. 23.4. starting with very few customers, as more are added the server utilization goes up nearly in proportion, so the curve is a straight line, up to about 15 customers in this case. Then, as further customers are added, server utilization does not go up so fast. As yet more customers are added, the corresponding increase in server utilization is even slower. A very large number of customers is needed to make ρ approach 1.

An interesting question is at what point we would consider the server to be 'saturated' or handling as many customers as reasonably possible. Theoretically, of course, we can continue adding customers indefinitely, since the system is self-limiting and will always be stable, although, of course, queueing time will steadily increase as we add more customers. We can identify a saturation point, though, and how we do this is shown in Fig. 23.4. We take the straight-line section of the queueing time curve and project it backwards. Where this line intercepts the line $T_Q/T_S = 1$ is the saturation point. These two straight lines form an 'ideal' performance characteristic, which could be achieved only with constant service-times and regular arrivals.

Number waiting or being served

Since we know the average waiting time, Little's law can be used to get the average number of customers waiting, i.e.

$$L_W = \lambda T_W = K - \lambda(T_A + T_S) \tag{23.8}$$

Similarly, we can get the average number of customers queueing, i.e. either waiting or being served. Using Little's law, this is

$$L_Q = \lambda T_Q = K - \lambda T_A \tag{23.9}$$

If we are interested in more detail about the distribution of the number waiting or in the system, then we will need the probability of a particular number being in the system. Note that a customer is considered to be 'in the system' if that customer is either waiting or being served, but not if the customer is in 'operating' or 'think' mode. These probabilities can be fairly easily calculated.

$$p_n = \text{Prob}(n \text{ waiting or being served}) = \frac{K!}{(K-n)!} \frac{p_0}{z^n} \text{ for } n = 0, \ldots, K \tag{23.10}$$

where p_0, which is the probability that all customers are in the operating or think mode, is given by

$$p_0 = \frac{1}{\sum_{n=0}^{K} \frac{K!}{(K-n)!} \frac{1}{z^n}} \tag{23.11}$$

If all the values of p_n are required for $n = 0, 1, \ldots, K$ then the relationship between successive p_n values is useful. This is

$$\frac{p_n}{p_{n-1}} = \frac{K - n + 1}{z} \tag{23.12}$$

Calculating the p_n requires p_0 to be evaluated. This can be done by first setting p_0 to an arbitrary value, usually 1, and then using Eq. (23.12) to calculate the relative values of the p_n. We can then use the fact that

$$\sum_{n=0}^{K} p_n = 1 \tag{23.13}$$

to scale all the relative values to their correct absolute value.

Offered traffic

In order to discuss the effect of the number of customers, and to define a way of measuring the 'efficiency' of this queueing system, we need the idea of offered traffic. Suppose there are 10 customers, each with an average operating time of 20 min. If the average service time T_S were zero then each customer would request the service every 20 min. The arrival rate would be 10 customers every 20 min, or 0.5 customers/min. The offered traffic rate is the arrival rate that would be achieved if service took zero time.

$$\lambda_O = \text{offered traffic rate} = \frac{K}{T_A} \qquad (23.14)$$

Effect of number of customers

Now let us investigate the effect of the population size. A particular offered traffic rate can be produced by many combinations of K and T_A. A small number of customers with short operating times will have the same offered traffic rate as a large number of customers with long operating times. In Fig. 23.5 we plot the average time-in-system against offered traffic, for different numbers of customers. With small numbers of customers there is little chance of customers simultaneously requiring service, so little contention for the server occurs and the average time-in-system stays low. As the number of customers increases and the operating time correspondingly increases to maintain the same offered traffic, there is more chance of contention for the server, so waits occur and average time-

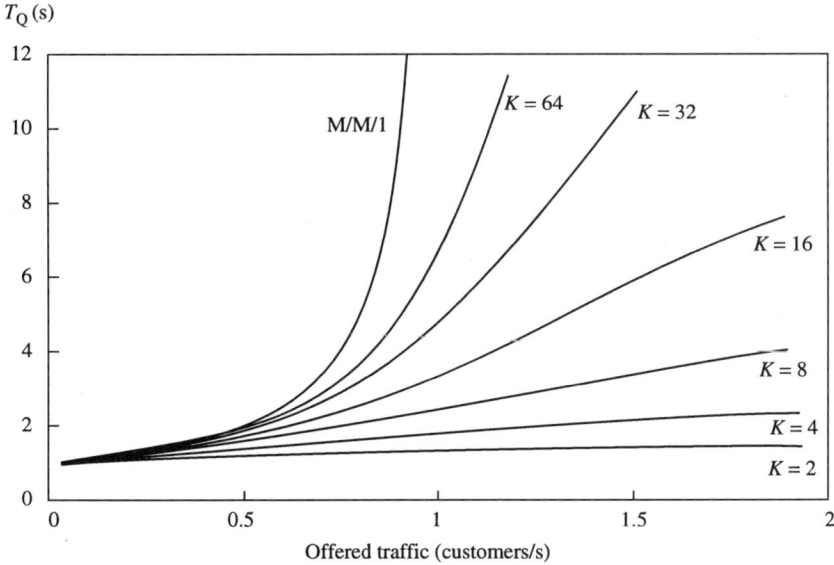

Figure 23.5. Time-in-system versus offered traffic for various numbers of customers.

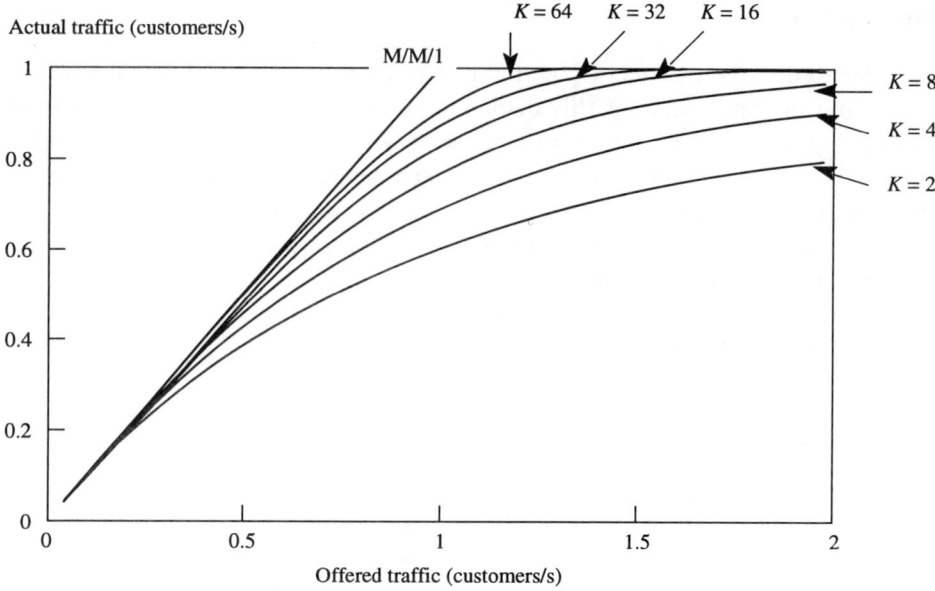

Figure 23.6. Actual versus offered traffic for various numbers of customers.

in-system goes up. The logical conclusion of this process is an infinite number of customers, and an M/M/1 queueing system. The performance curve for M/M/1 is shown in Fig. 23.5 for comparison.

In Fig. 23.6 we show the actual traffic versus offered traffic. For small numbers of customers actual traffic grows very slowly as offered traffic increases. For larger numbers of customers actual traffic grows faster. The M/M/1 curve is shown for comparison, where actual traffic equals offered traffic until the server's throughput capacity is reached.

Efficiency

We can now use the notion of offered traffic to define a measure of efficiency. This is very simply the ratio of the rate at which customers are processed to the rate at which customers would ideally be processed, as in

$$\text{efficiency} = \frac{\text{actual traffic}}{\text{offered traffic}} \times 100\% \tag{23.15}$$

which can also be written

$$\text{efficiency} = \frac{\lambda T_A}{K} \times 100\% \tag{23.16}$$

Equation (23.16) has a useful interpretation. Using Little's law, since T_A is the average operating time, λT_A is the average number of customers in the operating state. If we think of the customers as machines being tended by the server, then Eq. (23.16) is the percentage of machines that is, on average, in the operating state

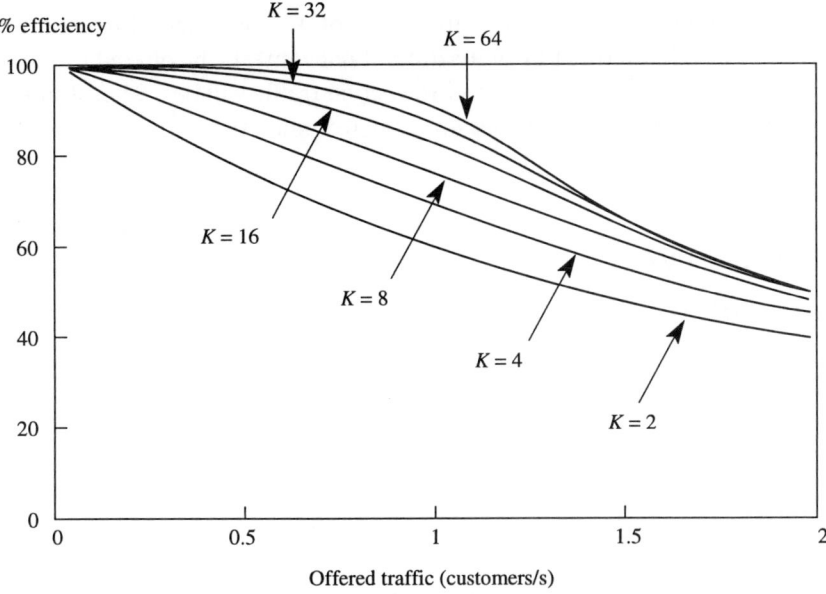

Figure 23.7. Efficiency versus offered traffic.

as opposed to waiting to be fixed or being fixed. This is a very reasonable definition of efficiency. Figure 23.7 shows how efficiency varies with offered traffic and
the number of customers in the system.

Latent demand

It is a well-known phenomenon that if a service is made quicker then the demand
for the service goes up, and waiting times therefore do not necessarily go down.
This increase in demand is called the 'latent demand'. Figure 23.8 shows an

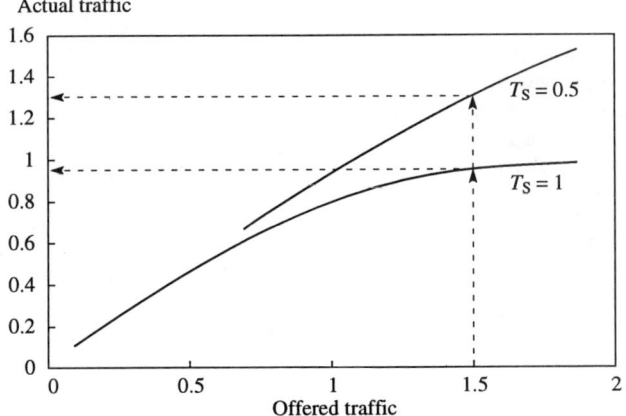

Figure 23.8. Latent demand.

example of latent demand, for a population size of 12 customers. Suppose we reduce T_S from 1 to 0.5 min (the time units are not relevant). If the offered traffic is 1.5 customers per minute, then with $T_S = 1$ the actual traffic will be 0.95 customers per minute. With the improved service time of 0.5 min, the actual traffic will increase to 1.32 customers per minute.

Programs for M/M/1/K/K

Prerequisite routines

In order to use the M/M/1/K/K routines, the following set of 'include file' statements would be used in Turbo-Pascal.

```
{$N+,E+ }
{$I \qthprog\qthtypes.pas    }
{$I \qthprog\qtherror.pas    }
{$I \qthprog\pratio.pas      }
{$I \qthprog\mm1fcclc.pas    }
```

MM1FcClc—general calculations for M/M/1/K/K

This routine calculates most of the statistics usually of interest for the single repairman model. Efficiency is not provided as an output from this routine, but is easily calculated using Eq. (23.16).

```
{-------------------------------------------------------}
{> MM1FcClc - Calculations for M/M/1 with finite        }
{             number of customers.                      }
{ Inputs:                                               }
{     TA          mean operating time                   }
{     TS          mean service time                     }
{     K           number of customers                   }
{ Outputs:                                              }
{     SVRAT       service ratio                         }
{     LAMBDA      arrival rate                          }
{     RHO         server utilisation                    }
{     TW          average waiting time                  }
{     TQ          average time in system                }
{     LW          average customers waiting             }
{     LQ          average customers in system           }
{ Copyright Mike Tanner 1993                            }
{-------------------------------------------------------}
Procedure MM1FcClc(TA,TS:QTHreal;K:integer;
                   Var SVRAT,LAMBDA,RHO:QTHreal;
                   Var TW,TQ,LW,LQ:QTHreal);
Var RK,RN,PRT:QTHreal;
Var N:integer;
begin
   RK:=K;       { type conversion  }
```

```
{-Calculate Poisson ratio function-------------}
SVRAT:=TA/TS;
PRT:=Pratio(K,SVRAT);
{-Arrival rate and server utilisation----------}
LAMBDA:=PRT/TS;
RHO:=PRT;
{-Average waiting and queueing time------------}
TW:=RK/LAMBDA-(TA+TS);
TQ:=TW+TS;
{-Average number waiting and in system---------}
LW:=LAMBDA*TW;
LQ:=LAMBDA*TQ;
end;
```

Example 23A

Until a few years ago it was common for a number of computer terminals to share a single relatively low-speed communications link to a computer centre. Suppose six terminals share a link, that it takes 0.7 seconds to transmit the input and output messages of a transaction, and the computer time for a transaction is negligible. If terminal users spend on average 10 seconds thinking between transactions, what is the transaction rate, line utilization, and transaction response time? Also, calculate results when the number of terminals is increased to 10, and then when a faster line is installed so that $T_S = 0.47$ seconds.

Using the single repairman model, we have $K = 6$, $T_A = 10$, and $T_S = 0.7$. Using Eqs (23.2–23.4) we can calculate the arrival rate and server utilization. Equation 23.7 then tells us the average time-in-system, which is the transaction response time. Table 23.1 gives the results.

Table 23.1. Results for Example 23A

Number of terminals	Average service-time	Transactions per s	Line utilization (%)	Transaction response time (s)
6.00	0.70	0.55	38.20	1.00
10.00	0.70	0.88	61.30	1.42
10.00	0.47	0.93	43.70	0.75

Example 23B—The Cotswolde Tearoom

The Cotswolde is a small tearoom in a tourist area. At times it is very busy, and its seven tables are always occupied. It takes on average 3.5 min for the waitress to serve each table. Once served, people stay on average 23 min, and when they leave their place is immediately taken by another party. How long do people wait to be served, what is the rate at which tables are served, and what is the utilization of the waitress? If the owner builds a small extension, and can now accommodate a total of 10 tables, calculate the result for the new situation.

The customers in this case are the tables! Using Eqs (23.1), (23.2), (23.4) and (23.5) we can calculate the results. Notice that it is waiting time, not time-in-system, we are concerned with. Q-Calc can also be used to calculate the results, which are given in Table 23.2.

The rate of serving tables has gone up from 12.7 to 15.7 per hour, an increase of nearly 24 per cent. Compare this to the increase in the number of tables from 7 to 10, an increase of 43 per cent. The average waiting time has more than doubled.

Table 23.2. Results for Example 23B

Number of tables	Tables served per min (per h)	Waitress utilization (%)	Average wait to be served (mins)
7.00	0.21 (12.7)	74.23	4.50
10.00	0.26 (15.7)	91.55	9.73

24
Multiple servers with finite number of customers (M/M/m/K/K), i.e. multiple repairmen model

Introduction

The multiple repairmen model is an obvious generalization of the single repairman case described in the previous chapter. Figure 24.1 illustrates the queueing situation. The single repairman model was treated separately in order to discuss the effects of a finite customer population and the definition of efficiency. Those ideas carry over directly to the multiple repairmen case. We shall see that the calculations are done slightly differently, since for a single repairman we can make use of the Poisson ratio function, whereas with multiple repairmen the probability distribution of the number of customers in the system has to be explicitly calculated to get the performance statistics we usually want. With the multiple repairman case

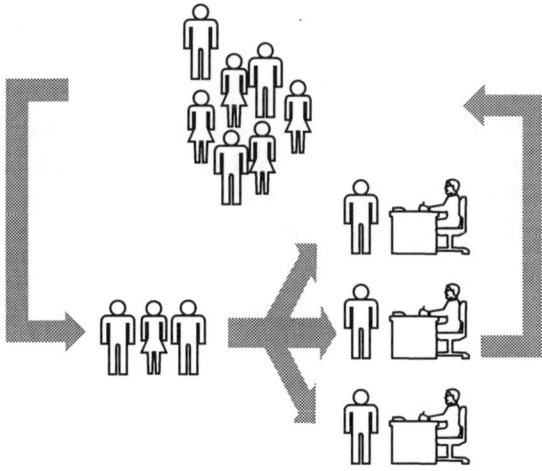

Figure 24.1. Multiple repairmen queueing model.

we can investigate the scaling effect, and also look at how to optimize the number of repairmen.

Assumptions, parameters, and initial calculations

There are a fixed number K of customers or machines that go through a cycle of 'operating' (i.e. not needing service), then requiring service, and then returning to operating mode. Service is provided by m servers, or repairmen. Customers are served first-come-first-served. Operating time is exponentially distributed with mean T_A, and service-time is exponentially distributed with mean T_S.

First we calculate the service ratio, which is the ratio of operating time to service-time.

$$z = \text{service ratio} = \frac{T_A}{T_S} \tag{24.1}$$

Some authors define the service ratio as T_S/T_A.

Probability distribution of number-in-system

With the single repairman case we used the Poisson ratio function, and the calculation of queueing statistics was otherwise quite simple. With the multiple repairmen case we have to calculate explicitly the probabilities of there being a particular number of customers 'in the system', by which we mean either being served or waiting for service. These probabilities are defined by

$$p_n = \text{Prob}(n \text{ waiting or being served}) \tag{24.2}$$

$$p_n = \binom{K}{n} \frac{p_0}{z^n} \text{ for } n = 1, 2, \ldots, m \tag{24.3}$$

$$p_n = \frac{n!}{m! m^{n-m}} \binom{K}{n} \frac{p_0}{z^n} \text{ for } n = m + 1, \ldots, K \tag{24.4}$$

These formulae make use of p_0, which is the probability that all the repairmen or servers are idle. The value of p_0 is defined by Eq (24.5). In practice, in a computer program, we would provisionally set $p_0 = 1$ and use Eqs (24.6) and (24.7) to calculate the p_n relative to p_0. Equation (24.5) would then be used to obtain the correct value of p_0 and the values of p_n would be adjusted accordingly.

$$p_0 = \left(1 + \sum_{r=1}^{K} \frac{p_r}{p_0}\right)^{-1} \tag{24.5}$$

$$\frac{p_n}{p_{n-1}} = \frac{K - n + 1}{n} \cdot \frac{1}{z} \text{ for } n \leq m \tag{24.6}$$

$$\frac{p_n}{p_{n-1}} = \frac{K - n + 1}{m} \cdot \frac{1}{z} \text{ for } n \geq m + 1 \tag{24.7}$$

Number waiting

The average number of customers or machines waiting is calculated from first principles, using Eq. (24.8). The value of L_W is the first statistic to be derived since it features in the formulae for most of the other statistics.

$$L_W = \text{average number waiting} = \sum_{n=m+1}^{K} (n-m)p_n \qquad (24.8)$$

Waiting time and time-in-system

Average waiting time is calculated by means of Eq. (24.9). If we want to know the probability that an arriving customer, or machine requiring service, has to wait for a server, then that is calculated directly from the p_n as in Eq. (24.10). The average time-in-system is calculated from the fundamental relationship in Eq. (24.11).

$$T_W = \text{average waiting time} = \frac{L_W(T_A + T_S)}{K - L_W} \qquad (24.9)$$

$$\text{Prob(arriving customer must wait)} = \sum_{n=m}^{K} p_n \qquad (24.10)$$

$$T_Q = \text{average time-in-system} = T_W + T_S \qquad (24.11)$$

Arrival rate and server utilization

Having obtained the average waiting time T_W, we can calculate the arrival rate. The formula for this is Eq. (24.12), which can be interpreted as follows. Each customer or machine goes repeatedly through a cycle of operating, waiting for service, being served, and then returning to operating mode. The length of this cycle is $T_A + T_W + T_S$, so each customer will 'arrive' for service once per cycle time. Since there are K customers, the arrival rate is clearly

$$\lambda = \text{arrival rate} = \frac{K}{T_A + T_W + T_S} \qquad (24.12)$$

Traffic intensity and server utilization are calculated in the usual way by

$$u = \text{traffic intensity} = \lambda T_S \qquad (24.13)$$

$$\rho = \text{server utilization} = \frac{u}{m} = \frac{\lambda T_S}{m} \qquad (24.14)$$

Number-in-system

The average number of customers in the system is derived using Little's law, as in

$$L_Q = \text{average number in system} = \lambda T_Q \qquad (24.15)$$

Probabilities for specific customers

Sometimes we want to know the probability, or proportion of time, that a specific machine or customer is in each of the operating, waiting, and being served states. Since we know the average time a machine spends in each state, we simply take appropriate ratios of average times to get the probabilities we want. Equations (24.16)–(24.18) define those probabilities. Note that in Eq. (24.16) we give the probability of not operating rather than operating.

$$\text{Prob(specific machine not operating)} = \frac{T_Q}{T_Q + T_A} \qquad (24.16)$$

$$\text{Prob(specific machine waiting for service)} = \frac{T_W}{T_Q + T_A} \qquad (24.17)$$

$$\text{Prob(specific machine being serviced)} = \frac{T_S}{T_Q + T_A} \qquad (24.18)$$

Offered traffic and efficiency

Chapter 23 discussed offered traffic and efficiency for the single repairman case. The same ideas and definitions apply in the multiple repairmen case, so that

$$\lambda_O = \text{offered traffic rate} = \frac{K}{T_A} \qquad (24.19)$$

$$\text{efficiency} = \frac{\lambda}{\lambda_O} \times 100\% = \frac{\lambda T_A}{K} \times 100\% \qquad (24.20)$$

Scaling effect

For the multiple repairmen model, as with most queueing systems, servers are used more efficiently when serving a larger population of customers than a smaller population. For example, suppose we have machines with an average operating time $T_A = 10$ and an average service-time $T_S = 1$. We want to have enough repairmen so that the average waiting time does not exceed $T_W = 0.5$. Figure 24.2 shows the number of repairmen required as the number of machines increases. Also shown is the utilization of the repairmen.

Clearly the repairmen are utilized more heavily with a larger number of machines to look after. This scaling effect is partly offset by a 'granularity' effect, since when the addition of one further machine necessitates another repairman the utilization drops back until further machines are added to make full use of the additional repairman.

Optimization

The multiple repairmen model lends itself to some straightforward optimization. If we have a set number of machines, what is the optimum number of repairmen? Suppose each machine is producing goods that are worth V_O per unit time. From

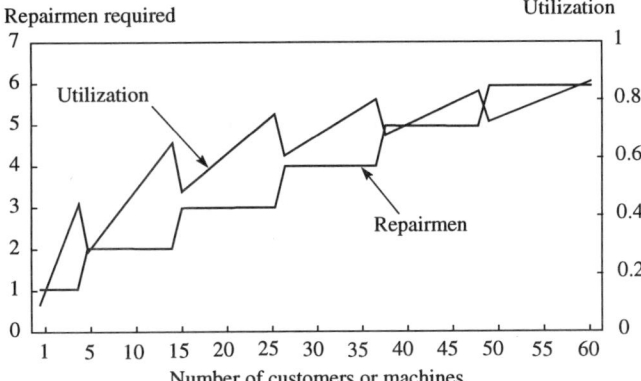

Figure 24.2. Scaling effect for multiple repairmen model.

Eq. 24.16, the value of lost output from all K machines is given by

$$\text{value of lost output} = K\frac{T_Q}{T_Q + T_A}V_O \tag{24.21}$$

Suppose each repairman costs R per unit time. If we want to find out the number of repairmen that minimizes total costs, then we have to find m so as to minimize (24.22)

$$\text{costs} = mR + K\frac{T_Q}{T_Q + T_A}V_O \tag{24.22}$$

While this is not easy algebraically, it is a simple matter to write a program that uses the subroutine supplied in this chapter to calculate the costs for a range of values for m, and find the optimum value. As an example, suppose $T_S = 1$, $T_A = 10$ and $K = 40$. The value of lost output while a machine is not operating is $V_O = 20$, and the cost of a repairman is $R = 25$. Figure 24.3 shows costs versus number of repairmen, and the optimum is easily identified as five repairmen.

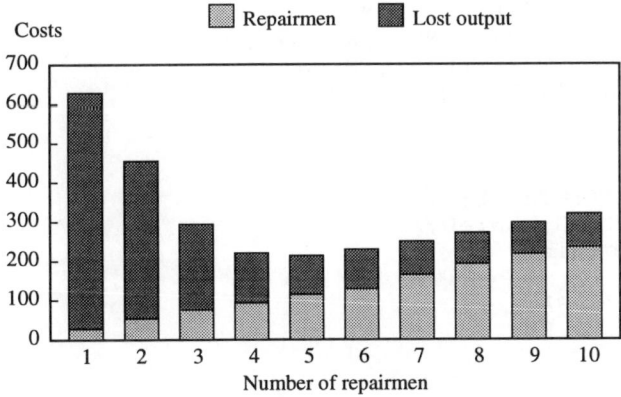

Figure 24.3. Optimization of a multiple repairmen situation.

Programs for M/M/m/K/K

Prerequisite routines

In order to use the M/M/m/K/K routines, the following set of 'include file' statements would be used in Turbo-Pacal.

```
{$N+,E+ }
{$I \qthprog\qthtypes.pas   }
{$I \qthprog\qtherror.pas   }
{$I \qthprog\mmmfcclc.pas   }
```

MMmFcClc—general calculations for M/M/m/K/K

This routine calculates most of the statistics usually of interest for the multiple repairmen model. Efficiency is not provided as an output from this routine, but is easily calculated using Eq. (24.20).

```
{----------------------------------------------------------}
{> MMmFcClc - Calculations for M/M/m/K/K                   }
{ Inputs:                                                  }
{     TA            mean operating time                    }
{     TS            mean service time                      }
{     K             number of customers                    }
{     M             number of servers                      }
{ Outputs:                                                 }
{     SVRAT         service ratio                          }
{     LAMBDA        arrival rate                           }
{     RHO           server utilisation                     }
{     PWAIT         Prob(arriving customer must wait)      }
{     PCIS          Prob(particular customer in system)}
{     TW            average waiting time                   }
{     TQ            average time in system                 }
{     LW            average number waiting                 }
{     LQ            average number in system               }
{   Copyright Mike Tanner 1993                             }
{----------------------------------------------------------}
Procedure MMmFcClc(TA,TS:QTHreal;K,M:integer;
                   Var SVRAT,LAMBDA,RHO:QTHreal;
                   Var PWAIT,PCIS:QTHreal;
                   Var TW,TQ,LW,LQ:QTHreal);
Var RK,RN,RM,P,P0:QTHreal;
Var N:integer;
begin
   RK:=K; RM:=M; SVRAT:=TA/TS;
   P:=1; P0:=1; PWAIT:=0; LW:=0;
   For N:=1 to K do begin
     RN:=N;
     P:=P*(RK-RN+1)/SVRAT;
     If N<=M then P:=P/RN
             else P:=P/RM;
     P0:=P0+P;
```

```
    If N>M then LW:=LW+P*(RN-RM);
    If N>=M then PWAIT:=PWAIT+P;
  end;
  P0:=1/P0;
  LW:=LW*P0;
  PWAIT:=PWAIT*P0;
  TW:=LW*(TA+TS)/(RK-LW);
  TQ:=TW+TS;
  PCIS:=TQ/(TQ+TA);
  LAMBDA:=RK/(TA+TW+TS);
  RHO:=LAMBDA*TS/RM;
  LQ:=LAMBDA*TQ;
end;
```

Example 24A—The Cotswolde TeaRoom II

Recall Example 23B, where the enterprising owner increased the number of tables from 7 to 10 (a 46 per cent increase) but the average wait to be served more than doubled and the number of customers served went up by only 24 per cent. Recognizing that one waitress is no longer enough, a second is hired. What now is the average wait to be served, the rate of serving customers, and the utilization of the waitresses? If a further extension to the tearoom is possible, how many more tables could be installed while keeping the average wait no more than the original 4.5 min?

The parameters are as for Example 23B, so that $T_S = 3.5$ min and $T_A = 25$ min. Doing the calculations directly from the formulae in this chapter would be tedious, so either the MMmFcCalc routine or QCalc could be used. The results, together with those from Example 23B, are given in Table 24.1. The final row of Table 24.1 shows that the number of tables can be increased to 16 for two waitresses, while keeping the waiting time no worse than the original situation.

Table 24.1. Results for Example 24A

Number of tables	Number of waitresses	Tables served per min (per h)	Waitress utilization (%)	Average wait to be served (min)
7	1	0.21 (12.7)	74.23	4.50
10	1	0.26 (15.7)	91.55	9.73
10	2	0.34 (20.3)	59.21	1.06
16	2	0.49 (29.5)	86.11	4.02

Part Eight
Some important topics

25
Server sharing

Introduction

In most queueing models we envisage that a server is dedicated to a single customer until that customer's service is complete. In pre-emptive priority queues we allow the server to be interrupted when a customer with higher priority than the one being served arrives. Even in this case, at the highest customer priority level customers are dealt with serially with the server dedicated to a single customer. There are cases when other types of server sharing are appropriate. One is where a number of jobs are being processed together on a computer system, all with the same priority level. The cpu power must be shared out between the jobs in some way, and to process the jobs completely serially would not give good performance in a time-shared system. Another case is when large data files are being sent over a telecommunications link, when it is often desirable to share the capacity of the link in a more dynamic way than just transmitting each file completely before sending the next file. Kleinrock (1976) gives a extensive discussion of cpu scheduling algorithms, which can become very complex. In this book we shall limit ourselves to a simple but representative algorithm, the 'round robin' (RR) processor-sharing algorithm.

M/G/1 processor sharing

Time-sharing computer systems use algorithms of various complexity to allocate the cpu time between the users or tasks currently wanting service. The basic algorithm is the round robin, where each user in turn is given access to the cpu for a fixed length of time. Once a time-slice has finished, the task rejoins the queue of tasks waiting for further service. The system is illustrated in Fig. 25.1. Of course a task that completes part-way through a time-slice will relinquish control. As long as the time-slice is small in proportion to the total cpu time required for each task, then most time-slices will be used in full, and we can safely ignore the complication of incompletely used time-slices.

Cycled arrivals

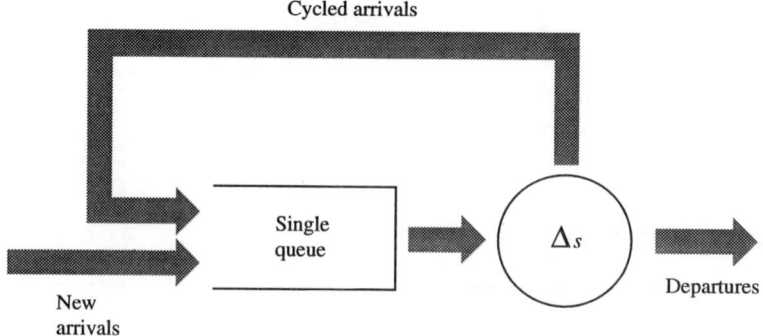

Figure 25.1. Round-robbin method for allocating cpu time-slices.

New tasks arrive at the cpu at a rate λ, and the average cpu time required by a task is T_S. No specific assumptions are made about the distribution of the cpu time needed. The length of a time-slice is Δs. Kleinrock showed that by letting Δs tend to zero, a simple analytic model is obtained that is a good approximation to a round-robin system with a finite time-slice. Details of the model are given by Kleinrock (1976), and we give the main results below. We assume that $\lambda T_S < 1$ so that the cpu is not overloaded and the system is stable.

$$\rho = \text{CPU utilization} = \lambda T_S \tag{25.1}$$

$$p_n = \text{Prob}(n \text{ tasks in the system}) = (1 - \rho)\rho^n \tag{25.2}$$

$$L_Q = \text{average number of tasks in system} = \frac{\rho}{1 - \rho} \tag{25.3}$$

$$T_Q = \text{average time-in-system} = \frac{T_S}{1 - \rho} \tag{25.4}$$

Equations (25.1)–(25.4) are the same as for an M/M/1 system, which may seem surprising since we allow the service-time to have a general distribution. Now let us consider the average time-in-system given that the service-time has a specific value t. We write this

$$E[\text{time in system} \mid \text{service time} = t] = \frac{t}{1 - \rho} \tag{25.5}$$

Equation (25.5) shows that the average time a task spends in the system is proportional to the amount of cpu time that task needs. This is saying something more than Eq. (25.4) which simply gives the average time-in-system regardless of the actual service-time for a specific customer. A task that needs twice as long as another will spend twice as long in the system. Another important property is seen by comparing eq. (25.4) with Eq. (9.1), the equivalent formula for M/G/1. Equation (9.1) shows that the M/G/1 the average time-in-system depends on the variance (equivalently the coefficient of variation) of the service-time, which means it depends on the particular distribution of service-time. In contrast,

Eq. (25.4) shows that for processor sharing it is only the average service-time that matters, not its distribution.

Example 25A

A computer time-sharing system supports many active users. On average four tasks per second need to be processed by the cpu. Each task takes on average 100 milliseconds of cpu time. What is the cpu utilization, and how long on average does it take for a task to be executed? How long will a specific task that requires 2 seconds of cpu time take to be executed, and how long for a specific task that needs 30 milliseconds of cpu time?

We have $\lambda = 4$ tasks/sec and $T_S = 0.100$ seconds. So $\rho = \lambda T_S = 0.4$, and from Eq. (25.4) we get

$$T_Q = \frac{T_S}{1 - \rho} = \frac{0.1}{1 - 0.4} = 0.167 \text{ s}$$

The task requiring 2 seconds will take $2/(1 - 0.4) = 3.33$ seconds. The task requiring 30 milliseconds will take $0.030/(1 - 0.4) = 50$ milliseconds. (Compare this with an M/M/1 system, where the average wait time per task would be $\rho T_S(1 - \rho) = 0.4 \times 0.1/(1 - 0.4) = 0.067$ secs, or 67 ms. The 2 second task would take 2.067 seconds, and the 30 ms task would take 107 ms.)

26
Networks of queues

Introduction

Customers often have to pass through a series of individual services. The road network is an obvious case, where we could regard each junction as a server. The term 'service centre' is used when discussing networks of queues. A service centre may consist of a single server, a multiple server, perhaps an unlimited number of servers. The theory of queue networks has been developed rapidly in recent years, particularly for its application to computer system performance. Much of the theory and the algorithms for finding solutions are really the province of the specialist. However, some of the simplest kinds of network are well within the scope of the non-specialist, and methods for their solution are given in this chapter.

Open networks

An open network of queues is one in which customers enter the system from óutside, follow a path through one or several service centres, and then leave the system. The path that a customer follows may be fixed, or may vary in a probabilistic way. Intuitively it seems obvious that the average rate at which customers enter the network must be the same as the average rate of departures from the network. The overall flow of customers is therefore not affected by the state of the network itself. We could think of the network as a single service centre with a complicated set of rules for serving customers.

Jackson networks

A Jackson network is a particular type of open network. Suppose there are n nodes or service centres in the network. Figure 26.1 illustrates a Jackson network with seven nodes. From outside the network customers may arrive at any of the nodes. The arrivals to node i are a Poisson, or random, arrival stream with average rate λ_i. Once customers have been served at node i, they may proceed to any of the

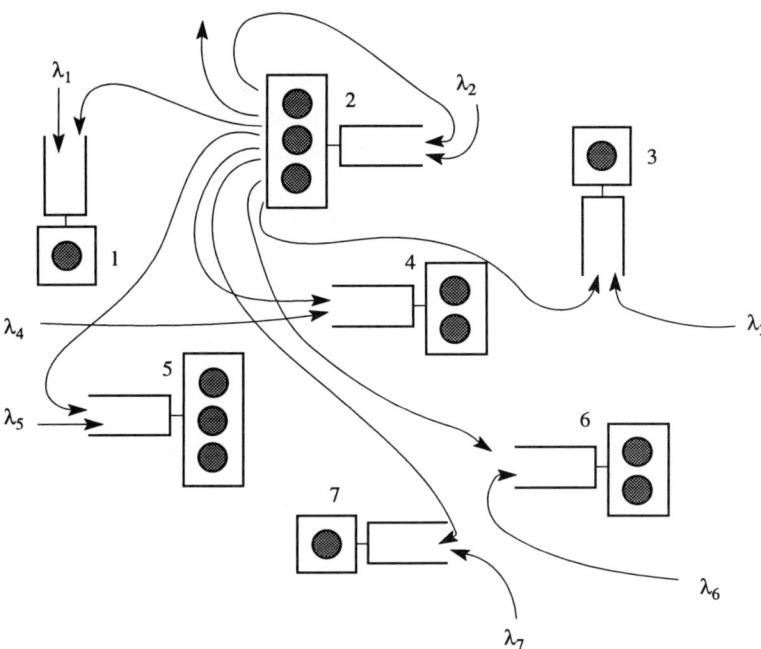

Figure 26.1. A 7-node Jackson network.

other nodes, rejoin the queue for another service at node i, or depart from the system. Figure 26.1 shows the arrivals from outside the system to each node, and the possible paths of a customer on leaving node 2.

Suppose the probability that a customer leaving node i goes to node j is p_{ij}. Instead of visiting a further node a customer may decide to leave the system. The probability of leaving after node i is given by Eq. 26.1.

$$\text{Prob(customer leaves after node } i) = 1 - \sum_{j=1}^{n} p_{ij} \qquad (26.1)$$

The total arrival rate at a particular node is the arrivals from outside to that node, plus arrival from every node in the network. So we have

$$\Omega_i = \text{arrival rate to node } i = \lambda_i + \sum_{j=1}^{n} p_{ji}\Omega_j \qquad (26.2)$$

Once the λ_i and p_{ij} are specified, Eq. (26.2) is a set of simultaneous linear equations that we can solve to get the total arrival rates at each node Ω_i.

So far we have said nothing about the nature of nodes or service centres themselves. The key characteristic of a Jackson network is that each service centre has a fixed number of servers and that the service-time at each centre is exponentially distributed. Each service centre is therefore an M/M/m system. Jackson's theorem says that under these conditions, each service centre can be treated independently. In other words, once we have worked out the arrival rates Ω_i, we just have to apply the M/M/m formulae to each node. A little thought will show that the arrival rate at a node is unaffected by the service-times or waiting times at other nodes. Arrival rates are determined by the arrival rates from outside the system, and the probabilities p_{ij}.

Suppose we have found the total arrival rates at each node, and applied M/M/m to each node. How do we work out the total time that a customer, on average, spends in the queueing network? Let us define

$$T_{Q(i)} = \text{time-in-system at node } i \qquad (26.3)$$

$T_{Q(i)}$ is the value of T_Q from applying M/M/m to node i, and is the average time a customer spends at node i in order to get one service. Having completed service at node i the customer, unless he or she departs, then proceeds to another node or the same node. The customer is then in exactly the same position as a customer arriving at that node from outside the system. So we denote by $T_{N(i)}$ the average time spent in the network by customers arriving at node i, and can write

$$T_{N(i)} = T_{Q(i)} + \sum_{j=1}^{n} p_{ij} T_{Q(j)} \qquad (26.4)$$

Equation (26.4) is another set of linear simultaneous equations that can be solved to get the $T_{N(i)}$. Another important aspect of Jackson's theorem is that if $P(k_1, k_2, \ldots, k_n)$ is the probability that there are k_1 customers at node 1, k_2 customers at node 2 and so on, then

$$P(k_1, k_2, \ldots, k_n) = P_1(k_1) \times P_2(k_2) \times \ldots \times P_n(k_n) \qquad (26.5)$$

where $P_i(k_i)$ is the probability of there being k_i customers at node i, treating node i as just an isolated M/M/m queue.

A simple feedback system

One of the simplest Jackson networks is shown in Fig. 26.2. A single server processes customers who after being served leave the system with probability $(1 - p)$, or rejoin the queue for another service with probability p. The arrival rate is λ.

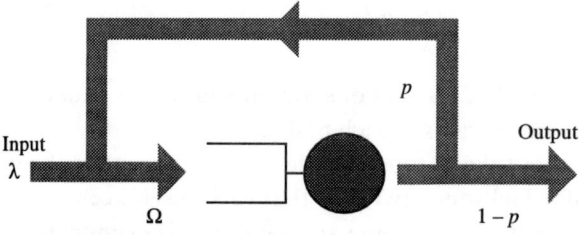

Figure 26.2. Simple feedback system.

We know that

$$\Omega = \lambda + p\Omega \tag{26.6}$$

so that

$$\Omega = \frac{\lambda}{1 - p} \tag{26.7}$$

Statistics on waiting times, etc. are obtained using the M/M/1 formulae. We might also be interested in the average number of times a customer is served. We can either observe that the number of visits to the server is a geometric series with parameter p, or note that

$$\text{average visits per customer} = \frac{\Omega}{\lambda} = \frac{1}{1 - p} \tag{26.8}$$

(This does not hold for each node in a multiple-node Jackson network.)

Example 26A—The helter-skelter

A popular ride at the fairground is the helter-skelter. Suppose each child occupies the ride for an average of 45 seconds, exponentially distributed. After a ride, 60 per cent of children decide to have another go. Children arrive at the helter-skelter at an average rate of one every 2 min. What is the utilization of the helter-skelter?

The calculations are very simple

$$p = 0.6, \ \lambda = 0.5, \ T_S = 0.75 \tag{26.9}$$

so that

$$\Omega = \frac{\lambda}{1 - p} = \frac{0.5}{0.4} = 1.25 \tag{26.10}$$

$$\rho = \Omega T_S = 1.25 \times 0.75 \approx 0.94 \tag{26.11}$$

This is a very busy ride, since the reader can verify using the M/M/1 formulae that children would wait on average just over 11 min for each ride. The 'time-in-system' would be 12 min, and since each child would have on average 2.5 rides, each child will be spending 30 min at the helter-skelter.

Closed networks

A closed network is one in which a fixed number of customers circulate through a set of service centres. We have already met some simple closed networks in Chapters 23 and 24 with the 'repairman' situation. With those simple networks there are only two service centres. One service centre is the repairman. The other service centre is less obvious, but the operating or think time can be represented as a service centre with an unlimited number of servers. The 'service-time' is in fact the operating time.

This way of looking at the repairman model is illustrated in Fig. 26.3.

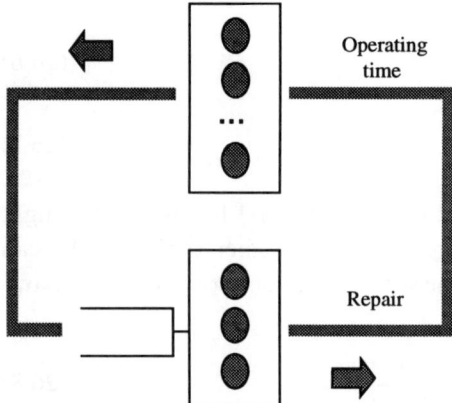

Figure 26.3. Repairman model as a queueing network.

Central server model

The central server model is a simple form of closed queueing network that has particular relevance to computer systems, and is illustrated in Fig. 26.4. In this chapter we shall describe the central server model and Buzen's algorithm for solving the model. Chapter 28 contains several examples of the use of the central server model, combined with other models, to analyse computer system performance.

There are K circulating programs. Each program performs some cpu activity and then an i/o operation, back to the cpu, and so on. The processing capacity or

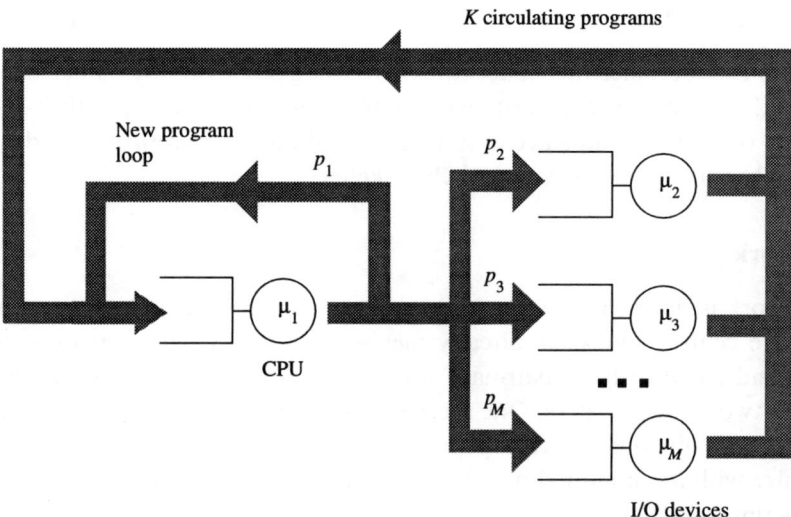

Figure 26.4. Central server model.

throughput of the cpu and each disk is denoted by $\mu_i = 1/T_{S(i)}$, i.e. capacity is the reciprocal of average service-time. Which of the M i/o devices is selected on each circuit depends on the probabilities $p_2, p_3, \ldots, p_M$. On leaving the cpu, a program may immediately loop back to the cpu, with probability p_1. The interpretation of this 'new program loop' may not be obvious to the reader, and it would be perfectly sensible for now just to accept that p_1 is defined by Eq. (26.12) without struggling for an intuitive meaning for p_1. There are K circulating programs, and when a program takes the new program loop, it represents one of the programs actually terminating and immediately being replaced by a new program. The purpose of this aspect of the model is to represent the throughput of the system. Each job or program will perform $1/p_1$ visits to the cpu and $(1/p_1) - 1$ visits to the pool of i/o devices. So we have (not strictly, but an extremely close approximation)

$$\frac{1}{p_1} \approx \frac{\text{total cpu time per job}}{\text{cpu time per i/o instruction}} = \frac{\text{cpu instructions per job}}{\text{cpu instructions per i/o}} \qquad (26.12)$$

So p_1 depends essentially on the total work required per job. Long jobs will give a small value for p_1 and short jobs will give a large value. The other parameters of the system are more easily interpreted. The probabilities $p_2, p_3, \ldots, p_M$ determine what proportion of i/o requests go to each i/o device. Note that we require

$$\sum_{i=1}^{M} p_i = 1 \qquad (26.13)$$

Sometimes it is helpful to define the probabilities that an i/o goes to a particular device, i.e.

$$f_j = \text{Prob(i/o goes to device } j) \text{ for } j = 2, 3, \ldots, M \qquad (26.14)$$

so that

$$\sum_{j=2}^{M} f_j = 1 \text{ and } p_j = (1 - p_1)f_j \text{ for } j = 2, 3, \ldots, M \qquad (26.15)$$

The μ_i represent the service rates of each server in the system. Service rate is the reciprocal of service-time, and service rate is sometimes more convenient and natural to work with. For disk devices we often talk about the i/os per second that a particular device can deliver. For the cpu, service rate is

$$\mu_1 = \text{cpu service rate} = \frac{\text{cpu instructions per i/o}}{\text{cpu speed in instructions per second}} \qquad (26.16)$$

If the queueing discipline at node 1, the cpu, is FCFS then we have to assume that cpu service-time is exponentially distributed. Alternatively, if the discipline is processor-sharing, then the service-time can have a general distribution. For all the other nodes, the i/o devices, we assume service-times to be exponentially distributed and the queueing discipline to be FCFS.

The key characteristic we want to find is the throughput. As there are a fixed number of jobs or customers circulating, longer waiting times at the servers will slow down the rate at which a job circulates, and so reduce the total throughput of the system. The throughput is calculated using Buzen's algorithm, which we describe next.

Buzen's algorithm

How many of the K jobs in the system are at a particular node i, either waiting for service or being served? We shall use the notation

$$k_i = \text{number of jobs at node } i \tag{26.17}$$

The number of jobs at a particular node is not independent of what is going on in the rest of the system, since the k_i must add up to K. Buzen showed that the probability of the jobs being spread around the servers in a particular way is

$$\text{Prob}(k_1, k_2, \ldots, k_m) = \frac{1}{G(K)} \prod_{i=2}^{M} x_i^{k_i} \tag{26.18}$$

where

$$x_i = \mu_1 \frac{p_i}{\mu_i} \tag{26.19}$$

We shall see in a moment how to calculate $G(K)$, but assuming we have done that we can calculate the server utilizations, the system throughput, and the average number of jobs at each node. The formulae for these are

$$\rho_1 = \text{cpu utilization} = \frac{G(K-1)}{G(K)} \tag{26.20}$$

$$\rho_i = \text{i/o device utilizations} = \rho_1 x_i = \mu_1 \rho_1 \frac{p_i}{\mu_i} \text{ for } i = 2, 3, \ldots, M \tag{26.21}$$

$$\lambda = \text{system throughput} = \mu_1 p_1 \rho_1 \text{ jobs per unit time} \tag{26.22}$$

$$E(k_i) = \text{average jobs at node } i = \sum_{n=1}^{K} x_i^n \frac{G(K-n)}{G(K)} \text{ for } i = 1, 2, \ldots, M \tag{26.23}$$

The vital part of the algorithm is the calculation of the function $G(k)$ for $k = 0, 1, 2, \ldots, M$. In order to do this Buzen defines a function $g(k, m)$. The definition of this function can be found in Allen (1978), which also gives the references to Buzen's original papers. What interests us here is the following properties of $g(k,m)$, which will allow us to calculate the values $G(k)$.

$$g(k, 1) = 1 \text{ for } k = 0, 1, \ldots, K \tag{26.24}$$

$$g(0, m) = 1 \text{ for } m = 1, 2, \ldots, M \tag{26.25}$$

$$g(k, m) = g(k, m - 1) + x_m g(k - 1, m) \tag{26.26}$$

$$G(k) = g(k, M) \text{ for } k = 0, 1, \ldots, K \tag{26.27}$$

Equation (26.26) gives a recurrence relation that we can use to calculate the value of $g(k, m)$ starting from the boundary conditions in Eqs (26.24) and (26.25). In this way we can get the values of $G(k)$ using Eq. (26.27), and in particular

$$G(K) = g(K, M) \tag{26.28}$$

The routine BuzenCS performs the above calculations.

Product-form networks

The machine repairman model and the central server model are simple examples of closed queueing networks having what is called a 'product-form' solution. Mathematically, this means that if s_i is the state (e.g. number of customers present) of node i in a network with J service centres, then

$$\text{Prob}(s_1, s_2, \ldots, s_J) = \frac{\text{Prob}_1(s_1) \times \text{Prob}_2(s_2) \times \ldots \times \text{Prob}_J(s_J)}{G(N)} \tag{26.29}$$

where $\text{Prob}(s_1, s_2, \ldots, s_J)$ is the complete state of the whole network, $\text{Prob}_i(s_i)$ is the state of the single service centre, i, N is the number of customers in the network, and $G(N)$ is a 'normalizing' constant to make the probabilities of every possible state of the network add up to 1. Put more intuitively, this means that each service centre can be analysed separately. The repairman and central server models have this property, but how general can a closed network of queues be while still retaining this 'product-form solution' property? This is an important question, since if a network does not have a product-form solution it is extremely difficult to find a solution at all. In order to have a product-form solution, only certain queue disciplines can be allowed at each service centre. These are FCFS, processor-sharing and infinite servers (i.e. unlimited number of servers so there is no queueing).

A mixed network is one in which some classes of customers each with a fixed number of customers who circulate around the network as in a closed network, but also 'open' classes of customers who arrive from outside the network, circulate, and then depart from the network. Mixed networks can also have a product-form solution, as indeed can open networks. (A Jackson network has a product-form solution.)

A number of algorithms have been developed for solving product-form networks. An excellent description of the algorithms and their strengths and weaknesses can be found in Lavenberg (1983). Solving a medium- or large-scale product-form network presents challenges in balancing numerical analysis difficulties, storing very large arrays of intermediate results, and the lengthy computations required. These techniques are beyond the scope of this book, and readers will generally encounter the algorithms incorporated into performance analysis software such as BEST/1 or RESQ.

There has been a great deal of research in recent years into the theory of product-form networks, prompted by its application to the performance of computer systems and data transmission networks. Much of the work has been concerned with approximate solutions for larger networks, approximate methods to solve networks with other queueing disciplines, and methods of handling some of the characteristics of computers and telecommunications networks that are not so easily represented in product-form networks. Readers wishing to pursue this area should consult Lavenberg (1983), Lazowska *et al.* (1984), King (1990) and the extensive references given in those books.

Programs for queue networks

In order to use the routines for solving queueing networks, a program must contain the following statements.

```
{$N+,E+,R+ }
{$I \qthprog\QTHTypes.pas }
{$I \qthprog\QTHError.pas }
{$I \qthprog\GJElim.pas   }
{$I \qthprog\Pratio.pas   }
{$I \qthprog\MMmCalc.pas  }
{$I \qthprog\Jackson.pas  }
{$I \qthprog\BuzenCS.pas  }
```

GJElim—solve simultaneous equations

Solving a set of simultaneous linear equations is a classic numerical analysis ·problem. The routine given here uses a technique called Gauss–Jordan elimination. There are other more sophisticated methods, but Gauss–Jordan is straightforward and reasonably robust. The algorithm is adapted and simplified from an algorithm given in Press *et al.* (1989), which readers are urged to consult for discussion of this and many other numerical analysis methods.

The subroutine is preceded by some type definitions that the calling program will need to define the arrays required. The constant MAXSC is the maximum number of service centres allowed in a Jackson network. Readers will probably want to increase this value, and if a large number of centres is needed it may be necessary to make the arrays dynamically allocated.

```
Const MAXSC=10;    { maximum service centres   }
Type AryRSCbySC=array[1..MAXSC,1..MAXSC] of QTHreal;
Type AryRSC=array[1..MAXSC] of QTHreal;
Type AryISC=array[1..MAXSC] of integer;
```

```
{----------------------------------------------------}
{> GJElim -- Gauss-Jordan elimination to solve a    }
{>          set of linear equations.                }
{ Inputs:                                           }
{   A,B    matrix and vector defining equations     }
{   N      number of equations and unknowns         }
{ Outputs:                                          }
{   X      solution vector                          }
{ Copyright Mike Tanner 1993                        }
{----------------------------------------------------}
Procedure GJElim(Var A:AryRSCbySC;Var B,X:AryRSC;
                 N:integer);
Var BIG,Z,PIV:real;I,COL,ROW,J,K,L,LL:integer;
Var INDXC,INDXR,IPIV:^AryISC;G:^AryRSCbySC;
begin
   { Set up working arrays------------------------}
   New(INDXC);  New(INDXR);  New(IPIV); New(G);
   For I:=1 to N do begin
      X[I]:=B[I];
      For J:=1 to N do G^[I,J]:=A[I,J];
   end;
   {-Solve the equations--------------------------}
   For J:=1 to N do IPIV^[J]:=0;
   For I:=1 to N do begin
      {-Find largest element for pivoting----------}
      BIG:=0.0;
      For J:=1 to N do
         If IPIV^[J] <> 1
         then For K:=1 to N do
                 If IPIV^[K] =0
                 then If Abs(G^[J,K]) >=BIG
                      then begin
                             BIG:=Abs(G^[J,K]);
                             ROW:=J; COL:=K
                           end
                 else If IPIV^[K] > 1 then
                      QTHError('singular matrix');
      IPIV^[COL]:=IPIV^[COL]+1;
      If ROW <> COL
      then begin
              For L:=1 to N do begin
                 Z:=G^[ROW,L];
                 G^[ROW,L]:=G^[COL,L];
                 G^[COL,L]:=Z
              end;
              Z:=X[ROW];
              X[ROW]:=X[COL]; X[COL]:=Z
           end;
      INDXR^[I]:=ROW; INDXC^[I]:=COL;
      If G^[COL,COL] =0.0
      then QTHError('GJElim-b singular matrix');
      {-Pivot on the selected element----------}
      PIV:=1.0/G^[COL,COL];
      G^[COL,COL]:=1.0;
      For L:=1 to N do G^[COL,L]:=G^[COL,L]*PIV;
      X[COL]:=X[COL]*PIV;
```

```
     For LL:=1 to N do
        If LL <> COL
        then begin
                 Z:=G^[LL,COL];
                 G^[LL,COL]:=0.0;
                 For L:=1 to N do
                    G^[LL,L]:=G^[LL,L]-G^[COL,L]*Z;
                 X[LL]:=X[LL]-X[COL]*Z
              end
   end;
   Dispose(IPIV);  Dispose(INDXR);  Dispose(INDXC)
end;
```

Jackson—Calculations for a Jackson network

This routine performs a calculation for a Jackson network, using the GJElim and
MMmCalc routines. The outputs LQ[] and TW[] are calculated by applying
MMmCalc to each service centre. Readers can easily modify this routine to
provide any of the statistics for each service centre that MMmCalc provides.

```
{-------------------------------------------------------}
{> Jackson -- Calculations for a Jackson network    }
{  Inputs:                                          }
{    LAMBDA[.] arrival rates from outside network   }
{              to each node in the network          }
{    P[.,.]    prob that customer goes from node    }
{              i to node j                          }
{    TS[.]     service time at each service centre  }
{    M[.]      servers at each service centre       }
{    N         number of service centres            }
{  Outputs:                                         }
{    TAR[.]    total arrival rate to each node      }
{    LQ[.]     number of customers at each node     }
{    TW[.]     waiting time at each node            }
{    PD[.]     prob of departure from each node     }
{    VALID     FALSE = service centre overloaded    }
{  Copyright Mike Tanner 1993                       }
{-------------------------------------------------------}
Procedure Jackson(Var LAMBDA:AryRSC;
                  Var P:AryRSCbySC;Var TS:AryRSC;
                  Var M:AryISC;N:integer;
                  Var TAR,LQ,TW,PD:AryRSC;
                  Var VALID:boolean);
Var I,J,K:integer;A:AryRSCbySC;B:AryRSC;VSW:boolean;
Var U,RHO,TQ,SDVTQ,SDVTW,PZW,LW,SDVLW:QTHreal;
begin
   {-Set up equations to find arrival rates-------}
   For I:=1 to N do begin
      B[I]:=-LAMBDA[I];
      A[I,I]:=P[I,I]-1;
      For J:=1 to N do If I<>J then A[I,J]:=P[J,I]
   end;
   GJElim(A,B,TAR,N);
```

```
{-Probability of departure after each node------}
For I:=1 to N do begin
   PD[I]:=1;
   For J:=1 to N do PD[I]:=PD[I]-P[I,J];
   If PD[I]<0 then QTHError('Jackson-P[] error');
end;
{-Apply the M/M/m model to each node-----------}
VALID:=true;
For I:=1 to N do begin
   MMmCalc(TAR[I],TS[I],M[I],U,RHO,TQ,SDVTQ,TW[I],
          SDVTW,PZW,LQ[I],LW,SDVLW,VSW);
   VALID:=VALID and VSW;
end;
end;
```

BuzenCS—Buzen's algorithm for central server model

This routine solves the central server model using Buzen's algorithm. The routine is preceded by constant and type definitions for the arrays the calling program will need to define. The maximum number of service centres is defined by MAXM, which is a different symbolic constant from that used with the Jackson routine. For large numbers of customers floating point overflow may occur, in which case the reader can either resort to double or extended precision variables or conclude that one of the more numerically robust algorithms in Lavenberg (1983) should be used. The BuzenCS routine is intended for fairly small numbers of service centres and customers.

```
Const MAXM=20;    { max service centres                 }
Const MAXK=50;    { maximum of jobs/tasks in system }
Type buzenvecm=array[0..MAXM] of QTHreal;
Type buzenveck=array[0..MAXK] of QTHreal;
{-----------------------------------------------------}
{> BuzenCS -- Buzen algorithm for solving the        }
{>            central-server model.                   }
{ Inputs:                                             }
{    M         number of service centres ( 1 plus     }
{              the number of i/o devices)             }
{    K         number of jobs or tasks in system      }
{    P[]       the p(i)                               }
{    SR[]      service rate for each service centre   }
{ Outputs:                                            }
{    G[]       G(k) for k=0,1,...,K                   }
{    RHO[]     utilisation of each service centre     }
{    TPT       throughput in jobs per unit time       }
{  Copyright Mike Tanner 1993                         }
{-----------------------------------------------------}
Procedure BuzenCS(M,K:integer;Var P:buzenvecm;
                  Var SR:buzenvecm;Var G:buzenveck;
                  Var RHO:buzenvecm;Var TPT:QTHreal);
Var X:buzenvecm;I,J:integer;
Var GG:array[0..MAXK,1..MAXM] of double;
Label L4;
```

```
begin
   X[1]:=1;
   For I:=2 to M do X[I]:=SR[1]*P[I]/SR[I];
   For I:=0 to K do GG[I,1]:=1;
   For J:=1 to M do GG[0,J]:=1;
   I:=1;
L4:For J:=2 to M do GG[I,J]:=GG[I,J-1]+X[J]*GG[I-1,J];
   I:=I+1;
   If I<=K then Goto L4;
   For I:=0 to K do G[I]:=GG[I,M];
   RHO[1]:=G[K-1]/G[K];
   For I:=2 to M do RHO[I]:=SR[1]*RHO[1]*P[I]/SR[I];
   TPT:=SR[1]*P[1]*RHO[1];
end;
```

Example 26B—A small database server

A small database server on a LAN handles requests from a large number of users. Each request on average involves 10 disk i/o operations and 0.6 s of cpu time. The server has three physical disk devices, each with a throughput capacity of 10 i/os per second. The server software allows a maximum of eight requests to be in progress simultaneously. Requests arrive fast enough that there are always eight in progress. What is the throughput of the server in requests per second? If the cpu of the server were doubled in speed, what then would be the throughput? If instead the disks were upgraded to process 15 i/os per second, what would be the throughput? Also calculate the throughput if both cpu and disks were upgraded.

We shall model this as a central server, and use Buzen's algorithm to solve it. There are three disks and the cpu, so $M = 4$. The disks have $\mu_2 = \mu_3 = \mu_4 = 10$ i/os per second (or 15 for the upgraded disks). Since each job will perform $(1/p1) - 1$ i/os, we know that $10 = (1/p_1) - 1$, so $p_1 = 1/11$. The 0.6 s (0.3 for the faster cpu) of cpu required by each task is split into $1/p_1 = 11$ slices, so that the cpu throughput is $\mu_1 = 11/0.6$ slices per second (11/0.3) for the faster cpu. Now we have all we need to apply Buzen's algorithm. In all cases $K = 8$ tasks are allowed to be circulating in the server. The BuzenCS routine was used to obtain the results in Table 26.1.

Table 26.1. Results for Example 26B

CPU time per request	Disk i/o rate per s	CPU utilization	Disk utilization	Throughput requests per s
0.60	10.00	0.96	0.53	1.60
0.30	10.00	0.67	0.74	2.23
0.60	15.00	1.00	0.37	1.66
0.30	15.00	0.88	0.65	2.92

27
Introduction to simulation

Introduction

By simulation we mean working out the series of events that occurs when a system deals with a series of customers. This requires us to write a computer program that mimics the behaviour of the system, so obviously a good knowledge of the workings of the system is essential. For many analysts a great advantage of simulation is that it requires less (but not zero!) mathematical and statistical competency than queueing theory. Simulation has some more important advantages. Complex arrival patterns and scheduling rules can be represented, and system-wide effects for systems with multiple service centres can be handled relatively easily. On the other hand, a queueing theory analysis can provide much more insight into the performance characteristics of a system than simulation.

The actual programming and testing of a simulation is also a non-trivial task, since the structure of a simulation program is akin to a real-time operation system, and not the straightforward structure that is typical of application programs. For this reason, specialized simulation languages have been developed. On the other hand, with a suitable subroutine library, it is perfectly possible to write significant simulations using general-purpose languages such as PL/I or Pascal. It is usually said that simulation is expensive in terms of computer time, since a simulation must be repeated a number of times for each set of parameters, and each run is a 'number crunching' sort of job. But with the cost of modern computing power this is a minor consideration.

Basic ideas of simulation

To see what simulation involves we shall simulate a single server queue. First of all, when will each customer arrive? For our simulation we shall generate random inter-arrival times by throwing a single die. Each throw of the die is the time before the next customer arrives. The results are given in Table 27.1, where the first column shows the inter-arrival times and the second column shows the arrival times. The second column is just the cumulative sum of the first column.

Table 27.1. Simulated arrival and service-times

Inter-arrival time I	Arrival time A	Service-time S
2.00	2.00	5.00
3.00	5.00	3.00
1.00	6.00	1.00
6.00	12.00	4.00
3.00	15.00	2.00
4.00	19.00	3.00
1.00	20.00	6.00
etc.	etc.	etc.

Next we need to assign a specific service-time to each customer. We shall use the die again to do this. In Table 27.1 the third column contains the service-times. (Astute readers will notice that the average inter-arrival time is 3.5, which is also the average service-time. This system is not stable, since the server utilization will be 100 per cent and for stability we need utilization to be strictly less than 100 per cent. We shall ignore the implications of this in return for the simplicity of using a die to generate inter-arrival and service-times.)

Now we have all the details about the customers that we need, and we can proceed to work out the timing of events in the system. For our simple example this is easy, but in a real simulation of a complex system the analysis of how the system works will be a substantial and challenging piece of work. In Table 27.2 we have added three more columns. Two of these columns are used to hold the times at which each customer begins and ends service. A customer cannot begin service until that customer has arrived, and until the previous customer has ended service. So for customer k we have the simple relationships

$$B[k] = \max(A[k], E[k-1]) \text{ where } E[0] = 0 \tag{27.1}$$

$$E[k] = B[k] + S[k] \tag{27.2}$$

The final column in Table 27.2 holds the waiting time for each customer. This is calculated in the obvious way

$$W[k] = E[k] - A[k] \tag{27.3}$$

Table 27.2. Calculating start and finish service-times, and the waiting time for each customer

Inter-arrival time I	Arrival time A	Service-time S	Begin service B	End service E	Waiting time W
2.00	2.00	5.00	2.00	7.00	0.00
3.00	5.00	3.00	7.00	10.00	2.00
1.00	6.00	1.00	10.00	11.00	4.00
6.00	12.00	4.00	12.00	16.00	0.00
3.00	15.00	2.00	16.00	18.00	1.00
4.00	19.00	3.00	19.00	21.00	0.00
1.00	20.00	6.00	21.00	27.00	1.00
etc.	etc.	etc.	etc.	etc.	etc.

Using these simple relationships we can calculate the times of all relevant events in the system, and observe the values of waiting times, queue lengths, busy-period lengths, or whatever characteristic interests us.

Random numbers

In the example above we used a die to generate random inter-arrival times and service-times. Within a simulation program how do we get random numbers? Researchers have given much attention to this subject, and simple techniques have been devised to generate sequences of 'pseudo-random' numbers. Most algorithms for generating a sequence of random numbers are of the form

$$X_n = (bX_{n-1} + c) \text{ modulus } m \tag{27.1}$$

where X_n is the nth number in the sequence while b, c and m are appropriately chosen constants. For a discussion of suitable choices of these constants see Lavenberg (1983) or specialized books on simulation. Equation (27.1) will produce numbers that are non-negative integers in the range 0 to m, so by using X_n/m we can get values in the range 0 to 1.

Pseudo-random number streams can be subjected to a number of tests to see if they have characteristics of a genuinely random sequence. The obvious test is that the numbers should be evenly distributed over the range of values, usually 0 to 1, required. Another test is that there should be no correlation between value separated in the sequence by a fixed lag, i.e. there should be no serial correlation. This test includes the case of trends and cycles. Other, more esoteric, tests can be applied to exclude various patterns in the sequence.

Simulation requires a random number generator that has been properly designed and tested. The actual programming of a generator may be quite trivial, but the choice of method and parameters is a specialized job. Special-purpose simulation languages can be expected to have random number generators that can be trusted. However, many general-purpose language compilers include a random number generator. Some of these may be well designed, but some may be intended just for use in computer games. For a computer game, it does not matter too much if the sequence of numbers would fail proper statistical tests.

Converting random numbers to particular distributions

Assuming we have a reputable random number generator that supplies us with values in the range 0 to 1, we somehow have to convert these into, say, exponentially distributed values, or some other distribution that we want to simulate. Several techniques have been developed for doing this. The basic method is to invert the cumulative probability distribution mathematically, but less obvious methods turn out to be better in some ways for some distributions. Again see Lavenberg (1983) for a detailed description and algorithms for specific distributions.

Analysis of simulation results

Simulation is often undertaken by analysts with limited statistical training. Because of this it is sometimes forgotten that the result of a simulation run is just one experimental observation, and an observation of an abstract model rather than the real system. Like all experiments, a simulation needs to be properly designed to answer the question of concern, and to be repeated a number of times (with different sequences of random numbers of course!) to provide a sample of results for analysis.

Simulation of an M/M/1 system

In order to illustrate some ideas about simulation, an M/M/1 queueing system was simulated. We certainly do not need to simulate an M/M/1 system, since we can readily calculate all the interesting characteristics from the formulae given in Chapter 7. This also means we can compare the simulation results with the exact theoretical calculations. The system being simulated had $T_S = 1$ and $\lambda = 0.7$.

Figure 27.1 shows a graph constructed from the trace of the simulation program. The horizontal axis is time, and the vertical axis shows the queue size (number waiting plus number being served) when a new customer arrives. This number does not include the arriving customer. The curve looks quite erratic, and it is worth reminding ourselves that even a stable queueing system will exhibit this kind of erratic behaviour.

Next observe that the extract selected starts at time 300, not 0. When we start up a simulation program the simulated system will be empty, and it will require quite a lot of customers to be processed before the system achieves a 'steady state'. So we need to run the simulated system for a 'warm-up' period before we can start collecting statistics such as waiting time. How long should the warm-up time be? The answer of course is long enough for a steady state to be reached. There are

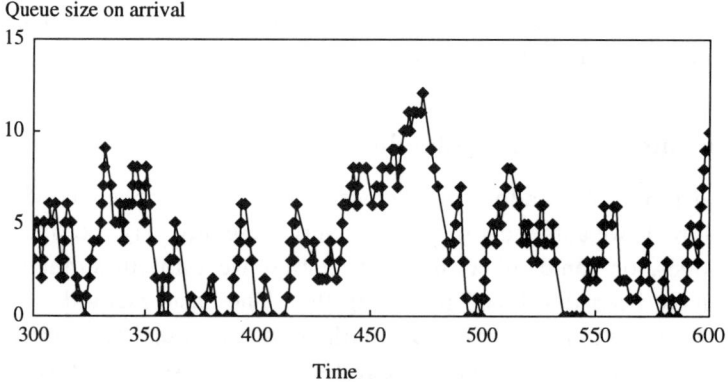

Figure 27.1. Extract from a simulation run.

formal statistical tests that could be used to decide when a steady state is reached, but in practice knowledge of the real system being simulated will often suggest a suitable warm-up time. There are also techniques to speed up the reaching of a steady state. We could set up the first few customers to arrive at time zero, so there is a queue of customers at the start. Or we could have a faster arrival rate for, say, half the warm-up period. These are refinements; the analyst still has to check somehow that the system reached steady state before statistics-gathering starts.

On the other hand, maybe we are not actually interested in the steady-state behaviour. We may want to understand how the system performs when starting 'cold', or dealing with short-term high arrival rates. Some real systems may never achieve a steady state. For example, a shop or store may not have enough customers during the day to make the concept of an average arrival rate meaningful. Most of the queueing theory in this book is about steady-state behaviour, which cannot be applied to such situations. Simulation, on the other hand, can cope with overloaded and unstable queueing systems, with changing arrival rates. Analysis of the results of simulating unstable systems needs great care, but at least is possible.

Now look at Fig. 27.2. This shows average time-in-system as measured for a series of separate simulation runs. For each level of utilization 10 separate simulation runs were done, each run for a given utilization using a different sequence of random numbers. Also shown is the exact theoretical result. The important point is that a single run produces one point, which is not in itself

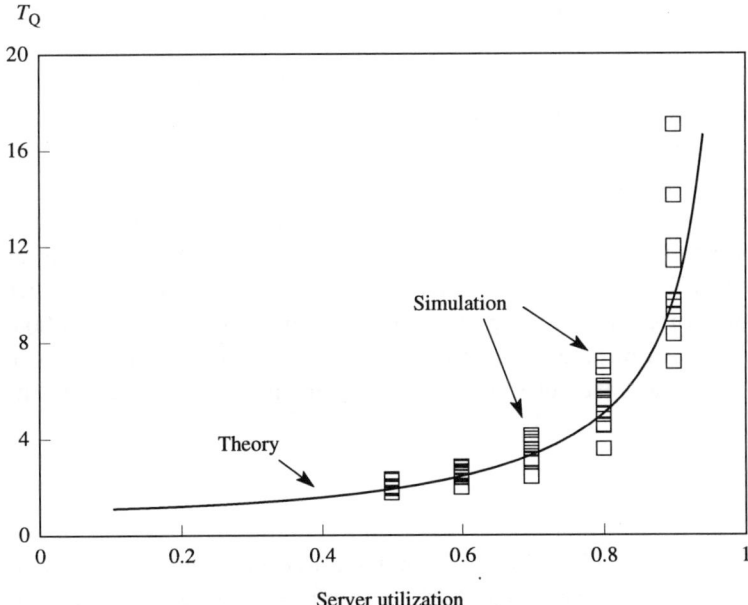

Figure 27.2. T_Q from a number of simulation runs.

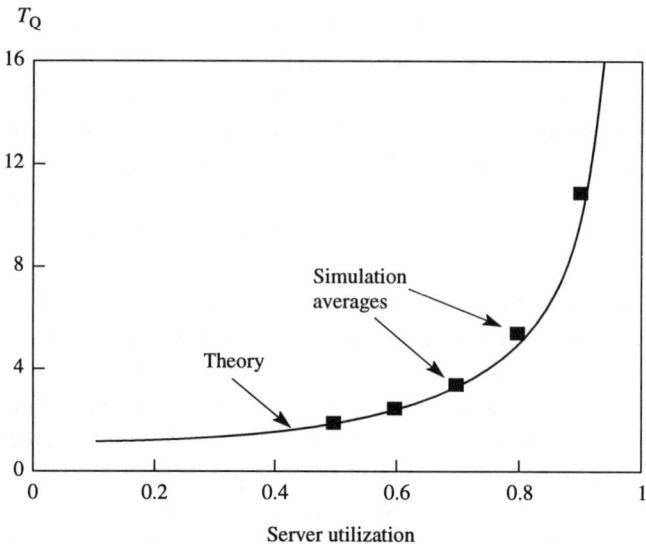

Figure 27.3. T_Q from an average of 10 simulation runs.

reliable. It is necessary to do several runs and take the average of the results. Figure 27.3 shows the averages of the 10 runs at each utilization value. The match between simulation and exact theory is clearly much better. Note also that the variance or spread of results is greater for higher utilizations. This is to be expected, since queueing theory tells us that the variance of time in system increases with server utilization. The need to take averages over a number of runs is an elementary point, but it is not uncommon for analysts to attach too much significance to a single simulation run.

One question we have not addressed so far is how long each simulation run should be. We know we need a warm-up time to get a steady state, but we have seen in Fig. 27.1 how the queue size varies even when a steady state exists statistically. The simple, but not necessarily helpful, answer is that enough customers need to be processed so that the variance of, for example, the average waiting time for one simulation run is not too high. Since we are going to take the average of the averages for several runs, there is a trade-off between doing fewer but longer runs and more but shorter runs. Doing more and shorter runs has the advantage that, if a distinct random number stream is used for each run, statistically more reliable results are obtained. However, fewer and longer runs tend to be more convenient in practice.

Exercise 27A

Construct a table like Table 27.2, and perform a simulation using a single die to generate inter-arrival and service-times. Add a column to represent the waiting-line size when the customer arrives.

Part Nine
Applications to computer systems and telecommunications

28
Applications to computer systems

Introduction

The purpose of this chapter is to show how straightforward queueing models can be used to analyse the performance of different aspects of computer systems. It is assumed that readers are familiar with the basic operation of disk storage drives and paging mechanisms.

Disk devices all operate in a broadly similar way, so although the examples here are based on IBM mainframe-style disks, the principles of the analysis can be applied to most disk systems. Engineers designing new disk devices are constantly trying to improve the base technology, and combining technologies in various ways to achieve a particular balance between cost, access speed and data capacity Bohl (1981) gives a good description of how disk storage works. Extensive discussion of the performance of modern large-scale disk systems is given in Houtekamer and Artis (1991). We shall look at the performance of a single disk drive using an $M/G/1$ model, and the impact of the number of paths between the disk devices and the processors using basic probability calculations.

Next we look at two examples of using the central server model, which is described in Chapter 26. The first example shows how the effect of paging on throughput can be modelled. The second example makes use of the disk performance model developed in this chapter, and combines this with the central server model. Finally we look at modelling transaction processing system, making use of the $M/M/m$ queueing model and combining it with the central server model of paging.

Many aspects of computer system performance are not covered here, for example the use of cache storage for disks, and multiprocessor cpus. Some performance analysis problems require advanced queueing-network techniques, which are beyond the scope of the non-specialist. On the other hand, a lot of valuable analysis can be done with the models presented in this book, and the author's aim is to encourage readers to make more use of 'intermediate' queueing theory.

Disk-drive performance

The principles of disk storage are straightforward. Data is stored magnetically on a revolving disk. A read/write head is mounted on an arm that can be extended or retracted to position the head over a particular 'track' of data. Data is stored in concentric circles on the disk. A schematic of a typical disk is shown in Fig. 28.1.

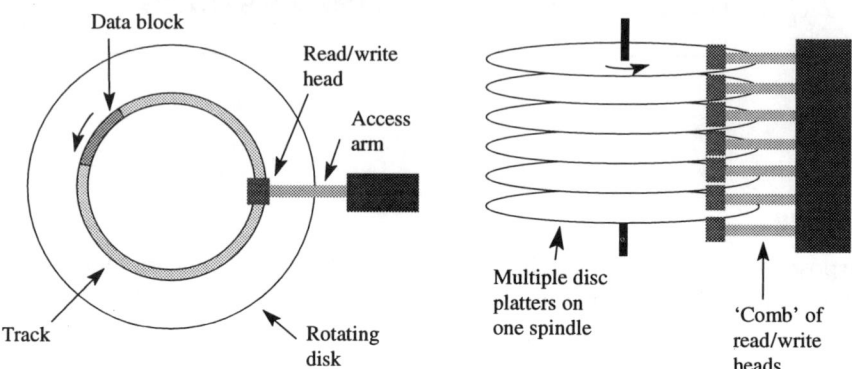

Figure 28.1. Schematic of a typical disk device.

In order to access a particular data position, the arm carrying the head must first be positioned over the relevant track, and then the system must wait for the disk to rotate and bring the particular data position under the head. The data can then be read or written. A number of individual disks are usually mounted on a single axle. The arms carrying a read/write for each disk surface are combined into a single unit, called the 'comb', that is extended and retracted as a unit. With the comb positioned at a particular track, the corresponding tracks on each disk surface form a 'cylinder' of data that can be accessed without moving the heads.

Moving the read/write head to a particular track is called 'seeking', and we speak of seek-distance (the number of cylinders traversed), and seek-time (the time taken to traverse the cylinders). Once the read/write head is positioned at the required cylinder, we have to wait for the data we want to read or the location at which we want to write some data to appear under the head. This, unsurprisingly, is called 'rotational delay'. Sometimes the terms 'latency' is used: less obvious but shorter. Since disks rotate at constant speed, and we assume there is no relationship between the time the seek completes and the position of the data we want, the probability distribution of rotational delay is easy to establish. Latency will be uniformly distributed between zero and the time for a complete disk rotation. (The distribution of latency may depend on the software in use, since some data access methods insist on starting the search for a required block at 'record 0' which is at a fixed physical position of the track. This could increase the latency by one revolution.)

Let us now build a queueing-theory model of a single disk drive. We shall assume that all blocks of data on the disk are the same length. To start with we shall assume 'random seeks' i.e. disk accesses are randomly spread over the disk. Later we shall use 'empirical seeks', i.e. make use of the kind of access pattern observed on real systems. The service-time for a disk read or write will therefore be made up from seek-time, rotational delay, and read/write time. The parameters we shall need to define are listed below, together with the values assumed for the examples. The parameters do not represent any specific disk device, but are not untypical of some products of the later 1980s.

$$N = \text{number of disk cylinders} = 885 \tag{28.1}$$

$$R = \text{disk revolutions per second} = 60 \tag{28.2}$$

$$B = \text{data blocks per track} = 11 \tag{28.3}$$

Seek-distance for random seeks

The number of cylinders traversed in a seek is called the seek-distance. This can vary from zero, when the seek starts and ends on the same cylinder, to $N - 1$ when the access arm must traverse the entire disk. Assuming that the access arm is equally likely to start on any cylinder and equally likely to finish on any cylinder, how shall we work out the probability of a seek of any particular distance? We can enumerate all the possible seeks that may occur. This is done in Table 28.1. The first column specifies the cylinder on which the seek starts, and each remaining column corresponds to a cylinder on which the seek may finish.

Table 28.1. Possible seeks

Start on cylinder	Finish on cylinder							
	1.00	2.00	3.00	4.00	5.00	...	$N-1$	N
1.00	0.00	1.00	2.00	3.00	4.00		$N-2$	$N-1$
2.00	1.00	0.00	1.00	2.00	3.00		$N-3$	$N-2$
3.00	2.00	1.00	0.00	1.00	2.00		$N-4$	$N-3$
...								
$N-1$	$N-2$	$N-3$	$N-4$	$N-5$	$N-6$		0.00	1.00
N	$N-1$	$N-2$	$N-3$	$N-4$	$N-5$		1.00	0.00

The first thing to note is that the number of possible seeks is N^2. How many of these are zero cylinders? The only zero entries are on the main diagonal from top left to bottom right. There are N zero entries, and so

$$p_0 = \text{Prob(seek distance} = 0) = \frac{N}{N^2} = \frac{1}{N} \tag{28.4}$$

How many possible seeks are there of 1 cylinder? There are two diagonal bands of 1s in the table, each containing $N - 1$ entries. So there are $2(N - 1)$ ways in which a seek of 1 cylinder can occur. In general, a seek of k cylinders can occur in $2(N - k)$ ways. So we have

$$p_k = \text{Prob(seek distance} = k) = \frac{2(N - k)}{N^2} \text{ for } k > 0 \tag{28.5}$$

Now that we know the probability distribution of seek-distance, we can work out the average seek.

$$\text{average seek} = \sum_{k=0}^{N-1} k p_k = \sum_{k=1}^{N-1} \frac{2k(N - k)}{N^2} \tag{28.6}$$

After some simple algebra we get Eq. (28.7) as the formula for the average seek. Some readers may find this result counter-intuitive, since they expect the answer to be $N/2$.

$$\text{average seek} = \frac{N^2 - 1}{3N} \approx \frac{N}{3} \tag{28.7}$$

Seek-time for random seeks

Next we need to know how long a seek of a particular number of cylinders takes. A zero seek will take zero time. Houtekamer and Artis (1991) suggest that a good approximation is

$$t_{\text{SEEK}} = \text{seek-time} = t_{\text{BASE}} + t_{\text{INCR}} \sqrt{\text{seek-distance}} \tag{28.8}$$

In other words, seek-time is a linear function of the square root of seek-distance. The values of t_{BASE} and t_{INCR} can be found by solving the simultaneous equations

$$t_{\text{S1}} = t_{\text{BASE}} + t_{\text{INCR}} \sqrt{1} \tag{28.9}$$

$$t_{\text{SMAX}} = t_{\text{BASE}} + t_{\text{INCR}} \sqrt{N - 1} \tag{28.10}$$

where t_{S1} is the time for a seek of 1 cylinder, and t_{SMAX} is the time for a maximum seek of $N - 1$ cylinders. The values for our hypothetical disk are

$$t_{\text{S1}} = 3 \text{ ms}, \, t_{\text{SMAX}} = 30 \text{ ms} \tag{28.11}$$

Simple rearrangement gives

$$t_{\text{INCR}} = \frac{t_{\text{SMAX}} - t_{\text{S1}}}{\sqrt{N - 1} - 1} \tag{28.12}$$

$$t_{\text{BASE}} = t_{\text{S1}} - t_{\text{INCR}} \tag{28.13}$$

Latency (rotational delay)

Latency is uniformly distributed between zero and the time for one complete revolution of the disk. Formulae for the mean and variance of a uniform distribution are given in Appendix 1. Using these we get

$$\text{average latency} = \frac{a+b}{2} = \frac{1}{2R} \text{ secs} = \frac{1000}{120} \text{ ms} = 8.333 \text{ ms} \qquad (28.14)$$

$$\text{variance of latency} = \frac{(b-a)^2}{12} = \frac{1}{12R^2} = \frac{1}{12}\left(\frac{1000}{60}\right)^2 = 23.148 \qquad (28.15)$$

$$\text{coeff of variation squared for latency} = \frac{(2R)^2}{12R^2} = \frac{1}{3} \qquad (28.16)$$

Read/write time

The layout of blocks of data on a track can be fairly complicated, but for our purposes we shall assume that all data blocks are of the same length and that read/write time (including the time to reconnect to the channel) is therefore a constant given by

$$\text{read/write time} = \frac{1}{BR} \text{ secs} = \frac{1000}{(11)(60)} \text{ ms} = 1.515 \text{ ms} \qquad (28.17)$$

Disk service-time for random seeks

Assuming random seeks, the mean and variance of the time for each component of a disk operation are listed in Table 28.2, together with the total. Seek-time is obviously the largest component. Using the mean and coefficient of variation squared for the i/o time, the M/G/1 formula (Eq. (9.1)) can be used to find average time-in-system for a range of i/o rates. In Fig. 28.2 the curve labelled 'random seeks' shows the results. These were calculated using the MG1Calc routine.

Table 28.2. Components of i/o time for random seeks

Activity	Mean	Variance	C2
Seek	16.97	38.29	0.13
Latency	8.33	23.15	0.33
Read/write	1.52	0.00	0.00
Total	26.82	61.44	
Standard development		7.84	
C2		0.09	

Disk service-time for observed seeks

In practice, to assume random seeks would be very pessimistic. Measurements of real systems show that a large proportion of i/os are to the same cylinder as the previous i/o to that disk. Consider a seek-distance distribution represented by the probability values in Table 28.3. Using this seek distribution, the program Seek-

I/O response (ms)

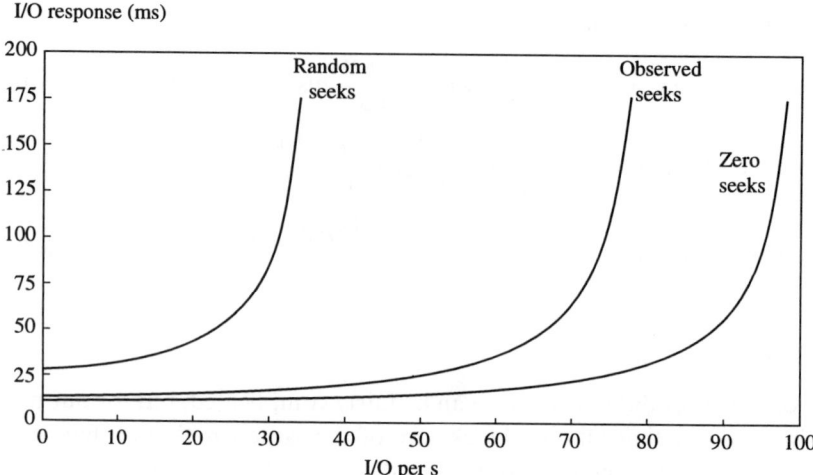

Figure 28.2. I/O response times for different seek distributions.

Table 28.3. Typical observed seek-distance distribution

Seek-distance	Probability
0.00	0.80
1.00	0.10
$k > 1$	$0.1/(N - 2)$

Calc was used to calculate the mean and variance of seek-time. The results are given in Table 28.4. Average seek-time is much smaller than for random seeks, although variance is much higher. Latency has become the biggest component of the time for an i/o operation. In Fig. 28.2 the curve labelled 'observed seeks' shows the corresponding performance as i/o rate increases.

Table 28.4. Components of i/o time for observed seeks

Activity	Mean	Variance	C2
Seek	2.37	42.51	7.55
Latency	8.33	23.15	0.33
Read/write	1.52	0.00	0.00
Total	12.22	65.66	
Standard development		8.10	
C2		0.44	

Disk service-time for zero seeks

A natural next step is to investigate performance when no seeks are required. This would correspond to providing a separate read/write head for each cylinder of the

Table 28.5. Components of i/o time for zero seeks

Activity	Mean	Variance	C2
Seek	0.00	0.00	
Latency	8.33	23.15	0.33
Read/write	1.52	0.00	0.00
Total	9.85	23.15	
Standard development		4.81	
C2		0.24	

disk. Table 28.5 shows the breakdown of i/o time now. Looking at Fig. 28.2, we can see that if we had a criteria of average disk response time of 25 ms, this would increase the i/o rate that could be handled from 50 i/os per second to 70 i/os per second. This would be a significant increase, but it is not obvious that the increased cost of such a disk would be justified.

Rotational position sensing (RPS)

So far we have looked at just a single disk device. A mainframe complex or powerful modern data server may have a large number of disk devices. Connections between the disks, cpu and memory are outlined in Fig. 28.3. The disks are connected to memory via a 'channel'. The channel provides access to the main memory independent of the cpu. Some direct communication between cpu and channel is needed, so that the cpu can tell the channel to initiate an i/o operation and the channel can tell the cpu via an interrupt mechanism that an i/o has completed. This kind of arrangement can be found in most types of computer system, although the terminology may be different. Clearly the channel is a shared resource or server, and if the channel is busy transferring data for one disk, then another disk cannot be simultaneously transferring data (although the channel might be shared by interleaving several data transfers).

Recalling that a disk device goes through seek, latency and read/write, what is the impact on the channel for a single i/o operation? We shall assume that once the

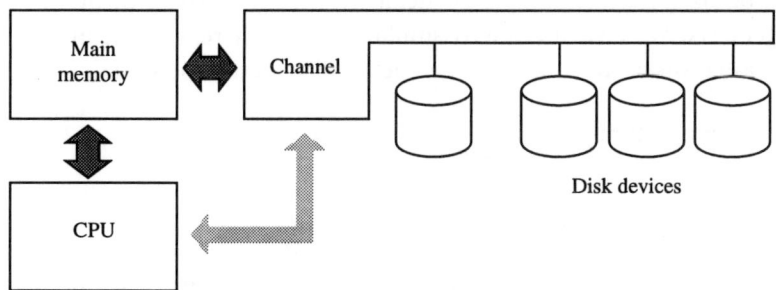

Figure 28.3. Connecting disks to main memory and cpu.

i/o operation has been comumunicated to the device, the disk does not require the channel during the seek. Similarly, the channel is not required during latency. It is only during read/write, when data transfer is actually happening, that the channel is needed. The disk device will usually have been instructed that a specific block of data from a particular track is wanted, so the disk knows when the required block is approaching the read/write head. At this point the disk will try to obtain control of the channel. If the channel is free then the read or write can proceed. If, however, the channel is busy, then the disk will have missed its chance, and will have to wait for a complete further rotation of the disk before trying again. This mode of operation is called 'rotational position sensing' or RPS, and a failure to obtain the channel when needed is called an 'RPS-miss'. What is the impact of RPS on i/o response time? We need to define some more parameters.

$$K = \text{number of disks} \tag{28.18}$$

$$\lambda_D = \text{i/o rate per disk} \tag{28.19}$$

$$T_S = \text{average disk service-time per i/o} \tag{28.20}$$

For simplicity we have assumed that the average i/o rate to each disk is the same. In practice this would obviously not be true, but the logic of the analysis is valid and can be extended straightforwardly for individual disk i/o rates. The average i/o service time is taken from Table 28.4. Now we can calculate the load on the channel.

$$\rho_C = \text{Prob(channel busy)} = K\lambda_D T_S \tag{28.21}$$

This is not, however, the probability of an RPS miss. If a disk is about to request control of the channel then by definition it is not already transferring data. It is only the load from the other disks that will prevent a disk getting control. So we have

$$p_{MISS} = \text{Prob(RPS miss)} = \frac{K-1}{K}\rho_C = (K-1)\lambda_D T_S \tag{28.22}$$

$$p_{REC} = \text{Prob(successful reconnect)} = 1 - p_{MISS} \tag{28.23}$$

How many attempts will be needed before the disk finally gets hold of the channel and the i/o can be completed? Example 2C dealt with the number of throws of a die needed to get a six. We have an exactly similar situation here, and the number of reconnection attempts needed follows a geometric distribution, so that

$$E(\text{reconnection attempts}) = \frac{1}{p_{REC}} \tag{28.24}$$

$$\text{Var(reconnection attempts)} = \frac{1 - p_{REC}}{p_{REC}^2} \tag{28.25}$$

The number of extra rotations caused by RPS misses is one less than the number of reconnection attempts. So the average number of extra rotations due to RPS is given by Eq. (28.26). Using the property of variance in Eq. (28.27), the variance of the extra rotations is the same as Eq. (28.25).

$$E(\text{extra rotations}) = \frac{1}{p_{\text{REC}}} - 1 = \frac{1 - p_{\text{REC}}}{p_{\text{REC}}} \tag{28.26}$$

$$\text{Var}(aX + b) = a^2 \, \text{Var}(X) \tag{28.27}$$

Now we have to convert the number of extra rotations into the time they take. To get the average time we obviously just multiply the average number of rotations by the time per rotation. For the variance we use Eq. (28.27). We shall refer to the time taken by extra rotations as 'RPS time'.

$$E(\text{RPS time}) = \frac{1 - p_{\text{REC}}}{R p_{\text{REC}}} \tag{28.28}$$

$$\text{Var}(\text{RPS time}) = \frac{1 - p_{\text{REC}}}{R^2 p_{\text{REC}}^2} \tag{28.29}$$

If we add this new component of response time to the components already listed in Table 28.4, then we can calculate the i/o response time for different combinations of i/o rate per disk and number of disks on the channel. The subroutine DiskPerf was used with a simple calling to produce the results that are plotted in Fig. 28.4.

With the type of disk and channel configurations considered so far, it is usual to limit the number of disks per channel so that channel utilization does not exceed about 30 per cent. This is because of the relatively large RPS delay times at higher channel utilizations. An obvious improvement is to provide more than one data transfer path between the disks and main memories. In fact, the number of data transfer paths depends not only on the number of channels, but also on the

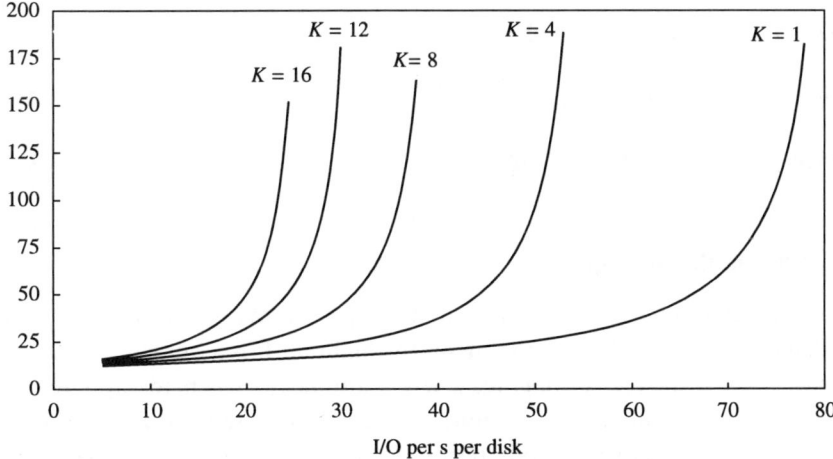

I/O response (ms)

I/O per s per disk

Figure 28.4. I/O response time with RPS delays (K = disks per channel).

Table 28.6. Probabilities of multiple paths all busy

Path utilization	Prob 2 paths all busy	Prob 4 paths all busy
0.20	0.04	0.00
0.40	0.16	0.03
0.60	0.36	0.13
0.80	0.64	0.41
0.90	0.81	0.66

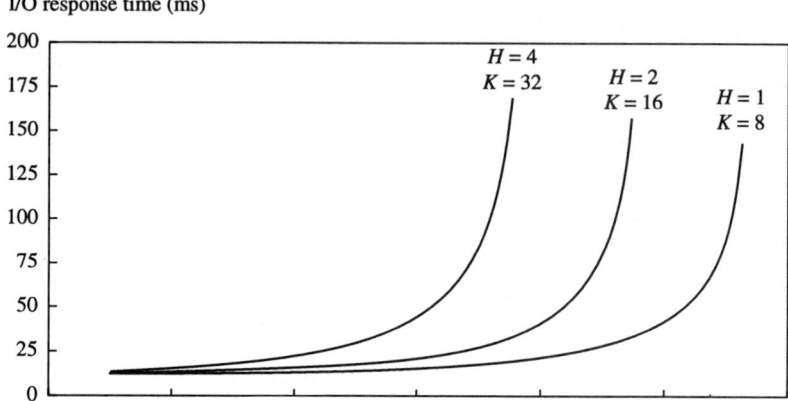

Figure 28.5. I/O response times for multiple paths.

sharing of disk control electronics. Readers should consult Houtekamer and Artis (1991) to understand the complexity of configurations that can occur. We shall consider what happens if multiple data paths are provided (Table 28.6, Fig. 28.5)

$$H = \text{number of paths or channels} \tag{28.30}$$

$$\rho_C = \text{Prob(channel busy)} = \frac{K\lambda_D T_S}{H} \tag{28.31}$$

$$\text{Prob(all channels busy)} = \rho_C^H \tag{28.32}$$

Central server model of paging

The central server model was described in Chapter 26. We shall now use this model to demonstrate the effect of paging on throughput. It is assumed the reader is familiar with the principles of paging, where main memory is made apparently larger than it really is. Only active sections or pages of programs are loaded into main memory, while inactive sections reside on disk until they are needed. The more programs that are competing for real memory, the more often a particular

Prob(page miss) $\times 10^5$

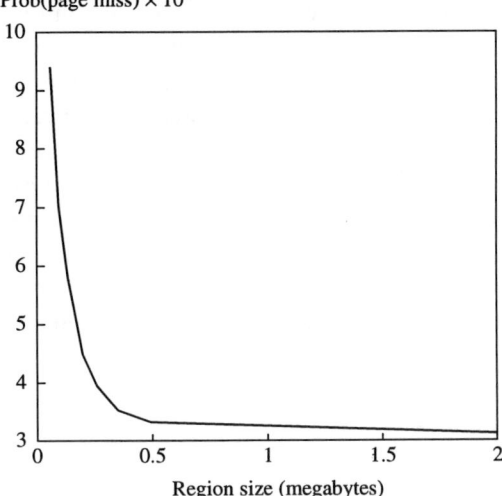

Region size (megabytes)

Figure 28.6. Page fault rate versus real memory region size.

program will have to wait until the section of apparent memory, or page, is brought in from disk. A typical characteristic is shown in Fig. 28.6, where the probability that a new page will be required when the next program instruction is attempted is plotted against the average amount of real storage allocated to each program. It is the shape of this curve that is typical; the particular values are based on an example in Lavenberg (1983). Lavenberg gives the parameters, analysis and results; here we shall show how the analysis is implemented in a computer program.

We shall use the central server model to look at how throughput varies as the multiprogramming level increases. For simplicity we assume that no disk i/o occurs except for paging. Also for simplicity we assume that when a page fault occurs it is necessary only to read in the required page from disk. In reality it will be necessary in a small proportion of cases to write out to disk the page in real memory selected to be replaced. The parameters we need to define are as follows, with the particular values chosen for the example.

$$A = \text{real memory size in megabytes} = 2 \tag{28.33}$$

$$N = \text{multiprogramming level} \tag{28.34}$$

$$\delta(A/N) = \text{Prob(page fault per instruction)} \tag{28.35}$$

$$L = \text{path length of a program} = 0.25 \times 10^6 \text{ instructions} \tag{28.36}$$

$$\alpha = \text{cpu speed (instructions/sec)} = 2 \times 10^6 \tag{28.37}$$

$$\beta = \text{disk speed (i/os per sec)} = 50 \tag{28.38}$$

$$D = \text{number of disks for paging} = 2 \tag{28.39}$$

In the central server model a program continues to execute instructions until a page fault occurs. The probability of a page fault is $(1 - p_1)$. On average, how many page faults will occur for each program before it completes? This takes us back to how many throws of a die are needed to get a six. (Readers could refer back to Example 2C in Chapter 2.) For a program, success equates to completing, which happens with probability p_1, while failure equates to a page fault, whose probability is $(1 - p_1)$. The average number of execution intervals per program is therefore $1/p_1$, and the average number of page faults per program is $[(1/p_1) - 1]$. Looking at page faults another way, we know that a program executes L instructions, and on each instruction the probability of a page fault is $\delta(A/N)$. Therefore we can write

$$\text{average page faults per program} = \frac{1}{p_1} - 1 = L\delta(A/N) \tag{28.40}$$

Some simple algebraic rearrangement of Eq. (28.40) gives

$$p_1 = \frac{1}{1 + L\delta(A/N)} \tag{28.41}$$

We also need to work out the speed of the cpu in programs per second, i.e. the number of program execution intervals per second. Since the speed of the cpu is α instructions per second, and each program is L instructions long, the cpu time needed by each program is L/α seconds, so the cpu can process α/L complete programs per second. Since each program has $1/p_1$ execution intervals

$$\text{cpu capacity} = \frac{\alpha}{Lp_1} \text{ execution intervals/sec} \tag{28.42}$$

Assuming that disk activity is evenly distributed over the disks, we can very easily calculate the other probabilities we need, i.e.

$$p_j = \frac{1 - p_1}{D} \text{ for } j = 2, \ldots, m \tag{28.43}$$

where $m = D + 1$. The service-time of each disk is assumed to be independent of the i/o rate, which means we are ignoring the effects of channel contention and RPS misses. Later in this chapter we shall show how to combine the central server model with models of channel contention and RPS. For now we want to keep the analysis simple, and it is not unreasonable to assume the disk configuration would be chosen to minimize channel effects for the paging devices. Putting the parameters we have into the central server model, we can plot throughput against the multiprogramming level. For the example we have chosen, the maximum cpu throughput *if no paging is required* is 8 tasks/s, since $\alpha = 2$ mips and $L = 0.25$ instructions.

Figure 28.7 shows the results, which were obtained using the CSPgTput routine. The shape of the curve is typical as the multiprogramming level in a paging system

Throughput (tasks/s)

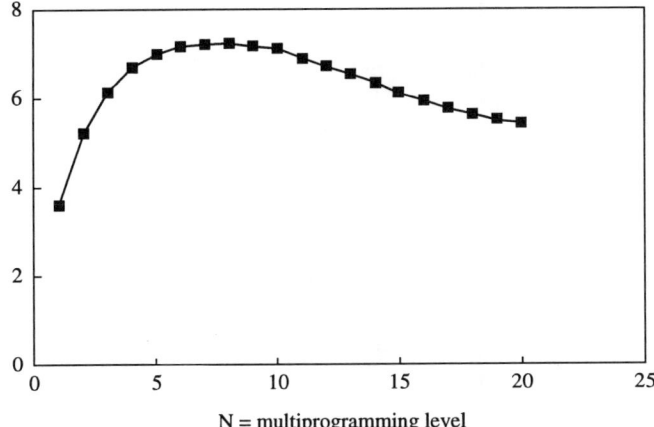

N = multiprogramming level

Figure 28.7. Throughput versus multiprogramming level.

increases. After reaching a maximum, throughput declines as real memory per program is reduced, necessitating more disk activity.

Central server model with disk performance model

In the previous section we used the central server model to investigate the effect of paging, assuming that the disk operated independently and there was no contention for the channel or paths. Now we shall look at how we can use the central server model and at the same time take account of channel/path contention and consequent RPS misses.

If we assume a particular disk service rate, then we can use the central server model to find out the system throughput and the i/o rate per disk. The i/o rate per disk can then be used with our disk performance model to find out what the effective disk service rate is at that i/o rate, taking account of channel contention and RPS misses. If that effective rate is different from the rate we assumed for the central server model, then we go back to the central server model with the new disk service rate. We continue iterating until the disk service rate has converged to a steady value (Fig. 28.8).

As an example we take

$$L = \text{cpu time per job} = 1 \text{ s} \tag{28.44}$$

$$\beta = \text{average cpu time per disk i/o} = 0.003 \text{ s} \tag{28.45}$$

$$D = \text{number of disks} = 16 \tag{28.46}$$

The disk characteristics we shall use are those in earlier examples. The initial disk service rate is taken from Table 28.4, which gives an average disk service-time of

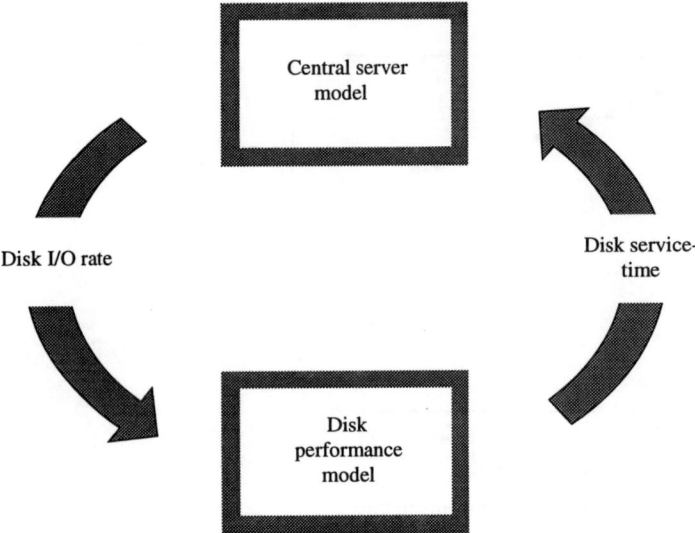

Figure 28.8. Combining central server and disk performance models.

12.2 ms, so the service rate is 1000/12.2 i/os per second. In order to set up the central server model we need to calculate p_1, which in this example is simply

$$p_1 = \frac{\beta}{L} \tag{28.47}$$

The calculations for a particular level of multiprogramming are carried out by the routine CSDkTput. The results for a range of N are shown in Fig. 28.9, which plots cpu utilization against multiprogramming level. The curve labelled 'variable'

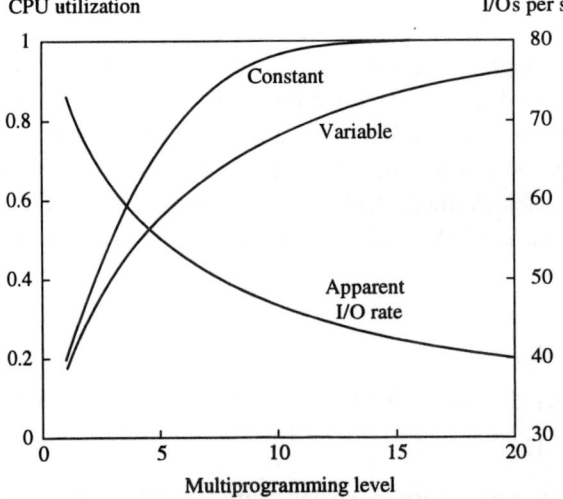

Figure 28.9. Results for combined central server and disk performance models.

is the result of our combined model. The curve labelled 'constant' is the result obtained if channel contention and RPS misses are ignored. This is obtained from the first pass through the central server model before adjusting the disk service rate. The 'apparent i/o rate' is the capacity of each disk device as perceived by the job's executing, taking account of RPS misses induced by channel contention. Ignoring channel contention obviously gives optimistic results.

Transaction processing

In a transaction processing system a number of transactions are processed in parallel. Some mechanism will be provided to limit the number of concurrently active transactions. At any instant in time there will be a number of active transactions contending for the resources of the computer system. Figure 28.10 illustrates the situation. How shall we model this? The idea of a limited number of regions, or other logical facilities for running transactions, suggests a multiserver queueing model. For simplicity we shall use M/M/m, although G/G/m could be used. However, adding another region is not quite the same as adding another server in the sense we usually mean when describing multiserver systems. The additional 'server' has to compete with the other servers for the cpu and disk capacity, and we must somehow model this resource contention.

As an illustration of how to model the resource contention, we shall use the paging model described earlier in this chapter. This means the transactions only need cpu to be processed: no disk i/o is required for data. We shall define the transaction in terms of the average number of instructions to be executed. The disk i/o

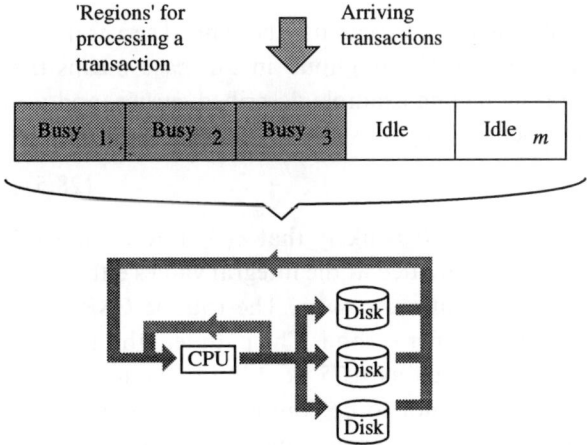

Figure 28.10. Transaction processing.

is purely for paging. The parameters we need to define are as follows.

$$\lambda = \text{transaction arrival rate per sec} \qquad (28.48)$$

$$L = \text{instructions executed per transaction} = 0.25 \times 10^6 \qquad (28.49)$$

$$\alpha = \text{cpu speed, instructions per sec} \qquad (28.50)$$

$$t = \text{service-time for a transaction} \qquad (28.51)$$

$$n = \text{number of concurrent transactions} \qquad (28.52)$$

$$h = \text{throughput in transactions per sec} \qquad (28.53)$$

The transaction service-time t is a service-time as though all the regions or servers were actually independent. As the number of concurrent transactions increases, t increases to reflect the contention for the cpu. Recall Fig. 28.7, which shows how the cpu throughput varies with the multiprogramming level. The multiprogramming level is the number of transactions that are concurrently in progress. Given λ, we need a method of finding n and t. We shall do this iteratively, so we use the notation

$$t_k, n_k, h_k \text{ are the } k\text{th successive estimates of } t, n, h \qquad (28.54)$$

Little's law tells us that the average number of concurrent transactions will be the average arrival rate times the average service-time. So we start by assuming the full power of the cpu is available to each transaction, so that

$$t_0 = \frac{L}{\alpha} \qquad (28.55)$$

Now we can work out the number of concurrent transactions, using Little's law, which gives

$$n_k = \lambda t_{k-1} \qquad (28.56)$$

With n_k, which is the multiprogramming level, we can use the central server model to calculate the throughput of the system. 'Throughput' in this case means the effective cpu rate. We do this as in the paging example described earlier, and use the CSPgTput routine. We can represent this by

$$h_k = \text{throughput}(n_k) \qquad (28.57)$$

meaning simply that h_k is a function of n_k. It is likely that n_k is not an integral value, in which case throughput must be evaluated at the integral values either side of n_k and interpolation used to get the required answer. The routine CSPgTput returns a throughput value of jobs or tasks per second. The path length of a task for CSPgTput is defined within that routine as 0.25×10^6 instructions. This happens to be the same as L, but they are not to be confused. The path length within CSPgTput is quite arbitrary as far as the current analysis is concerned. We need to know what it is only in order to convert the throughput value h_k into cpu instructions per second, i.e.

$$\text{effective cpu speed} = h_k \times 0.25 \times 10^6 \text{ instructions/sec} \tag{28.58}$$

Now we can form our next estimate of transaction service-time:

$$t_k = \frac{L \times n_k}{h_k \times 0.25 \times 10^6} \tag{28.59}$$

This completes one iteration, and if successful values of t are close enough the process can be terminated. Figure 28.11 shows how the number of active transactions and the apparent service-time for a transaction change as the arrival rate increases. The active transactions have to share the available cpu power between them, and we assume that the cpu power is shared in small time-slices equally between the transactions present. Figure 28.7 shows that the effective cpu power is not a constant, but changes with the multiprogramming level. In Fig. 28.11 we can see that the service-time is flat until the average number of concurrent transactions exceeds 1, after which apparent service-time increases as the cpu is shared between transactions.

The analysis so far has been the complicated part. The remainder of the analysis is simply applying the M/M/m queueing model to work out response time versus transaction arrival rate, for a given number of 'servers' or message-handling regions. The MMmCalc routine was used for this, and the results for $m = 4$ and $m = 15$ are shown in Fig. 28.12. For $m = 4$, the contention is for the 'servers'. The response time increases markedly in the familiar way as the arrival rate exceeds the capacity of the servers. Capacity of the system would be increased if five servers were provided instead of four. Now look at the graph for $m = 15$. Figure 28.7 shows that the effective cpu power would actually decrease once the multi-

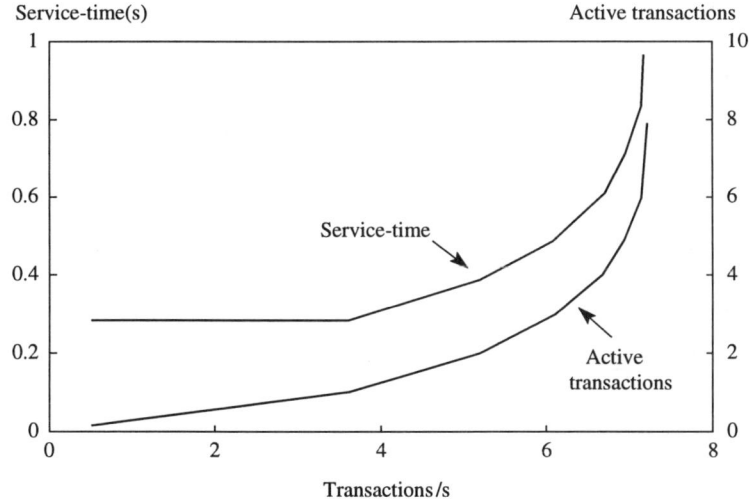

Figure 28.11. Service-time and number of active transactions versus arrival rate.

Response time (s)

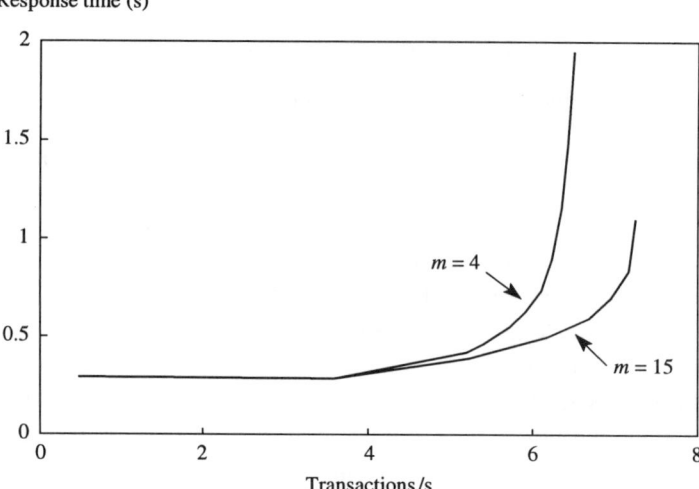

Figure 28.12. Response time versus arrival rate for different maximum concurrent transactions.

programming level exceeded about 10, so for $m = 15$ the response-time curve reflects the cpu approaching saturation, and the number of servers is not the bottleneck.

All models are approximations, and in our analysis of transaction processing we have simplified real systems. In particular we have used the average active transactions as a 'constant' multiprogramming level. Of course the number of active transactions will be fluctuating, and the cpu throughput is a decidedly non-linear function of the multiprogramming level. Even so, we have developed a model that shows how the parameters of the system interact, and helps us understand the behaviour of transaction processing systems.

Programs for computer system performance

In order to use the routines described in this chapter, a program must contain the following statements.

```
{$N+,E+,R+ }
{$I \qthprog\qthtypes.pas   }
{$I \qthprog\qtherror.pas   }
{$I \qthprog\qthfile.pas    }
{$I \qthprog\mg1calc.pas    }
{$I \qthprog\pratio.pas     }
{$I \qthprog\mmmcalc.pas    }
{$I \qthprog\diskperf.pas   }
{$I \qthprog\buzencs.pas    }
{$I \qthprog\getmrte.pas    }
{$I \qthprog\cspgtput.pas   }
{$I \qthprog\trantput.pas   }
```

The DiskPerf subroutine

The DiskPerf subroutine calculates the performance for multiple disks sharing
several channels or paths. The average i/o rate is assumed to be the same for each
disk. The disk characteristics assumed are for the IBM 3380 example, with empirical seek distribution. For a different type of disk or seek pattern the performance
parameters within the subroutine should be changed. These characteristics could
be supplied as parameters, but the DiskPerf routine will be used later in combination with other queueing models, so it simplifies the programming to keep the
number of parameters small. Readers who wish to make DiskPerf more general
should have no difficulty in doing so.

```
{----------------------------------------------------}
{> DiskPerf - Calculation of disk response time      }
{  Inputs:                                            }
{     K          no. of disk devices                 }
{     H          no. of channels or paths             }
{     LD         i/os per sec per disk                }
{  Outputs:                                           }
{     TQ         average disk response time           }
{     TS         average disk service time ms         }
{     RHOD       individual disk utilisation          }
{     RHOC       channel/path utilisation             }
{     PMISS      probability of an RPS miss           }
{     VALID      false if disk or channel             }
{                utilisation >=1                      }
{  Copyright Mike Tanner 1993                         }
{----------------------------------------------------}
Procedure DiskPerf(K,H:integer;LD:QTHreal;
                   Var TQ,TS,RHOD,RHOC,PMISS:QTHreal;
                   Var VALID:boolean);
{-Performance statistics for a single device-------}
Const SEEKMN=2.370;  SEEKVR=42.51;   { seek time  }
Const LTCYMN=8.333;  LTCYVR=23.15;   { latency    }
Const RDWRMN=1.515;  RDWRVR=0;       { read/write }
Const R=60;                     { disk revs per sec }
Var LAMBDA,TD,PREC,VR,C2,TQ90,TQ95:QTHreal;
Var TW,LQ,LW,Z,RK,RH:QTHreal;
Var RCATMN,RCATVR:QTHreal; { reconnection attempts }
Var RTMEMN,RTMEVR:QTHreal; { RPS time              }
begin
    RH:=H;  RK:=K;
    VALID:=true;
    {-TD is time channel busy per i/o--------------}
    TD:=RDWRMN;
    {-Channel utilisation and RPS miss prob--------}
    LAMBDA:=LD*RK;
    RHOC:=LAMBDA*TD/(RH*1000);{ channel utilisation }
    If RHOC>0.9 then begin VALID:=false; Exit; end;
    If K>H then PMISS:=Exp(RH*Ln((RK-1)*RHOC/RK))
           else PMISS:=0;
    PREC:=1-PMISS;              { prob of reconnect   }
```

```
{-Reconnection attempts and RPS time------------}
RCATMN:=1/PREC;   RCATVR:=(1-PREC)/Sqr(PREC);
RTMEMN:=1000*(RCATMN-1)/R;
RTMEVR:=RCATVR*Sqr(1000/R);
{-Add all the i/o components together-----------}
TS:=SEEKMN+LTCYMN+RDWRMN+RTMEMN;
VR:=SEEKVR+LTCYVR+RDWRVR+RTMEVR;
C2:=VR/Sqr(TD);
{-Calculate i/o response time-------------------}
MG1Calc(LD/1000,TS,C2,
        RHOD,TQ,TQ90,TQ95,TW,LQ,LW,VALID);
end;
```

The CSPgTput routine

The CSPgTput routine sets up the central server model for calculating throughput versus multiprogramming level for the example we have been considering. The routine GetMRte, described below, provides a representation of the page-fault rate versus region size shown in Fig. 28.6. The BuzenCS routine is used to solve the model. Many of the parameters of the example are fixed within the subroutine, but they could easily be made into parameters. Two disks are assumed, and each disk is given a speed of 50 i/os per second.

```
{--------------------------------------------------}
{> CSPgTput -- A central server model of paging    }
{  Inputs:                                          }
{     N            multiprogramming level           }
{  Outputs:                                         }
{     TPT          throughput in jobs/sec           }
{     RHO[]        utilisations of cpu, disks       }
{  Copyright Mike Tanner 1993                       }
{--------------------------------------------------}
Procedure CSPgTput(
            N:integer;              { no. of jobs/tasks }
            Var TPT:QTHreal;    { tput jobs/sec     }
            Var RHO:buzenvecm);{ utilisations       }
Var IPS,L,RD,RN,DELTA,A,REG:QTHreal;
Var I,J,D,M:integer;
Var P,SR:buzenvecm;G:buzenveck;
begin
   RN:=N;                     { type conversion        }
   {-Set job/task parameters----------------------}
   IPS:=2*1E6;                { instructions per sec   }
   L  :=0.25*1E6;             { ins per cpu slice       }
   D  :=2;    RD:=D;          { no. of paging devices   }
   M:=D+1;                    { no. of service centres  }
   A:=2.0;                    { megabytes main memory   }
   {-Set up and solve central server model---------}
   REG:=A/RN;                 { region size per job     }
   DELTA:=GetMrte(REG);  { page fault rate         }
   P[1]:=1/(1+L*DELTA);  { prob job completes      }
   For J:=2 to M do
```

```
begin
   P[J]:=(1-P[1])/RD;  { io evenly split         }
   SR[J]:=50;          { disk capacity ios/sec   }
end;
SR[1]:=IPS/(L*P[1]);   { cpu rate jobs/sec       }
BuzenCS(M,N,P,SR,G,RHO,TPT);
end;
```

The GetMRte routine

This routine represents the page-fault probability function $\delta(A/N)$. The function is defined by a series of points in the table MRT, and simple linear interpolation is used to get values of the function between the defined points. Figure 28.5 was constructed using the GetMRte routine.

```
{---------------------------------------------------}
{> GetMrte -- Page miss rate function for central   }
{>            server model of paging.               }
{  Input:                                           }
{     R           region size in megabytes          }
{  Returns:                                          }
{     Probability of a page fault                   }
{  Returns:                                          }
{  Copyright Mike Tanner 1993                       }
{---------------------------------------------------}
Function GetMrte(R:QTHreal):QTHreal;
Type missrte=record   { entry in miss ratio table  }
                RGSZ:QTHreal;
                MRAT:QTHreal;
             end;
Const MRTN=10;          { number of entries in ARF }
Const MRT:array[1..MRTN] of missrte=(
(RGSZ:0.01;MRAT:1.5*1E-4),(RGSZ:0.05;MRAT:1.0*1E-4),
(RGSZ:0.1; MRAT:7.0*1E-5),(RGSZ:0.15;MRAT:5.5*1E-5),
(RGSZ:0.20;MRAT:4.5*1E-5),(RGSZ:0.25;MRAT:4.0*1E-5),
(RGSZ:0.35;MRAT:3.5*1E-5),(RGSZ:0.5; MRAT:3.3*1E-5),
(RGSZ:1.0; MRAT:3.2*1E-5),(RGSZ:3.0; MRAT:3.0*1E-5));
Var I:integer;
begin
   If (R<=MRT[1].RGSZ) or (R>=MRT[MRTN].RGSZ)
   then QTHError('Region size out of range');
   For I:=2 to MRTN do
     If R<=MRT[I].RGSZ
     then begin
             GetMrte:=MRT[I-1].MRAT+
                  (R-MRT[I-1].RGSZ)
                  *(MRT[I].MRAT-MRT[I-1].MRAT)
                  /(MRT[I].RGSZ-MRT[I-1].RGSZ);
             Exit;
          end;
end;
```

The CSDkTput routine

The CSDkTput routine operates in the way illustrated in Fig. 28.8. A central server model is set up and solved using the BuzenCS routine. The BuzenCS routine tells us the disk utilization, which implies the disk i/o rate. The disk i/o rate is then input to the DiskPerf routine which calculates the effective disk service-time including RPS misses. This new disk service rate is then compared to the previous value, and if they are significantly different the new disk service-time is input to BuzenCS to recalculate the i/o rate. Iteration continues until the disk service-time converges to a stable value.

```
{------------------------------------------------------}
{> CSDkTput -- A central server model of disk and  }
{>            cpu activity for multiprogramming,    }
{>            with channel contention.              }
{  Inputs:                                          }
{      N        multiprogramming level              }
{  Outputs:                                         }
{      RHOA     cpu utilisation, no channel effect  }
{      RHOB     cpu utilisation, with channel effect }
{      RHOC     channel utilisation                 }
{      DSRT     effective disk service rate         }
{  Copyright Mike Tanner 1993                       }
{------------------------------------------------------}
Procedure CSDkTput(N:integer;
                   Var RHOA,RHOB,RHOC,DSRT:QTHreal);
Var BURST,RD,RN,TPT,DTP,L,NSR:QTHreal;
Var TQ,DS,RHOD,PMISS:QTHreal;J,D,M:integer;
Var LSW,VALID:boolean;
Var P,SR,RHO:buzenvecm;G:buzenveck;
Label L1;
begin
    L:=1;                 { total cpu time needed   }
    BURST:=0.003;         { length of cpu burst     }
    P[1]:=BURST/L;        {                         }
    D:=16; RD:=D;         { number of disks         }
    NSR:=1000/12.2;       { initial disk service rate}
    M:=D+1;
    RN:=N;
    {-Set up central server model------------------}
    LSW:=true;
    For J:=2 to M do P[J]:=(1-P[1])/RD;
    SR[1]:=1/(BURST);
    {-Solve central server model to get thruput-----}
L1:For J:=2 to M do SR[J]:=NSR; { disk ios per sec }
    BuzenCS(M,N,P,SR,G,RHO,TPT);
    DTP:=RHO[2]*SR[2];              { disk throughput }
    {-Record tput with initial disk service time----}
    If LSW then begin RHOA:=RHO[1]; LSW:=false; end;
    {-Use throughput to calculate disk response time}
    DiskPerf(D,1,DTP,TQ,DS,RHOD,RHOC,PMISS,VALID);
    { Adjust disk service rate---------------------}
    NSR:=1000/DS;         { new service rate         }
```

```
      If Abs(NSR-SR[2])>0.1 then Goto L1;
      RHOB:=RHO[1];              { return cpu utilisation   }
      DSRT:=SR[2];               { return disk service rate }
end;
```

The TranTput routine

The TranTput routine does the calculations for the analysis of transaction processing. An initial estimate of transaction service-time is made, which is used with the
arrival rate to calculate the number of active transactions. The CSPgTput routine
is then called to calculate the cpu throughout. Linear interpolation between integral numbers of active transactions is used. The throughput is then converted to an
effective cpu speed. A new estimate of transaction service-time is made, and the
process repeated until the estimated service-time has converged to a steady value.

```
{-------------------------------------------------------}
{> TranTPut - Transaction processing model.            }
{  Inputs:                                              }
{      LAMBDA transaction arrival rate                  }
{      PL0    path length per transaction i.e.          }
{             average instructions executed             }
{  Outputs:                                             }
{      NAT    number of active transactions             }
{      TS     average transaction service time          }
{      CPURHO cpu utilisation                           }
{  Copyright Mike Tanner 1993                           }
{-------------------------------------------------------}
Procedure TranTput(LAMBDA,PL0:QTHreal;
                    Var NAT,TS,CPURHO:QTHreal);
Var OLDTS,NATLR,NATHR,TPT,TPTL,TPTH,CPUIPS:QTHreal;
Var UTIL:buzenvecm;NATL,NATH:integer;
Const ALPHA=2*1E6;                      { cpu speed }
Label L1;
begin
    {-Set initial service time using full cpu power-}
    TS:=PL0/ALPHA; OLDTS:=0;
    {-Calculate active transactions----------------}
L1:NAT:=LAMBDA*TS;            { no. of active trans }
    NATL:=Trunc(NAT); NATLR:=NATL; { lower integer  }
    NATH:=NATL+1;      NATHR:=NATH; { higher integer }
    {-Use CSPgTput routine to calculate thruput-----}
    If NATL=0 then TPTL:=0
              else CSPgTput(NATL,TPTL,UTIL);;
    CSPgTput(NATH,TPTH,UTIL);
    TPT:=TPTL+(TPTII-TPTL)*(NAT-NATLR)/(NATHR-NATLR);
    {-Convert into cpu instruction rate, and util---}
    CPUIPS:=TPT*0.25*1E6;
    CPURHO:=CPUIPS/ALPHA;
    {-Calculate new service time--------------------}
    TS:=PL0*NAT/CPUIPS;
    If Abs(TS-OLDTS)<0.001 then Exit;
    OLDTS:=TS;
    Goto L1;
end;
```

29
Applications to voice networks

Introduction and terminology

A simple voice network is shown in Fig. 29.1. This network has two 'switches' or 'exchanges', linked by a number of circuits or lines. The question we are usually trying to answer is whether a circuit will be available when a call is attempted from X to Y. Generally we ignore the possibility that phone Y is busy when the call is attempted. If the network can carry the call, but Y is busy, then the call has not failed because of lack of network capacity. It is also possible, at least in theory, that a switch has insufficient capacity to connect a call even though a circuit is available to carry the call. In practice, switches these days are 'non-blocking' or at least have greater internal switching capacity than the number of phones and circuits that can be attached to them. In this book we shall not consider internal switch design. Interested readers should consult Bear (1988) and Schwartz (1987) for discussion of switch design.

The voice, or telephony, area has its own terminology. Lines or circuits connecting switches are often referred to as 'trunks'. The term 'blocking' is used for all the lines being busy, so that a call is 'blocked'. The probability of blocking is called the 'grade of service'. The term 'holding time' is used for service-time, with the

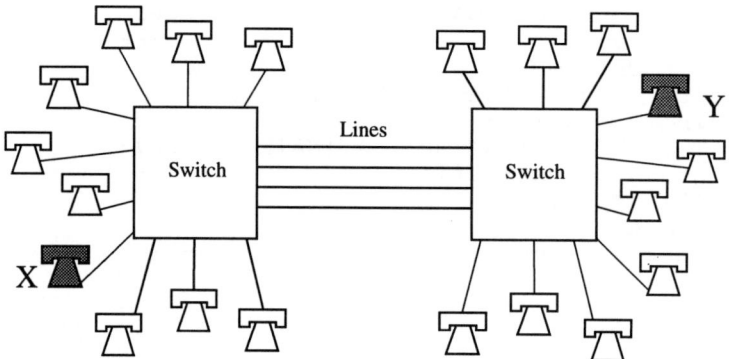

Figure 29.1. A basic voice network.

interpretation that a call holds on to the facilities required for the duration of a call.

The erlang

The erlang is a unit of traffic in voice systems, named after A.K. Erlang who conducted the early development of queueing theory in order to tackle design problems in telephone networks. The erlang can be defined in a number of equivalent ways. We shall define it as the average number of calls per holding time.

$$\text{traffic level} = \lambda T_S \text{ erlangs} \qquad (29.1)$$

This is identical to 'traffic intensity' which we encountered particularly when looking at multiserver systems. It is quite common, and perfectly correct, to describe the workload arriving at a queue as so many erlangs even for non-telephone applications.

Erlang-B (multiserver loss model)

We have already met one of the main queueing-theory models used in voice network design. This is the multiserver loss or $M/M/m/m$ model described in Chapter 19, where we saw that the probability that an arriving call (customer) is blocked (lost) is given by the Erlang-B function $E_B(m, u)$. Figure 19.2 illustrates the shape of the Erlang-B function, and Chapter 19 contains all the formulae and programs likely to be needed, including a discussion of the scaling effect for a multiserver loss system.

Limited traffic sources model (Engset)

The Erlang-B model assumes an infinite or very large number of potential 'customers' or telephones that may originate a call. The offered traffic rate is therefore unaffected by the number of lines (servers) provided to carry the traffic. What is the difference if the traffic is coming from a finite and not very large number of sources? The situation is depicted in Fig. 29.2, where K subscribers to telephones may make calls which then contend for one of the m lines available. We shall for simplicity assume that only outgoing calls are made, so that we do not have to consider incoming calls finding the required phone busy. Similarly we assume that calls are not made between phones attached to the switch.

First we need to define the offered traffic rate λ_0 in terms of the behaviour of each of the K sources. If a source is idle, then we assume that the time between calls is T_A, exponentially distributed. When a call is attempted and is successful in getting one of the m lines, then the call duration (holding time) is exponentially distributed with average T_S. If an attempted call is lost because all the lines are

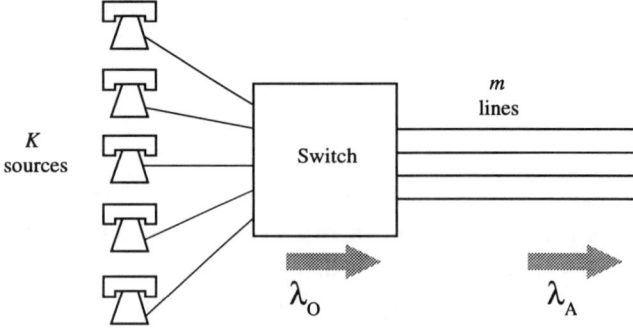

Figure 29.2. Limited traffic sources.

busy, then the source immediately goes back to idle state and will, after an average time T_A, attempt another call.

This is very like the machine repairman models of Chapters 23 and 24, except that in this case calls or customers are lost if all the servers are busy. The machine repairman model has a self-limiting effect, because customers queue, and as more of the fixed number of customers are queueing, the arrival rate decreases. What effect do we expect, intuitively, with the current finite-sources model? First there is an upper limit of the arrival rate from each source. The maximum average arrival rate will be $1/T_A$ calls per unit time for each source, and this will happen if practically all the attempted calls are lost. If at least some calls are successful, then while a successful call is in progress the originating source cannot attempt further calls. So if no calls are lost the minimum average arrival rate will be $1/(T_A + T_S)$ call per unit time. So the call rate is contained between limits, but on the other hand the average call rate will be higher if many calls are lost than if most calls are successful.

When we looked at the G/M/1 model in Chapter 14 we saw that there is a distinction between the average number of customers in the system and the average number of customers found in the system by a newly arriving customer. These two quantities are the same only for Poisson, or random, arrival patterns. In an analogous way, for our limited sources model, we shall need to distinguish between

$$P_B = \text{Prob(all servers are busy)} \tag{29.2}$$

$$P_L = \text{Prob(attempted call is lost)} \tag{29.3}$$

This difference arises because the arrivals do not follow a Poisson pattern. If all call attempts were successful and call duration was zero then the arrival pattern would be Poisson, but since the Poisson or random arrival pattern is suspended for the duration of a successful call, the offered traffic has a non-Poisson pattern. Because the call durations are exponentially distributed, the inter-arrival times will have a distribution that is a 'mixed-Erlangian' form, and traffic generated from a limited number of sources is 'smoother' than traffic from an infinite number of sources. This means that the arrival pattern will be somewhat more regular than a Poisson pattern. In the telephony world, P_B is sometimes known as the 'time congestion' and P_L as the 'call congestion'. Often in practice the difference is small, but may be important. In any case we must take account of the difference in our mathematical model. Let us define

$$\lambda_O = \text{offered call rate from all sources} \tag{29.4}$$

$$\lambda_A = \text{actual call rate carried} \tag{29.5}$$

The following relationships hold.

$$\lambda_O = \frac{K}{(1 - P_L)T_S + T_A} \text{ calls per unit time} \tag{29.6}$$

$$\text{offered traffic} = \lambda_O T_S = \frac{K}{(1 - P_L) + z} \text{ erlangs} \tag{29.7}$$

$$\lambda_A = \text{actual call rate} = (1 - P_L)\lambda_O \text{ calls per unit time} \tag{29.8}$$

$$\text{actual traffic} = \lambda_A T_S = \frac{(1 - P_L)K}{(1 - P_L) + z} \text{ erlangs} \tag{29.9}$$

The key parameter here is obviously P_L, and we calculate this from the formula

$$P_L = \text{Prob(arriving call is lost)} = \frac{\binom{K-1}{m} z^{-m}}{\sum_{n=0}^{m} \binom{K-1}{n} z^{-n}} = \frac{q_m}{\sum_{n=0}^{m} q_n} \tag{29.10}$$

where

$$q_n = \binom{K-1}{n} \frac{1}{z^n} \tag{29.11}$$

The following relationships are useful when coding an algorithm to calculate P_L:

$$\frac{q_n}{q_{n-1}} = \frac{K-n}{nz} \text{ for } n = 1, 2, \ldots, m \text{ and } q_0 = 1 \tag{29.12}$$

The distribution of the number of customers being served is usually of secondary interest to the customer loss rate. But should we need it, it is defined by

$$p_n = \text{Prob}(n \text{ customers being served}) \tag{29.13}$$

$$p_n = \binom{K}{n} \frac{p_0}{z^n} \text{ for } n = 1, 2, \ldots, m \tag{29.14}$$

$$p_0 = \left[\sum_{n=0}^{m} \binom{K}{n} \frac{1}{z^n} \right]^{-1} \tag{29.15}$$

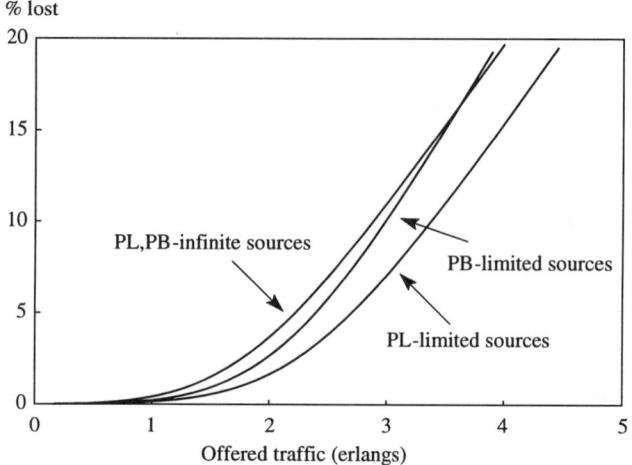

% lost

Figure 29.3. Comparing loss rates for limited and infinite sources.

The probability that all the servers are busy is therefore

$$P_B = p_m \tag{29.16}$$

and the interesting relationship between the probability of all servers being busy and the probability of a lost customer, i.e. between time congestion and call congestion, is

$$P_L \le P_B \tag{29.17}$$

This means that looking at the proportion of time all the servers are busy gives a pessimistic view of the service as seen by the customers or calls.

Erlang-B extended (some failed calls retried)

We return now to the Erlang-B model with an infinite set of sources. In that model we suppose a certain amount of traffic or rate of calls is offered to the available lines or servers, and a proportion of those calls is lost. The assumption is that those lost calls are never retried. In practice, of course, a proportion of calls would be retried. Let us assume that if a call attempt fails and is retried, the retry occurs a long time after the previous attempt. This means the retries just go to increase the level of offered traffic, and do not distort the random arrival of calls. If we define

$$\phi = \text{Prob(failed call attempt is retried)} \tag{29.18}$$

then we can write

$$\text{Prob(attempt fails, and is repeated)} = \phi P_L \tag{29.19}$$

$$\text{Prob(attempt succeeds, or fails and is not repeated)} = 1 - \phi P_L \tag{29.20}$$

We are back to the type of problem exemplified by throwing a die until you get a six (Chapter 2), so that

$$\text{Prob}(k \text{ attempts for a call}) = (\phi P_L)^{k-1}(1 - \phi P_L) \tag{29.21}$$

The average attempts made per call, that is a 'real' call from the original offered traffic, is given by

$$\text{average attempts per call} = \frac{1}{1 - \phi P_L} \tag{29.22}$$

Let us define

$$\lambda_R = \text{rate of call retries} \tag{29.23}$$

$$\lambda'_O = \text{composite offered traffic} = \lambda_O + \lambda_R = \frac{\lambda_O}{1 - \phi P_L} \tag{29.24}$$

We want to find the value of λ'_O, which is easy if we know P_L, but in order to calculate P_L we first have to know λ'_O. An iterative solution to the problem of finding P_L is straightforward. First we define

$$\lambda'_O(n) = n\text{th estimate of } \lambda'_O \tag{29.25}$$

$$P_L(n) = n\text{th estimate of } P_L \tag{29.26}$$

The algorithm is then

Step 1: Set $\lambda'_O(0) = \lambda_O$, and $n = 0$
Step 2: Use $\lambda'_O(n)$ in the Erlang-*B* formula to get $P_L(n)$
Step 3: Form next estimate $\lambda'_O(n+1) = \dfrac{\lambda_O}{1 - \phi P_L(n)}$
Step 4: If $\lambda'_O(n+1) \approx \lambda'_O(n)$ to sufficient accuracy, then the algorithm has finished. Otherwise set $n = n + 1$ and go back to step 2.

Once we have discovered what the composite offered traffic rate is, we simply use the Erlang-*B* formula to get any other queueing statistics that we are interested in. As an example we take 10 erlangs of 'real' traffic being offered to 12 lines or servers. Figure 29.4 shows the results for the range of possible values of ϕ.

The curve labelled 'carried traffic' shows how the composite traffic grows as call attempts are more and more likely to be retried. The other curve shows the probability that a particular call attempt will fail. This failure probability increases from just over 1 in 10 if calls are never retried to about 1 in 4 if nearly all calls are retried. If all failed call attempts are retried then in effect we have a multiserver queueing system rather than a loss system, although the service order is rather random.

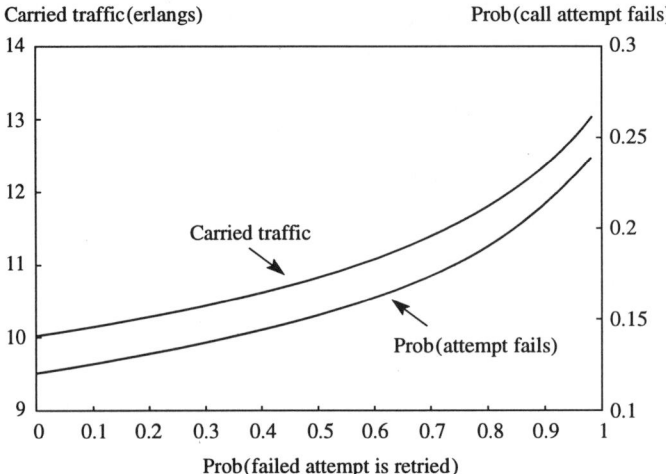

Figure 29.4. Carried traffic versus retry probability for 10 erlangs offered to 12 circuits.

Models for actual instead of offered traffic

In the Erlang-*B*, limited sources, and Erlang-*B* extended models we take the offered traffic and work out the proportion of calls that will be lost. The design question of how many lines should be provided is handled by increasing m until P_L is reduced to an acceptable figure. Target figures for P_L are to some extent arbitrary, but experience has established sensible values. Since P_L is generally of the order of 1 per cent, the difference between offered and actual traffic is small, and almost certainly less than the margin for error in the estimates of traffic to be dealt with. All the models mentioned can be modified to take actual rather than offered traffic as an input parameter. This involves an iterative approach, which of course is a nuisance when working with paper calculations. However, if a computer program is being used a few more lines of code to do the necessary iterations are of no consequence. The results in terms of answers to design questions may be trivial, but there will be instances when the difference is relevant (particularly in non-telephony applications), and it also removes a potential point of confusion when explaining queueing-theory-derived designs to non-technical people.

So how do we use an iterative approach to work with actual rather than offered traffic? We need to find the value of λ_O that gives rise to the value of λ_A that we want, taking account of the loss rate that will occur according to the particular queueing model we are using. As in many other cases in this book, a simple binary search procedure is effective. As some calls will be lost, a good lower estimate of λ_O to start with is obviously λ_A. An upper bound can be found empirically by trying multiples of λ_A. Programs that use this approach are given later in this chapter.

Links in series

Even in a modest-sized private voice network, a large proportion of calls will be routed over more than one link in series. A 'link' here means a group of circuits, with each circuit able to carry a single call at any one time. Suppose we have the situation depicted in Fig. 29.5, where calls from A to D traverse three links in

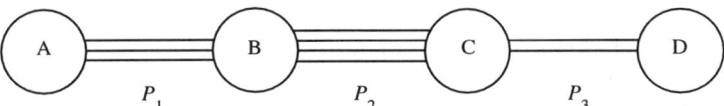

Figure 29.5. Voice links in series.

series. The probability of all the circuits in the link from A to B being busy is P_1, for the link B to C the probability is P_2, and for the link from C to D it is P_3. What is the probability that a call attempt from A to D will be blocked? Elementary probability and algebra give

$$\text{Prob(AD blocked)} = 1 - \text{Prob(AD not blocked)} \qquad (29.27)$$
$$= 1 - (1 - P_1)(1 - P_2)(1 - P_3) \qquad (29.28)$$
$$= (P_1 + P_2 + P_3) - (P_1P_2 + P_2P_3 + P_3P_1)$$
$$+ P_1P_2P_3 \qquad (29.29)$$

Now in a practical network, the probability of any particular link being blocked will be kept quite small, certainly under 0.1, say. So the terms in Eq. (29.29) that involve more than one blocking probability multiplied together, such as P_1P_2, are going to be very small and for practical purposes can be ignored. So we have

$$\text{Prob(AD blocked)} \approx P_1 + P_2 + P_3 \qquad (29.30)$$

In other words, the blocking probabilities of links in series can simply be added together. This is obviously a very convenient result.

Of course the above argument conveniently overlooks some complications. Presumably a number of calls will be made between A and D. This being so, the probabilities of each of the links being blocked are not independent. If the link BC is heavily loaded at a particular instant, this is likely to be because a number of calls are in progress between A and D, so links AB and CD are also likely to be heavily loaded. It is possible that link BC is blocked because of calls between B and C, with little traffic at that instant on link AB or CD. However, the blocking probabilities of the links are not independent. Nevertheless, in practical design work we do just add blocking probabilities together.

Capacity, routing and performance

A telephone network will usually provide more than one possible route between any two switches. Protection against failure of a link is one important factor in designing networks with alternative routes, although multiple routes often arise naturally from the need to carry traffic between any two switches in the network. Figure 29.6 shows a simple hypothetical network in which there are a number of possible routes between any particular pair of nodes, say A and G. If one particular route has limited capacity, for example link AF, then some of the traffic from A to G could be routed A–D–F–G or A–D–E–G or A–B–E–G, etc.

Real switches may not actually allow completely flexible routing, but assuming a reasonable degree of sophistication in the routing methods, how do we work out where the traffic will actually flow? This is not an easy problem to solve, since as traffic is diverted away from a congested link, other links will become congested, which in turn affects routing decisions. If we cannot be sure what traffic will flow over a link, how do we then decide the capacity of that link? In practice we might assume that all traffic goes via the physically shortest route, or perhaps the shortest route that uses only 'main' switching nodes. We can then work out the link capacities needed using the queueing theory discussed in this chapter. Such an approach means that the shortest route will generally be the least blocked route, so only a small proportion of traffic will overflow onto 'second-choice' routes. This assumes, of course, a traffic pattern that is known and is stable, i.e. does not vary dramatically with time of day. More complex situations may well require simulation to evaluate the performance of a network. The issues of routing and the topology of networks are outside the scope of this book, which focuses on queueing theory. Readers should refer to Schwartz (1987) and Kleinrock (1976) for discussion of routing in both voice and data networks.

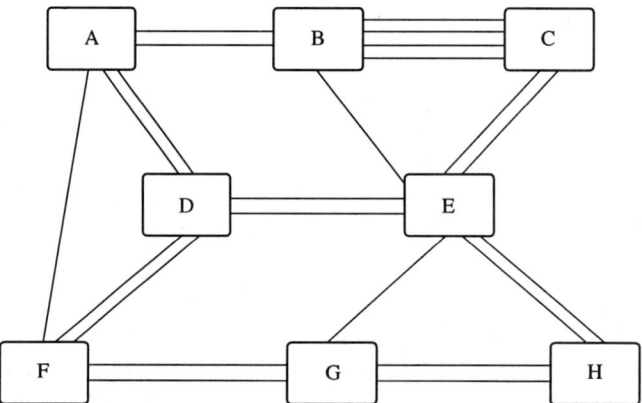

Figure 29.6. Multiple routes in a network.

Call-centres

An important aspect of voice networks is the 'call-centre'. When we make an enquiry about airline or train timetables, or buy something over the phone on a credit card, or book a theatre ticket, our call will be to a group of people answering calls usually in first-come-first-served order. If all of them are busy, our call is not normally lost, rather we are put in a waiting-line. We might simply have to listen to a ringing tone, or perhaps hear a message assuring us that our call will be dealt with shortly and then be subjected to some 'canned' music. The situation is multiple servers with a single queue, as described in Chapter 17. Managers of call-centres have to decide how many 'agents' or servers they will provide. If there are not enough servers, then waiting times will be long, and many customers will either give up or at least be irritated about the service they are getting even before they are answered. On the other hand, too many servers means unnecessary costs. Complications arise because of the difficulty in predicting the call-rate.

Target service levels for a call-centre are usually quoted as 'x per cent of calls should be answered within y seconds'. Equation (17.8) gives the distribution of waiting time, and Eq. (17.10) gives the percentiles of waiting time. These formulae provide a method of working out the number of servers needed to meet a service-level target of x per cent within y seconds. Note the use of the Erlang-C function $E_C(m, u)$ in Eq. (17.8). An alternative way of setting a target is to say that z per cent of calls should be answered immediately. The probability that a call is answered immediately is $1 - E_C(m,u)$

In practice, the modelling of a call-centre needs to take account of a number of factors. The call-rate will obviously vary during the day, and the stability of this pattern needs to be understood. The average call duration may not be a stable parameter, and could vary over the day. It is likely that there will be several categories of customers, or at least enquiries, each with its own call pattern and durations. It is not unusual to have more than one level of server, where a call is initially handled by a general pool of servers, but if it is lengthy or specialized or complex may be transferred to other groups of servers. Some of these factors take us away from random arrivals and exponential service, so the M/M/m model may give incorrect estimates of the service level or number of servers needed. In Chapter 21 we looked at G/G/m, and in particular the heavy-traffic approximation can be used. Some care is needed, since Fig. 21.2 shows that the heavy-traffic approximation can be quite pessimistic for high values of C^2_S and C^2_A. The Allen–Cunneen approximation in Eq. (21.3) is more accurate, but unfortunately gives only the average waiting time and not percentiles. The tables in Seelen *et al.* (1985) can be used, although this is less convenient than an analytic form that can be turned into a program.

Programs for voice networks

In order to use the subroutines from this chapter, the following statements should be included in a program.

```
{$I \qthprog\qtherror.pas  }
{$I \qthprog\qthtypes.pas  }
{$I \qthprog\engstclc.pas  }
{$I \qthprog\engstrte.pas  }
{$I \qthprog\pratio.pas    }
{$I \qthprog\mmmlsclc.pas  }
{$I \qthprog\mmmlsrte.pas  }
```

EngstClc—general calculations for limited sources model

This routine performs the calculations for the Engset, or limited sources, model.

```
{-----------------------------------------------------}
{> EngstClc -- Calculations for multiple servers     }
{               with finite number of sources.        }
{ Inputs:                                             }
{      TA        mean operating time                  }
{      TS        mean service time                    }
{      K         number of sources                    }
{      M         number of servers                    }
{      SHORTSW   T=do only calc needed by EngstRte    }
{ Outputs:                                            }
{      AROFF     offered arrival rate                 }
{      ARACT     actual arrival rate                  }
{      RHO       server utilisation                   }
{      PB        prob. all servers are busy           }
{      PL        prob. arriving customer is lost      }
{ Copyright Mike Tanner 1993                          }
{-----------------------------------------------------}
Procedure EngstClc(TA,TS:QTHreal;K,M:integer;
                SHORTSW:boolean;
                Var AROFF,ARACT,RHO,PB,PL:QTHreal);
Var RK,RN,RM,P,PO,Q,SQ,SVRAT:QTHreal;N:integer;
begin
    RK:=K; RM:=M;                 { type conversion  }
    SVRAT:=TA/TS;                 { service ratio     }
    {-Calculate prob an arriving customer is lost---}
    Q:=1; SQ:=1;
    For N:=1 to M do
    begin
        RN:=N;
        Q:=(Q/SVRAT)*(RK/RN-1);
        SQ:=SQ+Q;
    end;
    PL:=Q/SQ;
    {-Arrival rates and server utilisation---------}
    AROFF:=RK/(TA+TS*(1-PL));
    ARACT:=(1-PL)*AROFF;
```

```
RHO:=ARACT*TS/RK;
If SHORTSW then Exit;
{-Calculate P0 and prob all servers are busy----}
P:=1; P0:=1;
For N:=1 to M do
begin
    RN:=N;
    P:=P*(RK-RN+1)/(RN*SVRAT);
    P0:=P0+P;
end;
P0:=1/P0;
PB:=P*P0;
end;
```

EngstRte—calculate T_A for given traffic rate

The EngstRte routine calculates the operating time T_A for the sources in a finite source model such that a specified offered or actual traffic rate is achieved.

```
{-----------------------------------------------------}
{> EngsetRte -- Find TA for Engset to give a          }
{                specified offered or actual          }
{                traffic rate.                         }
{ Inputs:                                              }
{      SAR        1=offered, 2=actual                 }
{      LAMBDA     required arrival rate               }
{      TS         mean service time                   }
{      K          number of sources                   }
{      M          number of servers                   }
{ Outputs:                                             }
{      TA         "operating time" to give LAMBDA     }
{   Copyright Mike Tanner 1993                        }
{-----------------------------------------------------}
Procedure EngstRte(SAR:integer;
                   LAMBDA,TS:QTHreal;K,M:integer;
                   Var TA:QTHreal);
Const ACC=0.0001;              { accuracy required  }
Var RK,RHO,PB,PL:QTHreal;
Var AR,ARLOW,ARHGH,ARO,ARA,TALOW,TAHGH:QTHreal;
Label L1;
begin
    RK:=K;
    {-Check for sensible parameters----------------}
    If LAMBDA>(RK/TS)
    then QTHError('EngstRte - LAMBDA invalid');
    {-Form initial estimates. TALOW must be >0------}
    TAHGH:=TS;
    Repeat
        TAHGH:=TAHGH*2;
        EngstClc(TAHGH,TS,K,M,true,ARO,ARA,RHO,PB,PL);
        Case SAR of
           1:ARHGH:=ARO;
           2:ARHGH:=ARA;
        end;
```

```
    until ARHGH<LAMBDA;
    TALOW:=TS;
    Repeat
        TALOW:=TALOW/2;
        EngstClc(TALOW,TS,K,M,true,ARO,ARA,RHO,PB,PL);
        Case SAR of 1:ARLOW:=ARO; 2:ARLOW:=ARA; end;
    until ARLOW>LAMBDA;
    {-Binary search for value of TA to give LAMBDA--}
L1:TA:=(TALOW+TAHGH)/2;
    EngstClc(TA,TS,K,M,true,ARO,ARA,RHO,PB,PL);
    Case SAR of
        1:AR:=ARO;
        2:AR:=ARA;
    end;
    If Abs(AR-LAMBDA)<ACC then Exit;
    If AR>LAMBDA
    then begin TALOW:=TA; ARLOW:=AR; end
    else begin TAHGH:=TA; ARHGH:=AR; end;
    Goto L1;
end;
```

MMmLsRte—find offered rate to give specified actual rate for Erlang-B

The MMmLsRte routine calculates what the offered traffic rate must be in order
to obtain a specified actual traffic rate for the Erlang-*B* (multiserver loss) model.

```
{------------------------------------------------------}
{> MMmLsRte -- Calculate offered rate to achieve     }
{             a given actual rate for multiserver }
{             loss system.                          }
{ Inputs:                                             }
{     LAMBDA      required actual rate                }
{     TS          mean service time                   }
{     M           number of servers                   }
{ Outputs:                                            }
{     AROFF       offered arrival rate needed         }
{ Copyright Mike Tanner 1993                          }
{------------------------------------------------------}
Procedure MMmLsRte(LAMBDA,TS:QTHreal;M:integer;
                   Var AROFF:QTHreal);
Const ACC=0.0001;                { accuracy required  }
Var AR,ARLOW,ARHGH,FR,FRLOW,FRHGH:QTHreal;
Var U,RHO,PLOSS,ARVRTE,P0,LQ,SDVLQ:QTHreal;
Label L1;
begin
    {-Form initial low and high estimates----------}
    FRLOW:=LAMBDA;
    MMmLsClc(FRLOW,TS,M,
             U,RHO,PLOSS,ARLOW,P0,LQ,SDVLQ);
    FRHGH:=LAMBDA;
    Repeat
        FRHGH:=FRHGH*2;
        MMmLsClc(FRHGH,TS,M,
                 U,RHO,PLOSS,ARHGH,P0,LQ,SDVLQ);
```

```
      until ARHGH>LAMBDA;
      {-Binary search for offered rate required------}
L1:FR:=(FRLOW+FRHGH)/2;
      MMmLsClc(FR,TS,M,
              U,RHO,PLOSS,AR,P0,LQ,SDVLQ);
      If Abs(AR-LAMBDA)<ACC then Exit;
      If AR>LAMBDA
      then begin ARHGH:=AR; FRHGH:=FR; end
      else begin ARLOW:=AR; FRLOW:=FR; end;
      Goto L1;
end;
```

ErlgBEx—calculate offered call rate for Erlang-B including retries

The ErlgBEx routine calculates the offered call rate that results from some calls being retried with an Erlang-*B* (multiserver loss) model. The parameter LAMBDA is the fundamental offered call rate, excluding retries of failed call attempts.

```
{-----------------------------------------------------}
{> ErlgBEx -- Erlang B extended. Calculate the        }
{>            offered call rate including             }
{>            retries of lost calls.                  }
{ Inputs:                                             }
{     LAMBDA     underlying offered call rate         }
{     TS         mean service time                    }
{     M          number of servers                    }
{     PR         prob. failed call retried            }
{ Outputs:                                            }
{     AROFF      total offered call rate              }
{     PLOSS      prob. a call attempt fails           }
{     ACA        average attempts per call            }
{  Copyright Mike Tanner 1993                         }
{-----------------------------------------------------}
Procedure ErlgBEx(LAMBDA,TS:QTHreal;M:integer;
                  PR:QTHreal;
                  Var AROFF,PLOSS,ACA:QTHreal);
Const ACC=0.001;                   { accuracy required }
Var OLDAR,U,RHO,ARVRTE,P0,LQ,SDVLQ:QTHreal;
Label L1;
begin
      AROFF:=LAMBDA; OLDAR:=AROFF;
L1:MMmLsClc(AROFF,TS,M,
           U,RHO,PLOSS,ARVRTE,P0,LQ,SDVLQ);
      ACA:=1/(1-PR*PLOSS);
      AROFF:=LAMBDA*ACA;
      If Abs(AROFF-OLDAR)<ACC then Exit;
      OLDAR:=AROFF;
      Goto L1;
end;
```

30
Applications to data networks

Introduction

In this chapter we look at the components of data networks such as links, data switches, LANs and bridges. The essential performance characteristics are described and simple queueing-theory models are suggested. The objective is not to build performance models suitable for debating the internal design or configuration of network components. Instead we are seeking models that will give us a reasonable estimate of the contribution of each component to the performance, i.e. delay, across a complete network.

We start by looking at a single line supporting either an 'infinite' number of terminals or a small, finite, number. Then we look at networks made up from multiple switches and links, and examine how to convert user or application traffic into a flow of frames across links and through switches. This is the difficult part of network performance analysis, and the part that requires a good knowledge of network protocols. Once the frame flow has been established, queueing-theory models of switches and links can be used. Ethernet and token-ring LANs are described, and formulae for calculating LAN delay are presented. Then we look at LAN bridges, and suggest a slightly different model from that of a general data switch. Lastly we look at the throughput achievable between two terminals on a network, taking account of the 'windowing' flow-control method commonly used.

Several references are given to detailed books specifically covering network performance rather than, as here, network performance as an example of the use of queueing theory. Readers are encouraged to refer to Schwartz (1977, 1987), Keiser (1989) and Kleinrock (1976).

Polled line with interactive traffic

A common situation is a number of terminals sharing a communications link to a data centre, as illustrated in Fig. 30.1. The terminals may be attached to a terminal controller directly, or may be on a LAN to which the terminal controller, or gateway, is also attached. Another possibility is for the terminals to be spread over

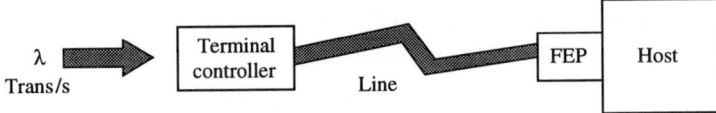

Figure 30.1. Single line supporting a number of terminals.

several locations or offices and attached in a 'multipoint' or 'multidrop' arrangement. When terminal controllers share a line, a 'polling' scheme is used. A central control point, the FEP in Fig. 30.1, sends a poll to each controller in turn inviting that controller to transmit pending messages. Even with a single controller on a line, polling may still be used. A detailed analysis of a polled line is a challenging task for queueing theory, and in practice simulation is widely used. In this section we shall see what can be achieved with a very simple queueing-theory approach. A more detailed approach to polling will be found in Schwartz (1987).

We shall assume that each transaction is made up of a single input message and a single output message. The detailed operation of a polled line, such as SDLC, is complicated and depends on many factors. In order to keep our model simple we shall ignore many of these complications and treat the line as if it is a single server where the 'customers' are the combined flow of input and output messages, as shown in Fig. 30.2. Note that we are assuming that data can flow in only one direction at a time, not both ways simultaneously. This is known as 'data half duplex' or 'two-way alternate' mode.

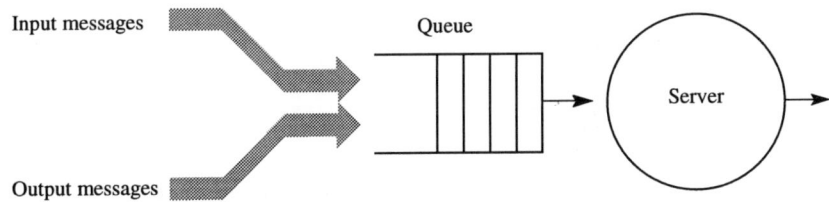

Figure 30.2. Simple model of a polled line.

This is how the line is modelled. The host processing will be represented as a simple random delay, not as another server with a queue. In general a mainframe or large data server can process many transactions in parallel, and we are more interested in the line than the host. Once we have constructed a queueing-theory model, we shall compare the results with simulation results to see how useful a simple model can be. We need to specify the following values.

$$T_{\text{IN}} = \text{average transmission time for an input message} \qquad (30.1)$$

$$T_{\text{OUT}} = \text{average transmission time for an output message} \qquad (30.2)$$

$$T_{\text{H}} = \text{average host processing time} \qquad (30.3)$$

$$\lambda = \text{transaction arrival rate} \qquad (30.4)$$

From these basic parameters we can calculate the following values, remembering
that λ is the transaction arrival rate and the message arrival rate is double this.

$$T_S = \text{average service-time per message} = \frac{T_{IN} + T_{OUT}}{2} \tag{30.5}$$

$$\rho = \text{line utilization} = \lambda(T_{IN} + T_{OUT}) \tag{30.6}$$

Since we are deliberately building a simple model, we shall use the M/M/1 formula
for the average wait per message.

$$T_W = \text{average wait per message} = \frac{\rho T_S}{1 - \rho} \tag{30.7}$$

Now we can construct the formula for response time. For each input message
there will be a wait-time and then a transmission time. When an input message
arrives at the host, host processing time will occur, and then for the output
message there will be a wait-time and a transmission time. Putting all this together,
we get

$$T_R = \text{average response time} \tag{30.8}$$

$$T_R = \text{input wait} + \text{input transmission} + \text{host processing}$$
$$+ \text{output wait} + \text{output transmission} \tag{30.9}$$

We can now write

$$T_R = T_W + T_{IN} + T_H + T_W + T_{OUT} \tag{30.10}$$

$$= \frac{2\rho T_S}{1 - \rho} + 2T_S + T_H \tag{30.11}$$

$$= \frac{T_{IN} + T_{OUT}}{1 - \rho} + T_H \tag{30.12}$$

Let us now put Eq. (30.12) to use by considering the following specific example.

    input message: constant 100 bytes
    output message: 400 to 1200 bytes uniformly distributed (average 800)
    host processing: 0.25 to 0.75 s uniformly distributed (average 0.5)
    line speed: 9600 bit/s

Applying Eq. (30.12) over a range of transaction rates, we get the curve labelled
'q-theory' in Fig. 30.3. Our simple model ignores many potentially important
factors in an actual system, such as an SDLC line attached to a communications
controller or FEP running a particular control program. These factors include
propagation delays in the line and modems. SDLC framing characters, NCP
scheduling parameters, and frame and message size constraints. A simulation
package developed by the author was used to analyse the example given here.
The package contains a detailed representation of SDLC operation, line and
modem propagation times, relevant SNA message protocols, and IBM's NCP line

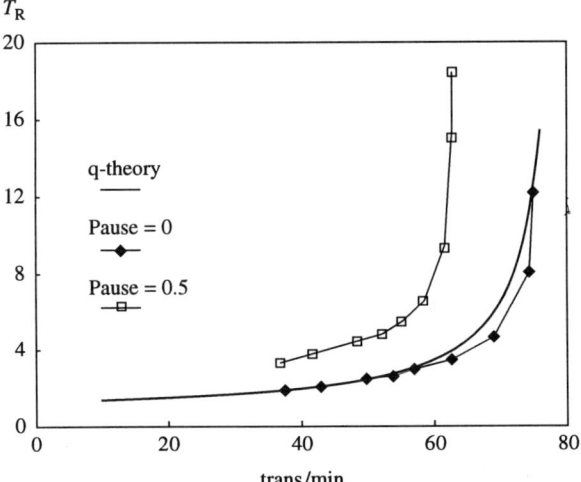

Figure 30.3. Simulation and theory at 9600 bit/s.

scheduling parameters. In particular, simulations were done using different values of the PAUSE parameter. PAUSE sets a minimum poll cycle for a line. NCP will not issue another poll to a terminal controller until at least the PAUSE time has elapsed since the previous poll was issued to that same terminal controller. PAUSE is intended to prevent excessively high polling rates consuming FEP processing power. As long as PAUSE is set to a reasonably low value, it will have no effect when the line load is moderate to high, since the 'natural' poll cycle will exceed the PAUSE value. When the line is idle, PAUSE will prevent continuous polling. It is apparent from Fig. 30.3 that for a PAUSE value of zero, our simple queueing theory model is perfectly adequate. However, if PAUSE is increased to 0.5, a somewhat high value for a 9600 bit/s line, our model differs significantly from the simulation results. It is the simulation results we should believe, since we know our queueing model just ignores the effect of PAUSE. If we repeat the exercise for a line speed of 64 000 bit/s, we get the results shown in Fig. 30.4. Again we see that with PAUSE = 0 the queueing model is quite reasonable. This is actually a bit surprising, since at 64 000 bit/s the message transmission times are small, and some of the things we have ignored such as propagation times are much more significant than at 9600 bit/s. We have looked at the effect of PAUSE as just one example of the complexities that exist in real systems. There are of course many such complicating factors in any system. However, the conclusion is that even elementary queueing theory models have something to offer the network performance analyst. It would be possible to construct a more sophisticated queueing-theory model of a polled line, taking account of more factors. Whether it is worth doing so depends on the availability of simulation tools.

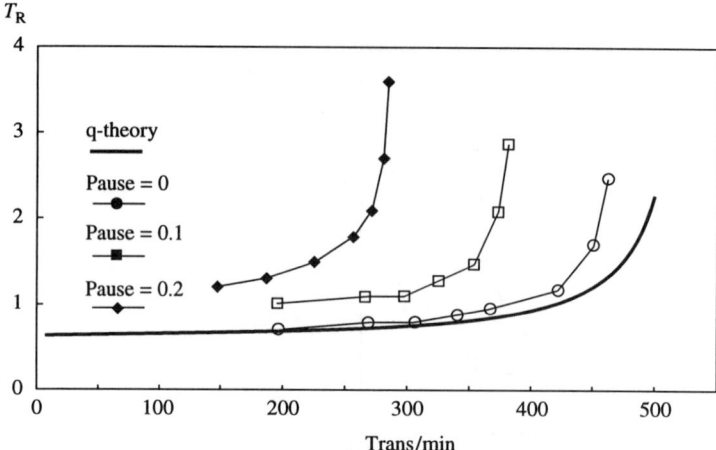

Figure 30.4. Simulation and theory at 64 000 bit/s.

Repairman model of a line

So far we have been comparing our simple model to simulation results, and concluding that simulation will probably be necessary, at least above 9600 bit/s, to get accurate performance estimates. However, the queueing model serves another valuable purpose. In Figs 30.3 and 30.4, the shape of the queueing theory curve is similar to the simulation curves even though the actual values are different. The queueing model is telling us something useful about the behaviour of the system, despite some inaccuracy. The non-linear increase in response time as the line saturates is clearly demonstrated by the queueing model.

The simple model of the previous section is based on the M/M/1 model. We might have gone to a little extra trouble and used M/G/1, but in either case we would be assuming an 'infinite' number of potential customers. This assumption would be reasonable if the real system consisted of a large number of terminals, say on a LAN complex, each making occasional enquiries of the mainframe system. On the other hand, we know that if the number of terminals is small the behaviour of the system will be different. The machine repairman model of Chapter 23 then becomes applicable, and we could represent the system as in Fig. 30.5.

We shall use the repairman or M/M/1/K/K model described in Chapter 23. In order to do this we have to treat the input and output messages of a transaction as a single composite service. Similarly we have to add the 'think' time and the host processing time to get an average 'operating' time. The 'think time' is the average time a terminal user waits, or thinks, after receiving an output message before initiating the next transaction by sending an input message. For our example we shall assume that the average think time is 10 s, with other parameters as in the

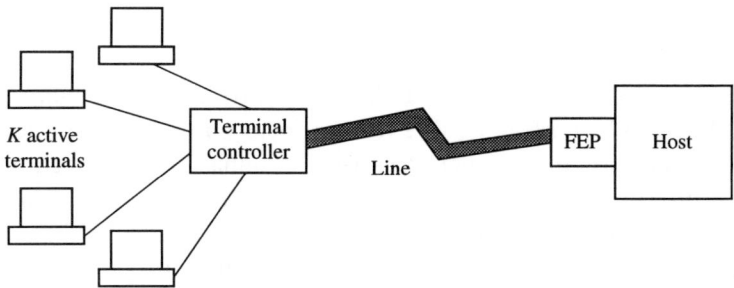

Figure 30.5. Line supporting limited number of terminals.

previous example. So we have

$$T_K = \text{average think time} = 10 \text{ s} \tag{30.13}$$

$$T_S = \text{average service-time} = T_{IN} + T_{OUT} \tag{30.14}$$

$$T_A = \text{average operating time} = T_K + T_H \tag{30.15}$$

$$K = \text{number of active terminals} \tag{30.16}$$

Now we can apply the formulae from Chapter 23 to calculate T_Q, the average time-in-system. Note that T_Q excludes the host processing time, which we have so far treated as part of the think time. So we have to add the host time to get response time. Using the MM1FcCLc routine we can easily perform the calculations for a range of numbers of active terminals. Figure 30.6 shows the results, together with simulation results. The simulations were done using PAUSE = 0.2, a typical value for 9600 bit/s, which could account for much of the difference between the simulation results and the repairman model. Even so, the two

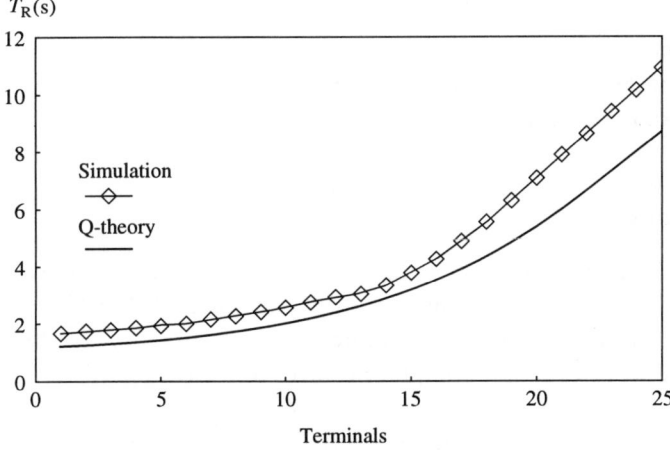

Figure 30.6. Results for repairman model of a line.

methods give similar results, and both indicate saturation at about 15 active terminals.

The intuitive ideas about system behaviour that we get from 'infinite source' models such as M/M/1 can be misleading. We tend to focus on transaction rates and assume that as we increase the number of users or terminals, the transaction rate will go up in proportion. As the transaction rate approaches the maximum capacity of the line we expect response times to get markedly bad. In theory adding one too many users will result in infinite response time. In practice we know that real systems do not behave like this, and so the naïve use of inifinite source models can damage the credibility of performance analysis based on queueing theory. Even when using simulation, it is not uncommon for an 'infinite' number of terminals to be assumed as the source of transactions. One reason for this over-reliance on infinite-source models is the difficulty of knowing what the think time is.

How can we use measurements of a real system to estimate think time? The hardest thing to measure may be the number of active terminals. A user may well sign on to a host application at the start of the day, remain signed on most of the day, but be active only for short periods. We could count the transactions for each terminal in successive 5 or 10 minute intervals, and consider the terminal as active in an interval only if the number of transactions exceeds some threshold. If we also have some way of measuring the average response time, then we can calculate the average think time as follows. We know that each user will generate a transaction on average every $(T_K + T_R)$ seconds, and there are K active users. So we have

$$\lambda = \text{transactions per second} = \frac{K}{T_K + T_R} \qquad (30.17)$$

Simple rearrangement of Eq. (30.17) gives

$$T_K = \text{average think time} = \frac{K}{\lambda} - T_R \qquad (30.18)$$

With a finite-source approach, we can use the notion of efficiency explained in Chapter 23 as a basis for describing performance versus, say, different line speeds. Calculating latent demand also becomes possible.

Wide-area data networks

So far we have looked at a single line supporting a number of terminals. There are many 'networks' that do consist of terminal controllers connected directly to a single computer centre, but a data network is more often like Fig. 30.7 or 30.8. Figure 30.7 depicts a network of major computer centres (CC) to which terminal sites (T) are connected via links and data switches (S). A terminal site might be a single terminal, several terminals attached to a terminal controller, or a LAN complex with gateways to the 'wide-area network' or WAN. The links between

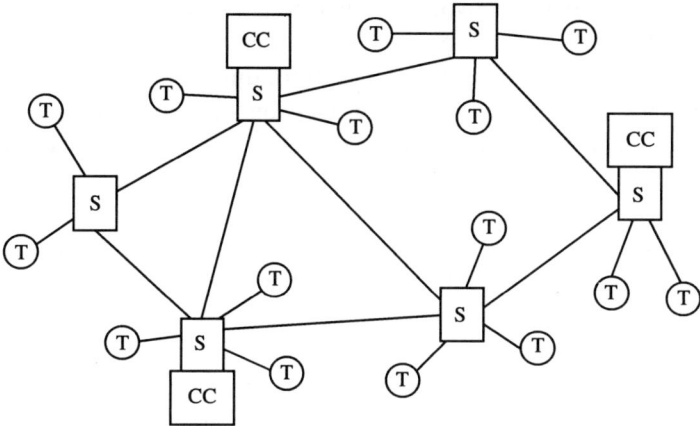

Figure 30.7. Data network supporting computer centres.

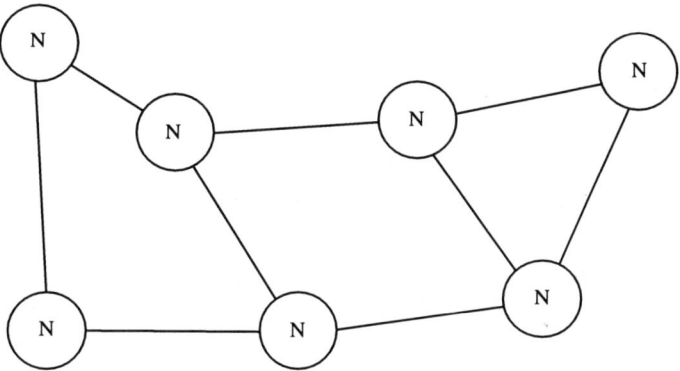

Figure 30.8. 'Distributed' data network.

switches may have various speeds, and a link may contain multiple circuits. Figure 30.8 shows a network where the distinction between a terminal site and a computer centre is not meaningful, and each node (N) has terminals attached either directly or via a LAN to a computer system that also acts as a switch. Different degrees of distribution may be found, and it is not the purpose of this book to explore the issues of network architecture and distributed and client/ server processing. We shall look at the contribution that queueing theory can make to understanding the performance of data networks.

Routing

As we saw in Chapter 29 on voice networks, we need to know the route that will be taken by traffic between any two end-points before we can work out the traffic carried by each link and switch in the network. If the routes are predefined then we can do this. However, a network with completely fixed routes has severe disadvantages. A network may use one of several approaches to routing. A typical method is for the route to be selected when a connection (session or virtual circuit) is created between two terminals. The route chosen is the 'best' at that time based on network transit time, or in some cases security or throughout criteria. This method is used with SNA and by many X25 implementations. There are also a number of X25 implementations that use completely dynamic routing, where each packet is independently routed by the perceived fastest available route. This is called dynamic adaptive routing, and the routing algorithms used to implement it have been refined and developed over many years, starting with the original ARPA network. Routing will not be discussed further in this book. Interested readers will find an extensive literature, and could start by referring to Schwartz (1987) and Kleinrock (1976). For our purposes we assume that the route between any two end-points in the network has been determined.

Data flow in a network

We need to look at the fundamental ideas used to transmit messages over a data network. The most important of these is packet-switching, as illustrated in Fig. 30.9. A limit is set on the length of a data unit that can be sent over the network. A data unit is called a 'packet' and a message that is longer than a maximum-size data unit is broken up into a sequence of packets. When the packets arrive at the destination terminal or computer system the original message is reconstructed. There are advantages to this approach for the design of terminal and switching equipment, since the maximum-size packet they have to deal with is defined.

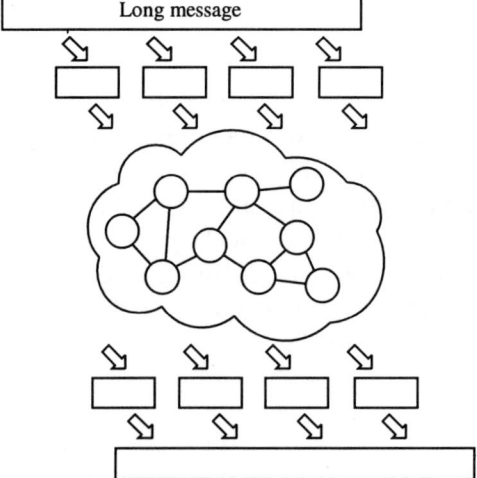

Figure 30.9. The packet-switching principle.

Also, the approach is analogous to the scheduling of jobs or tasks in a computer system, where each job is given a time-slice and then must wait for further time. Chapter 28 showed how this results in wait-time being proportional to service, an 'equitable' arrangement.

Now let us consider data flow between two 'terminals'. A terminal might be a user terminal or a program running in a mainframe or workstation, or even a network management or monitoring process in a switch. The explanation is easiest to understand if we think of a continuous stream of large units of data being sent, such as might occur when transferring a file. The data flow has to be considered at three levels. (Network protocols are usually discussed with reference to the OSI 7-layer model. However, it is simpler for performance purposes to use the less architecturally rigorous approach suggested here.) At the top level, which we could call the application level, the data flow will be perhaps very simple, as shown in Fig. 30.10. Each data unit M is a 'message' whose size will depend on the appli-

Figure 30.10. Application-level data flow.

cation. The size might be related to the record size in a file, or another criterion unrelated to the network. Note that we show only one direction of flow. There will undoubtedly be messages flowing in the reverse direction, such as confirmation that data has been received or application-level flow control messages.

The messages M will be split into packets for transmission over the network. We shall call this the 'network' level, and Fig. 30.11 shows the data units involved. The

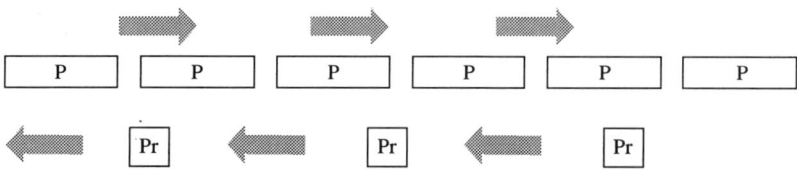

Figure 30.11. Network-level data flow.

data units P are the packets that make up the messages M. At the network level it is necessary to have some form of confirmation that packets have been received. Later we shall see the role that 'packet responses', labelled Pr, play in regulating the flow of data across a network. The frequency of packet responses varies. With X25 networks there is generally a packet response for every packet. The definition of X25 does not require this to be so, but most X25 implementations take this straightforward approach to packet responses. With SNA, the packet responses may come in different forms. With file transfers there will usually be 'pacing responses', or for interactive traffic there may be application-level responses. Either way there will typically be some response every 1 to 10 packets (or RUs in SNA terminology). Readers are expected to have a fairly good understanding of the network technology they are dealing with in order to apply queueing theory to a network, so a detailed exposition of network technologies would be out of place in this book (and too lengthy).

The network level of data flow takes place between one terminal and another across, typically, several links and switches. Now we shall look at what takes place on a particular circuit between two switches, or between a terminal and a switch. We have in mind here a wide-area link. The same principles apply to a local-area network, but there the flows may be simpler. The 'link-level' flow is shown in Fig. 30.12. First of all, the packets P may be split into the frames F, with more

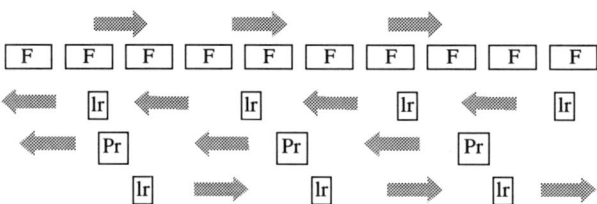

Figure 30.12. Link-level flow.

than one frame per packet. In fact this is quite likely with SNA, where large packet (RU) sizes may be used, but with X25 the smaller packet sizes almost always used will fit in a single link-level (HDLC) frame. The packet reponses, Pr, are small so they will need only a single frame. In both directions we now have link-level responses, labelled lr in Fig. 30.12. These are required for error control. The definition of HDLC/SDLC allows these link-level responses to be combined with data frames, but in practice separate response frames are used. Link-level responses occur typically every 7 data frames (or 128 data frames when satellite links are in use). They will occur more often if the flow of data is intermittent, since the sender will mark a data frame as 'final' if it has no more frames ready to send at that time, and this 'final' flag requests the receiver to acknowledge all the frames sent so far.

The explanation given so far demonstrates that performance analysis of a network requires a good knowledge of the protocols used in a network. Considerable work is needed to establish what is flowing over a link or through a switch before we start to build queueing theory models. This is an important motivation for using simulation to analyse network performance. It is not so much that queueing-theory models of the components of a network are particularly difficult to build, but a simulation package that contains sub-models of the various protocols does the hard work of breaking down application-level data flow into the lower level flows.

Model of a link

At the start of this chapter we saw how quite simple models of communications links can give useful accuracy. Figure 30.2 shows a model of a link that transmits data in only one direction at a time. Such links are generally confined to connecting individual terminals or simple terminal controllers to a network switch. Most links in a network are capable of simultaneous transmission of data in both directions. (Whether data can be transmitted simultaneously in both directions depends on both the circuit characteristics and the attached terminals/controller/switches. A common situation is where the circuit can handle data in both directions at the same time, but one of the attached pieces of equipment cannot. Note that the terms 'full-duplex' and 'half-duplex' are so ambiguous as to be almost useless.) For modelling purposes we treat each direction as a separate server, as shown in Fig. 30.13. Once we have worked out the data flow for the link, we then apply M/G/1 to each direction separately. If there are multiple circuits in a link, then we can either apply M/M/m to each direction or use the Allen–Cunneen approximation (Eq. (21.3)).

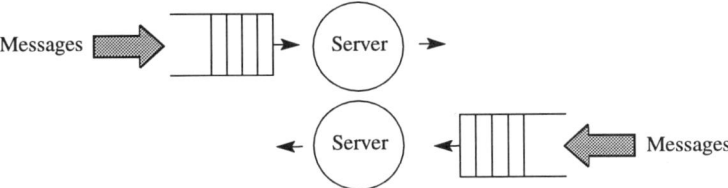

Figure 30.13. Model of link transmitting data in both directions simultaneously.

We are relying on the fact that transmission times for data packets are the dominant factor in performance. If the data packets are small or, more likely, circuits speeds are very fast, then propagation delays and detailed link protocol flows may become significant. Simulation is then indicated.

Model of a data switch or router

A simple representation of a data switch is shown in Fig. 30.14. Incoming packets arrive from attached circuits or from LAN connections or mainframe channel connections. We assume that link-level responses are handled by the line adapters, and do not impact the switch processor, which is primarily concerned with routing. Each packet is routed by the switch processor, after which the packet is placed on the queue for the appropriate link. The two questions of concern are the delay to each packet caused by the switch processor, and the number of packets present in the switch and requiring storage to accommodate them.

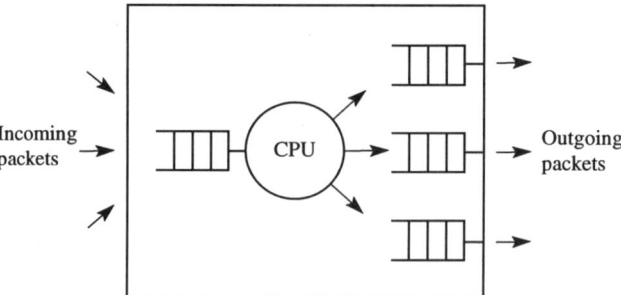

Figure 30.14. Model of a data switch.

The switch processor can be modelled just as an M/G/1 system (probably an M/D/1). This will also provide an estimate of the number of packets waiting for the processor. The queues for outgoing links are part of the link models, described in the previous section. The link models will provide estimates of the number of packets waiting for each line. We can therefore arrive at an estimate of the average number of packets present in the switch. The average is not what we require; we need to estimate some high percentile of the number of packets in the system, since if buffer storage for packets becomes full various emergency mechanisms will have to be invoked to prevent further traffic arriving until the backlog of packets has been reduced. A simple approach to this is to calculate the variance of the number of packets in each queue as well as the average. This can be done using the M/GX/1 formulae (Eqs (10.13) and (10.14)) via the routine MG1XCalc. This requires knowledge of the third moment of packet transmission time. Packets will not be of constant length, since some data packets will be smaller than the maximum, and there will be many non-data packets such as acknowledgement packets that will be very small. Another approach would be just to calculate the mean and variance of packet transmission time, and use the M/Ga/1 formulae (Eqs (12.8) and (12.9)) via the routine MGa1Calc. We can then calculate the mean and variance of the total packets in the switch by simply adding the means and using Eq. (2.26), which tells us that we can just add the variances also. A first approximation to the probability of a particular number of packets being exceeded can then be obtained using the Normal distribution. (Readers should consult any introductory statistics book for an explanation and tables of the Normal distribution.) A much more detailed approach to buffering in data switches will be found in Schwartz (1977). The question of buffer storage in switches is less of a problem than it used to be owing to the lower cost of memory (so that switches can be configured with generous amounts of buffer storage), and the sophistication of overall network flow-control mechanisms than can prevent congestion occurring.

The model shown in Fig. 30.14 is, of course, a great simplification of real switches. Many switches have multiple processors. This might mean the switch is really several switches, with each processor handling a subset of the links with an internal bus to route packets to another processor. Alternatively, the switch may act as a genuine multiprocessor system, with any processor able to service any link. In this case the M/M/m or G/G/m (Allen–Cunneen) formulae can be used. Our objective with these simple models is not to understand the fine details of switch design, but to represent the contribution that a switch makes to the delays imposed on messages or packets.

A data network as a network of queues

A data network is an obvious case of a network of queues. In Chapter 26 we looked briefly at networks of queues. In particular we looked at Jackson networks, in which each server in the network is an M/M/m system. If we depart from random arrivals and exponential service times we violate the assumptions of a Jackson network. On the other hand, if we just use M/M/m for each switch and link instead of G/G/m, then we are anyway making unnecessary approximation for the individual components. We may as well use M/G/m (in the form of the Allen–Cunneen G/G/m approximation) for each node and still treat the network as a Jackson network of queues. One alternative is to use simulation, and a number of simulation packages are available for networks of various kinds. (Within IBM the SNAPSHOT simulation package has been widely used for many years to analyse the performance of complex networks.) Another alternative is advanced queueing network theory. Simulation has generally been preferred by network performance analysts.

Local area networks

Local area networks, or LANs, typically span a building, complex of buildings, or campus. LANs operate at higher speeds (10 or 16 Mbit/s) than wide-area links (the accuracy of this statement continually changes as both wide- and local-area network technologies develop). LANs are relatively free of transmission errors and, together with the higher speeds, this makes different access protocols appropriate. We shall look at the performance of Ethernet and token-ring networks. (In this book the terms 'Ethernet', 'CSMA/CD' and 'IEEE802.3' may be considered equivalent, although in other contexts there are important technical and legal differences. Similarly, this book makes no distinction between 'token-ring' and 'IEEE802.5'.) For a detailed explanation of LAN technology readers are referred to Keiser (1989). Performance analysis for these and other LANs can be found in Keiser (1989), Schwartz (1987) and Kleinrock (1976).

Token-ring

In a token-ring the stations are connected in a unidirectional ring, with each station transmitting to the next station around the ring. Before a station can send

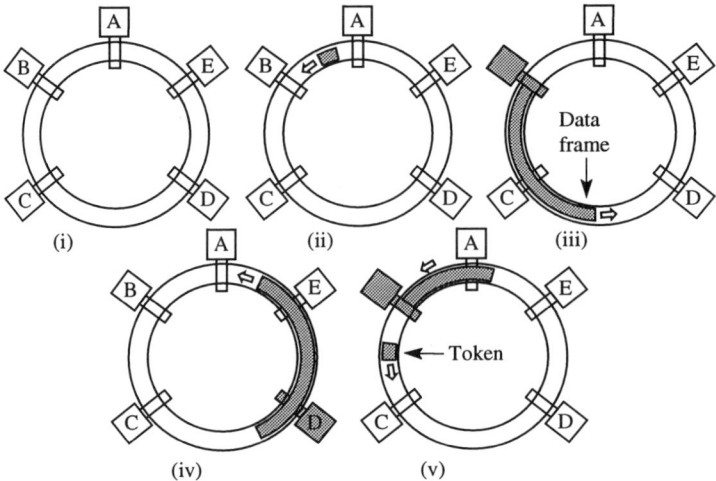

Figure 30.15. Token-ring operation.

data on the ring, it must wait for a token. Figure 30.15 illustrates the basic operation of a token ring. Diagram (i) shows the stations A to D connected to the ring.
Diagram (ii) shows the token, which is a special short frame, circulating round the
ring. If a station receives the token and has nothing to transmit, then the station
simply relays the bit-stream onwards to the next station. Suppose, however, that
station B wants to send a frame of data to station D. When the token reaches B the
token will be removed from the ring by B and replaced with the data frame, as
shown in diagram (iii). In fact the token is converted to a frame header by invert-
ing a particular bit in the token. The frame proceeds around the ring, and is
inspected by each station. If the destination address in the frame header is not that
of a particular station, e.g. C, then the station simply relays the bit-stream on to
the next station. When D examines the frame, as in diagram (iv), the destination
address will match D and D will copy the frame from the ring. D will also relay the
frame onwards, and will invert a bit in the frame trailer to indicate that the frame
has been received. When the frame again reaches B, as in (v), B will recognize the
source address in the frame, convert the frame header back into a token which is
relayed on, and remove the rest of the frame from the ring. The propagation delay
around the ring is made up of cable propagation time and 1 bit time per station.
The transmission speed of the ring may be 4Mbit/s or 16Mbit/s.

    How do we analyse the performance of a token-ring? There are two approaches.
We can simply regard the ring as a single server, convert the frame-length distribution into a distribution of service-time, and apply the M/G/1 formula (Eq. (9.1)).
This is the approach recommended when we are interested in the token-ring as
part of an overall system. The token-ring, or other LAN, is likely to be a minor
contribution to overall response-time, so a detailed approach would be unwar-
ranted. The second, detailed, approach is relevant when we want to understand

the effect of different ring configurations e.g. cable length, or when the token-ring is a significant part of the overall response. The above description of how a token-ring operates is the one we shall analyse, since this will demonstrate the approach needed. However, readers should realize that there are other factors that could be included in the model. There is a mode of operation called 'early token release' in which B would issue a free token immediately after transmitting the frame trailer. This improves throughput for high-speed rings that are physically large and therefore have a long propagation time, e.g. in excess of 1 km for 16 Mbit/s. Another factor is that one of the stations on a ring will be acting as a monitor or control station, responsible for taking care of various error conditions and for issuing a token to start the operation of the ring. The monitor station will impose a 24-bit delay on all transmissions, in addition to the physical propagation delay of the ring. This 24-bit delay ensures that a station does not start receiving a token before it has finished sending it, and makes the design of station logic simpler. For performance purposes it is just a component of ring propagation time.

The first value we require is the 'walk-time' or ring latency. This is made up of the propagation delay for the cable, and delays caused by the stations. Cable propagation time in theory depends on the media (shielded twisted pairs, unshielded twisted pairs, optical fibres), but is typically estimated as 5 μs/km. An active station inspects each token or message passing around the ring, and introduces a delay, usually assumed to be equivalent to one bit time. So

$$L = \text{ring latency} = \tau + \frac{Nb}{R} \qquad (30.19)$$

where

$$\tau = \text{propagation delay once round the ring} \qquad (30.20)$$

$$N = \text{number of active stations} \qquad (30.21)$$

$$b = \text{station latency in bits} = 1 \qquad (30.22)$$

$$R = \text{ring transmission speed in bit/s} \qquad (30.23)$$

We shall also need the ring utilization, which is given by

$$\rho = \text{ring utilization} = \lambda T_S \qquad (30.24)$$

The transfer delay, or time-in-system, is then given by

$$T_Q = T_S + \frac{\rho T_S(1 + C_S^2)}{2(1 - \rho)} + \frac{L[1 - (\rho/N)]}{2(1 - \rho)} + \frac{L}{2} \qquad (30.25)$$

The first two terms on the right-hand side are identical to the M/G/1 formula (Eq. (9.1)). The remaining terms involve the ring latency L. As long as ring latency is small compared to T_S, the extra terms in Eq. (30.25) can be ignored. Conversely, Eq. (30.25) demonstrates that as ring transmission speed increases and/or the physical cable-length of the ring increases, ring latency can become important if the average frame size is fairly small.

'Ethernet' or CSMA/CD

The 'Ethernet' or CSMA/CD LAN uses a contention protocol to regulate access to the transmission medium. A CSMA/CD LAN uses a coaxial cable bus as shown in Fig. 30.16. Other media can also be used. Unlike token-ring, each station can remain passive, i.e. just listening, unless it wishes to send a frame. Diagram (i) shows stations A to Z attached to the cable, but with no transmissions taking place. If station A wants to send a frame to station Z then A first 'listens' to see if another station is already transmitting. If no transmission is already in progress, then A just sends its frame. The frame propagates along the cable as shown in (ii). All other stations monitor the cable, and Z, recognizing the frame's destination address, copies the frame. However, since it takes a finite time for signals to travel along the cable, it is possible for a station to listen, find the cable free, and transmit even though another station at some distance along the cable has already started transmitting. This situation is shown in (iv), where both A and Z have started transmitting. (It is not relevant that Z may have been the destination for A's transmission.) In this case the frames or transmissions will 'collide' and the data in each frame will be corrupted. The stations will detect this collision, abort transmission, and then wait a random time before resending.

We have seen how with CSMA/CD, packets can 'collide' and need to be retransmitted. Some of the capacity of the LAN is therefore used up by unsuccessful transmissions. The 'carrier-sense' part of the protocol is intended to minimize packet collisions, and the 'collision-detect' mechanism seeks to minimize the wasted capacity by aborting a failed transmission as quickly as possible. Still, the

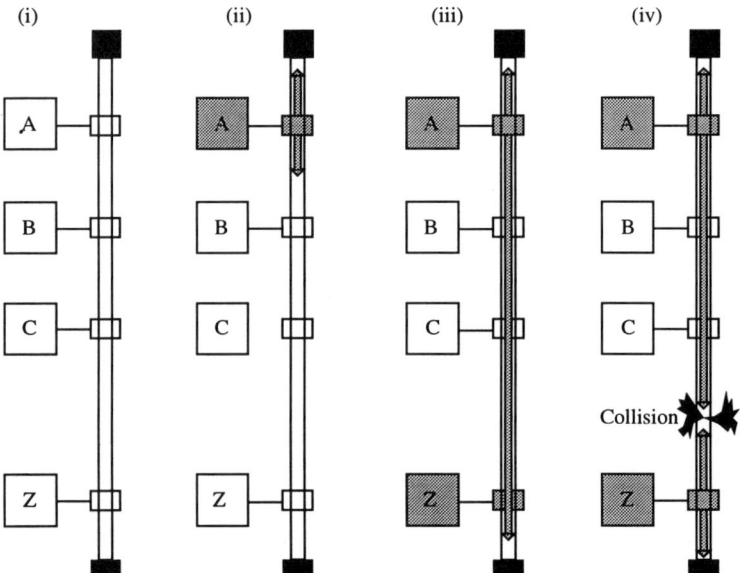

Figure 30.16. CSMA/CD operation.

effective capacity of the LAN will be significantly less than the nominal transmission capacity.

A detailed performance analysis of CSMA/CD was conducted by Lam and refined by Bux. Their results are discussed in some detail in Schwartz (1987), and here we present the main results of practical use. The maximum achievable value of ρ is given by

$$\rho_{MAX} = \text{maximum CSMA/CD throughput} = \frac{1}{1 + 6.44a} \tag{30.26}$$

where

$$a = \frac{\tau}{T_S} = \frac{\text{cable propagation time}}{\text{average packet transmission time}} \tag{30.27}$$

The average transfer time, or time-in-system, for a packet is given by the following fairly complicated formula, known as the Bux/Lam formula

$$\frac{T_Q}{T_S} = \rho G + 1 + 2ea - H + \frac{a}{2} \tag{30.28}$$

where

$$G = \frac{(1 + C_S^2) + (4e + 2)a + 5a^2 + 4e(2e - 1)a^2}{2\{1 - \rho[1 + (2e + 1)a]\}} \tag{30.29}$$

$$H = \frac{(1 - e^{-2a\rho})\left(\frac{2}{\rho} + \frac{2a}{e} - 6a\right)}{2[F^*(\lambda)e^{-(1+\rho a)} - (1 - e^{-2a\rho})]} \tag{30.30}$$

The only remaining factor in the above expressions to be explained is $F^*(\lambda)$, which is the Laplace transform of the frame-length distribution. Readers will recall that a guideline for this book is the avoidance of Laplace transforms if at all possible. All we need to do to make the above formula usable is to give the appropriate definitions of $F^*(\lambda)$ for the two particular cases usually of interest to us. First of all, if frames are a constant length, we have

$$F^*(\lambda) = e^{-\rho} \text{ and } C_S^2 = 0 \tag{30.31}$$

If frame lengths and therefore frame transmission times are exponentially distributed, then instead we have

$$F^*(\lambda) = \frac{1}{1 + \rho} \text{ and } C_S^2 = 1 \tag{30.32}$$

Do we need such a complex formula? If, as we did for token-ring, we just treat the LAN as an M/G/1 system, we get the result illustrated in Fig. 30.17. This is for exponentially distributed packet lengths with an average of 256 bytes. The horizontal axis is nominal utilization, which means the utilization for successful data transmissions, i.e. throughput. The curve labelled M/G/1 is obtained using Eq. (9.1), whereas the other curve shows the result of Eq. (30.28). The results are

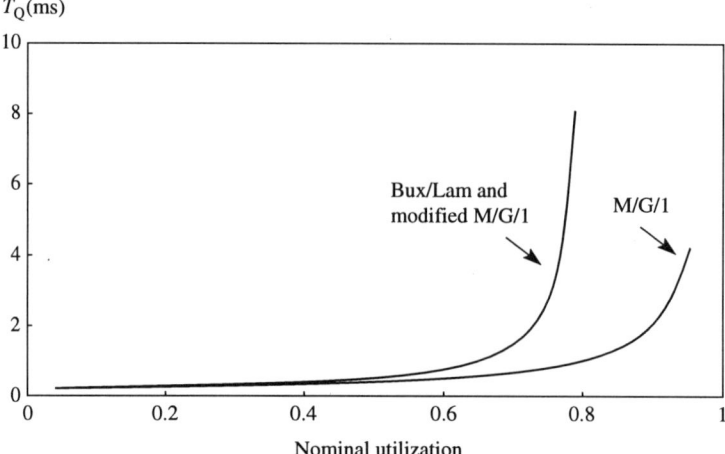

Figure 30.17. Using M/G/1 to approximate CSMA/D.

rather different, so maybe a simple M/G/1 approach will not work satisfactorily. However, we can get a good simple approximation using a modification to the M/G/1 approach. We know that the maximum throughput is ρ_{MAX}, and that time-in-system becomes unbounded as LAN utilization approaches ρ_{MAX}. If we use

$$\lambda_E = \text{effective packet transmission rate} = \frac{\lambda}{\rho_{MAX}} \qquad (30.33)$$

and then apply the M/G/1 formulae, we get results that for practical purposes are indistinguishable from the Bux/Lam formula (Eq. 30.28).

Comparison of token-ring and CSMA/CD performance

Many books on LAN and telecommunication network performance contain detailed comparisons of LAN access protocols with various physical con-figurations. The point of view taken is usually that of fundamental design, for example comparing CSMA/CD, earlier less sophisticated contention protocols such as ALOHA, token-ring and other varieties of controlled or polled access, under comparable conditions such as the same transmission speed. Readers of this book are likely to be faced with analysing 'Ethernet' running at 10 Mbit/s and token-ring running at either 4 Mbit/s (becoming less common) or 16 Mbit/s. We shall present here a simple comparison of these alternatives. Figure 30.18 shows the LAN transit times for 128-byte frames for the three types of LAN of interest. In each case the cable length was 1.5 km. For the 4 Mbit/s token ring the number of active stations was 20, and for the 16 Mbit/s rings there were 100 active stations. (The number of stations has only a very small effect for the same traffic rate.) For

T_Q(ms)

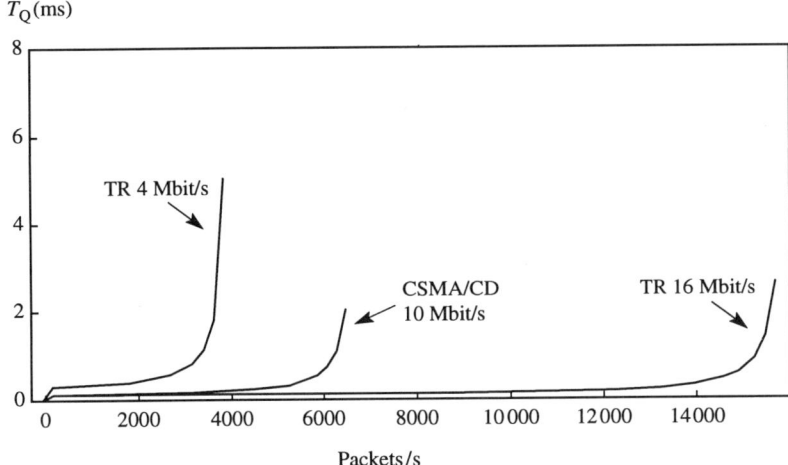

Packets/s

Figure 30.18. Comparative performance of 10 Mbits/s CSMA/CD with token-ring at 4 Mbit/s and 16 Mbit/s.

CSMA/CD the traffic was assumed to be generated from a large number of stations. The effective maximum capacity of the CSMA/CD LAN is given by Eq. (30.26) and is about 68 per cent in this case. For a token-ring the theoretical maximum capacity is nearly 100 per cent, but a more practical value to use is 80 per cent. In any case no sensible designer is going to plan for a LAN to be anything like 80 per cent utilized.

LAN bridges

The LANs we have looked at so far have been 'single segment' LANs. LAN segments can be interconnected. The most obvious way is with a local bridge, as shown in Fig. 30.19. (The figures here show rings, but could equally well be Ethernet buses, or a combination of the two.) A bridge inspects each frame on each

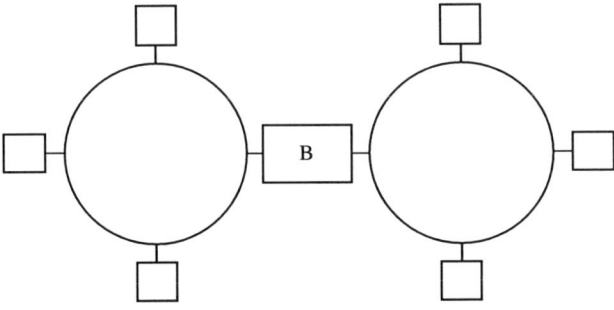

Figure 30.19. Local bridge between LAN segments.

LAN segment   LAN segment

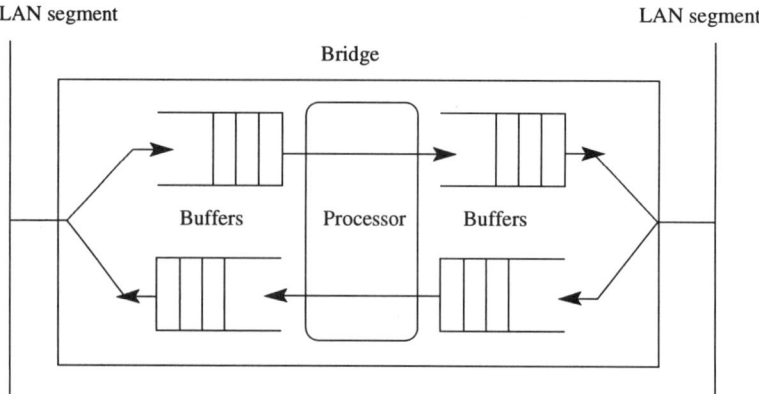

Figure 30.20. Representation of a LAN bridge.

LAN segment to which it is attached, and uses the addressing information to decide whether a frame should be copied onto the other LAN segment. The method by which this is done differs between token-ring and Ethernet, and readers are referred to Keiser (1989) for details of the addressing and routing schemes used.

The internal structure of a bridge can be represented as in Fig. 30.20, which is very similar to Fig 30.14 which shows a general data switch. A bridge may have two processors, one for each direction of packet transfer. Manufacturers provide bridge performance information in various ways. A common method is to quote the 'filtering rate' and the 'forwarding rate'. The filtering rate is the maximum rate at which the bridge is capable of inspecting packets to decide if they should be copied to the other side of the bridge. This is an important characteristic for Ethernet bridges, which must basically do an address table lookup, and sometimes a table update, per packet. Token-ring bridges with source routing do not require a table lookup. Both types of bridge need to copy the frame from one LAN adapter to the other, and may need to manipulate the frame (for example to change the frame format when bridging between token-ring and Ethernet, or perhaps to perform user-specified filtering). The 'forwarding' rate is the maximum rate at which the bridge can process frames that are to be copied across the bridge. (Readers should be careful in interpreting published performance figures for bridges, since the exact conditions under which performance is measured are often not explained adequately. See Rickert (1990).)

A reasonable model of bridge processing is to say that

$$t_P = \text{service-time for a packet} \tag{30.34}$$

$$= \alpha + \beta L + \theta \text{ for a forwarded packet} \tag{30.35}$$

$$= \alpha \text{ for a non-forwarded packet} \tag{30.36}$$

where L is the length of the frame in bytes, α is the time taken to decide if a frame should be forwarded, β is the time per byte required to copy the frame, and θ is the

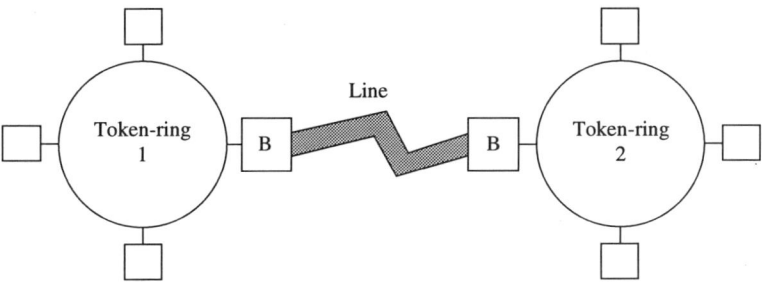

Figure 30.21. Remote bridge between LAN segments.

time needed to re-format the frame. If these parameters and the distribution of frame size are known, then the distribution of frame service-time in the bridge can be derived. This is a different approach to a general data switch, because bridges are usually high-performance devices designed to handle frames at something approaching the capacity of the attached LANs.

LANs can be bridged remotely, as shown in Fig. 30.21. This is sometimes called a 'split bridge' because in effect the bridge is split in two and a wide-area link inserted between the two halves of the bridge. Since WAN link speeds are usually limited to 2 Mbit/s, and LAN speeds are 4 Mbit/s or higher, it is obvious that the WAN link can be a bottleneck. If the WAN link speed is 64 kbit/s then bridge processing time will be insignificant compared to link delays. In this situation it is hardly worth modelling the bridge, and the link can be modelled as in Fig. 30.13. At 2 Mbit/s, then, it is still reasonable to ignore the bridge in calculating network transit times, although it would be sensible to check that a particular bridge has sufficient throughput.

Connection throughput achievable with flow control

Networks use various mechanisms to prevent congestion, and therefore excessive build-up of packet queues in switches. Once congestion occurs, emergency mechanisms can sometimes regain stable operation, but it is also possible for 'knock-on' effects to disrupt the whole of a complex network. One of the most important mechanisms to prevent congestion is flow control applied to each connection. A connection here may be an X25 virtual circuit, or an SNA session, or the equivalent connection between end-points in any other network architecture. The basic technique is to allow the sender to transmit only a limited number of packets before acknowledgements for the packets are received. (The term 'packet' is used throughout this section, but includes SNA RUs or any other data unit.) The number of packets that may be sent before waiting for acknowledgement is called the 'window-size'. Flow control also serves the purpose of preventing the sender transmitting packets into the network at a faster average rate than the receiver can handle them. (Other, more global, flow-control techniques will also be found in networks. SNA in particular contains a powerful set of global flow-control methods.)

What concerns us here is the throughput that can be achieved on a connection where a flow-control technique is operating. Simulation is well suited to this sort of question, which involves the interaction of widely separated components in a network. On the other hand, queueing theory can give us some useful insights into the effect of windowing methods. Schwartz (1987) gives a good description of the two queueing-theory approaches that can be used to analyse throughput. Both approaches make use of some theory that is slightly beyond what is covered in this book. Interested readers are recommended to study Schwartz (1987) for the derivation, but the results are presented here since they may prove useful.

We have to make a number of assumptions for the analysis to be possible. First we assume that the network path followed by a connection consists of M stages (or links), as shown in Fig. 30.22. The service-time for a packet on each link is assumed to be exponentially distributed, and service-times on each link to be independent. This last assumption may seem strange, since packets are likely to be fixed-length, and even if this is not so a packet does not change its length between one link and another. However, work by Kleinrock and others has shown this assumption to be justifiable since it produces usefully accurate estimates of performance. Refer to Schwartz (1987) and Kleinrock (1976) for discussion of this point. The next assumption is that the effective capacity of each link is θ packets per second. Effective capacity means the capacity available to an individual connection, i.e. the share of that link. For simplicity, readers can just think of a network with no traffic except the connection in question, and effective capacity then becomes the link capacity. Every link is assumed to have the same capacity. If the effective link capacities are different, then only the slowest links need to be considered, and the number of stages is reduced. Queueing, within a single connection, is not going to occur on the faster links. Remember that it is effective speed and capacity we are talking about, so a 2 Mbit/s link shared by many connections may have a lower effective capacity than a 64 kbit/s link shared by fewer connections. We also assume that delays in the switches are insignificant compared to

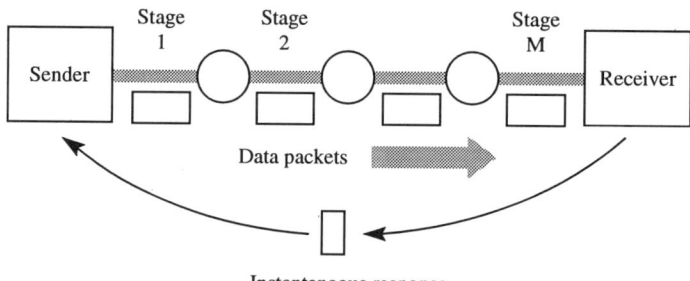

Figure 30.22. Connection via M network stages.

link delays. The last important assumption is that packet acknowledgements are returned instantly from receiver to sender. This is obviously artificial, although something approaching it may occur in practice if acknowledgements (which are small packets anyway) are handled at higher priority than data packets.

The parameters we need to define are few.

$$M = \text{number of stages} \qquad\qquad (30.37)$$

$$\theta = \text{capacity per connection for each link (packets/s)} \qquad (30.38)$$

$$N = \text{maximum unacknowledged packets allowed} \qquad\qquad (30.39)$$

The other 'parameter' we must specify is whether an unlimited source of data is available to the sender. If the traffic represents, say, a file transfer, then the sender will always have a data packet available whenever the flow control mechanism will allow a packet to be sent. We call this the 'heavy-traffic' case. If the traffic is intermittent, say for interactive transactions, then we call this the 'limited-traffic' case, and assume that data packets are supplied to the sender as random arrivals at an average rate θ. Now we can present the results for three variations of flow control, for heavy and limited traffic in each case. The routine ThruPut later in this chapter can perform all the following calculations.

Sliding window

The sliding-window method is the basic flow-control mechanism, and is used with X25. Every packet is separately acknowledged, so that a considerable volume of acknowledgement packets is generated. Figure 30.23 illustrates the method. The sender transmits packets, incrementing a counter PC each time a packet is sent. Packets can be sent only if PC is less than N, the window size. Whenever an acknowledgement is received the counter PC is decremented. (The actual packet formats in X25 allow either sliding window or acknowledgement at end of window to be used. It is the receiver that determines the method used, not the transmitter.)

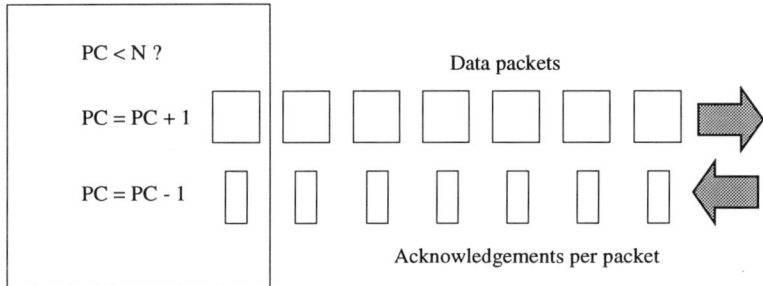

Figure 30.23. Flow control with sliding window.

For the heavy-traffic case, we have

$$\Omega_{\text{HEAVY}} = \text{throughput} = \frac{N\theta}{N + (M - 1)} \text{ packets/sec} \qquad (30.40)$$

$$D_{\text{HEAVY}} = \text{packet delay} = \frac{N + (M - 1)}{\theta} \text{ secs} \qquad (30.41)$$

In the limited-traffic case we have instead:

$$\Omega_{\text{LIM}} = \text{throughput} = \frac{N\theta}{N + M} \text{ packets/sec} \qquad (30.42)$$

$$D_{\text{LIM}} = \text{packet delay} = \frac{M(M + N)}{(M + 1)\theta} \text{ secs} \qquad (30.43)$$

Acknowledgement at end of window

In order to reduce the volume of acknowledgement packets it would seem a sensible scheme to send just one acknowledgement when all the packets in a window have been received. Figure 30.24 shows the method. The throughput will be less than for the sliding window, since there will be a pause in the sender's transmitting packets at the end of each window's worth of packets.

For the heavy-traffic case:

$$\Omega_{\text{HEAVY}} = \text{throughput} = \frac{\theta}{1 + (M - 1)f(N)} \text{ packets/sec} \qquad (30.44)$$

where

$$f(N) = \frac{1}{N} \sum_{r=1}^{N} \frac{1}{r} \qquad (30.45)$$

$$D_{\text{HEAVY}} = \text{packet delay} = \frac{1}{\theta} \left[M + 1 + \left(\frac{1 + N}{2} \right) \right] \text{ secs} \qquad (30.46)$$

The number of packets in transit along the connection will vary from 1 to N, so the factor $(1 + N)/2$ is the average number in transit. In the limited-traffic case

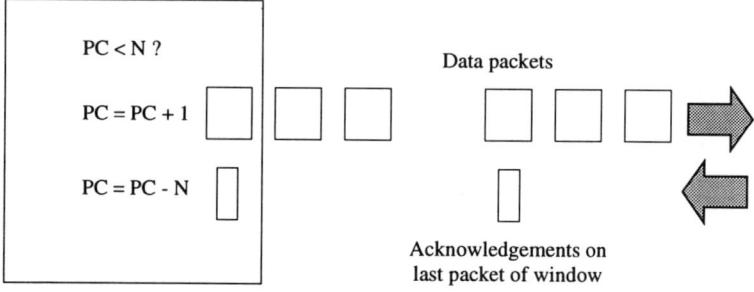

Figure 30.24. Flow control with acknowledgement at end of window.

we have:

$$\Omega_{\text{LIM}} = \text{throughput} = \frac{\theta}{1 + Mf(N)} \text{ packets/sec} \tag{30.47}$$

$$D_{\text{LIM}} = \text{ packet delay} = \frac{M}{(M+1)\theta}\left[M + \left(\frac{1+N}{2}\right)\right] \text{ secs} \tag{30.48}$$

Acknowledgement at start of window (SNA pacing)

Another scheme is to send an acknowledgement when the first packet of a window has been received, this acknowledgement permitting a further full window's worth of packets to be transmitted. If the window size is N, then this scheme is approximately equivalent to a window size of $2N - 1$ with acknowledgement at the end of the window. In practice it has the advantage that acknowledgements have more time to reach the sender before the sender's authority to transmit packets has been exhausted. Figure 30.25 illustrates the method, which is the way that session pacing is performed in SNA. (The analysis assumes acknowledgements are returned instantly: this will somewhat underestimate the relative performance of this scheme compared to acknowledgement at the end of the window.)

For the heavy-traffic situation we have:

$$\Omega_{\text{HEAVY}} = \text{throughout} = \frac{\theta}{1 + (M-1)g(N)} \text{ packets/sec} \tag{30.49}$$

$$g(N) = \frac{1}{N}\sum_{r=N}^{2N-1}\frac{1}{r} \tag{30.50}$$

$$D_{\text{HEAVY}} = \text{packet delay} = \frac{1}{\theta}\left[M - 1 + \left(\frac{3N-1}{2}\right)\right] \text{ secs} \tag{30.51}$$

Since the number of packets in transit may vary from N to $2N - 1$, the factor $(3N - 1)/2$ represents the average number of packets in transit. In the limited-

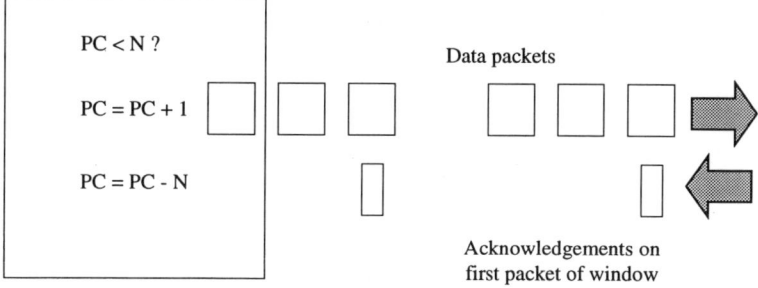

Figure 30.25. Flow control with acknowledgements at start of window.

traffic case the formulae become:

$$\Omega_{\text{LIM}} = \text{throughput} = \frac{\theta}{1 + Mg(N)} \text{ packets/sec} \tag{30.52}$$

$$D_{\text{LIM}} = \text{packet delay} = \frac{M}{(M+1)\theta}\left(M + \frac{3N-1}{2}\right) \text{ secs} \tag{30.53}$$

Comparison of connection throughput for flow-control methods

Now we compare the throughput and delay characteristics of the different flow-control methods. The basis for comparison is $M = 4$, with the capacity of each link assumed to be 1 packet/sec. Heavy traffic is assumed, although the results are similar for limited traffic. Figure 30.26 show the throughput of each scheme as the value of N increases. In all three methods throughput rises rapidly as N increases from 1 to M, and then continues to increase but more slowly. Large values of N, the window size, are needed to achieve throughput approaching the available capacity. Acknowledging at the end of the window give markedly lower throughput than either of the other two methods. Absolute values of throughput should be treated carefully, since we have assumed a number of things, notably that packet service-times are exponentially distributed and arrival patterns at each stage are random. In a real system the packet service-times would be more constant, and packet arrival patterns would be smoother than random. Intuitively, from looking at the behaviour of M/G/1 and G/M/1, we would expect the throughput curves to rise more steeply at first, and to be closer to the 'ideal' curve in Fig. 30.26. However, the relative values are instructive, and show the efficiency of each method and the effect of N.

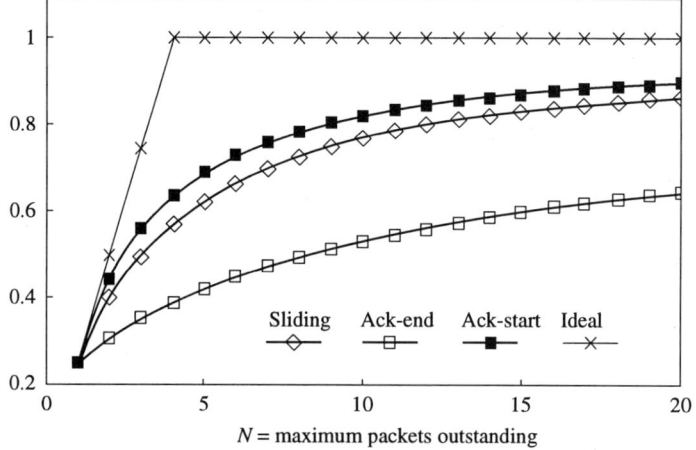

Figure 30.26. Connection throughput for flow-control schemes.

Packet delay/link delay

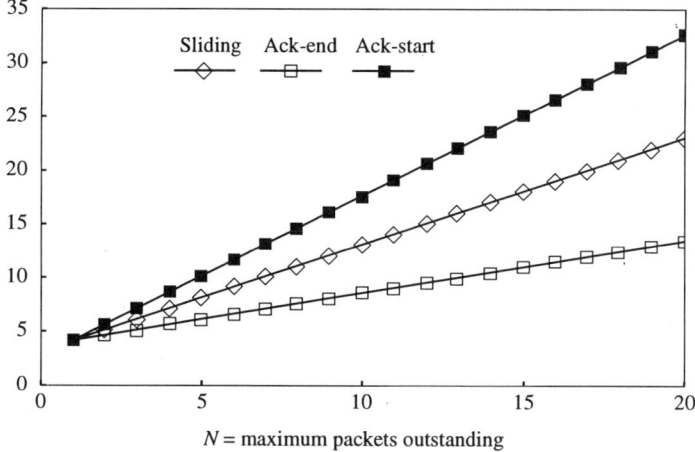

Figure 30.27. Connection packet delay for flow-control schemes.

Figure 30.27 shows the average packet delay for each method. The differences are easiest to understand if we regard the sliding window as the 'norm', which keeps roughly N packets in transit all the time. The ack-at-end method gives a lower packet delay because there are on average fewer packets, in fact $(1 + N)/2$, in transit, and so smaller queues at each stage. Conversely, with ack-at-start there are an average of $(2N - 1)/2$ packets in transit, and so larger queues are possible.

An interesting way of comparing both throughput and packet delay times for each of the methods is given in Schwartz (1987) for $M = 3$. The same approach, for $M = 4$, is shown in Fig. 30.28. The horizontal axis shows the throughput achieved, and the vertical axis shows the packet delay. The curve, for example the sliding-window curve, plots the performance as N increases. The trade-offs between throughput and packet delay is clearly shown. Note how the curves for sliding window and ack-at-start are nearly identical, except for the values of N to produce a given trade-off.

Note on flow control for modern high-speed networks

Telecommunications technology has been developing rapidly in recent years, and high-speed transmission facilities are now available using techniques known as 'frame relay', 'asynchronous transfer mode' (ATM), and other variations. At the time of writing it seems that ATM will become the dominant technology over the next few years.

With such high-speed transmission, the propagation times across a network are large compared to the transmission time for a packet. This means that any

Packet transit time/link delay

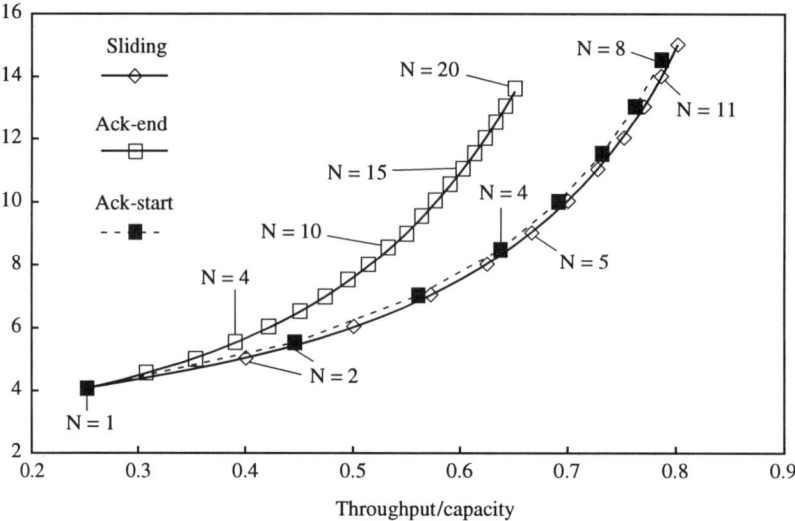

Figure 30.28. Connection throughput and packet transit for flow-control schemes.

flow-control technique that relies on the receiver sending some kind of 'now send me some more' message to the sender will be very inefficient. Figure 30.29 shows an alternative approach used in the 'rapid transport protocol' that is part of IBM's advanced peer-to-peer networking. Feedback is still required from receiver to sender, but instead of permits to send more data the feedback is about the actual rate at which data is being received from the network. The sender then adjusts the sending rate so as not to exceed either the allowed maximum or the actual receive rate.

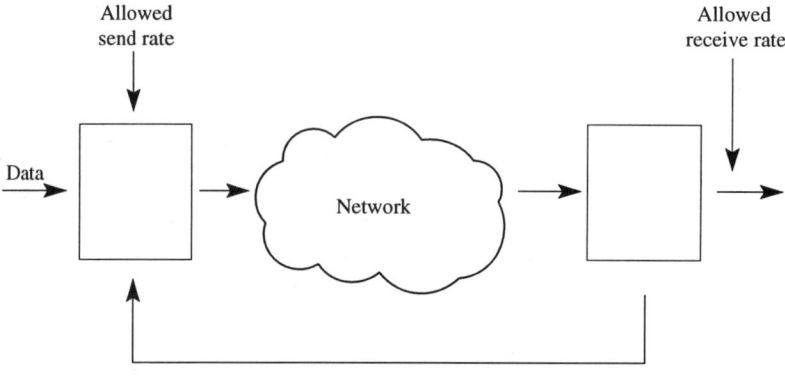

Figure 30.29. Flow control for a high-speed network.

Programs for data networks

In order to use the routines for data networks, a program must contain the following statements.

```
{$N+,E+ }
{$I \qthprog\QTHTypes.pas }
{$I \qthprog\QTHError.pas }
{$I \qthprog\CsmaCD.pas    }
{$I \qthprog\TokRng.pas    }
```

CsmaCD—calculations for a CSMA/CD LAN

This routine calculates the average delay per packet for a CSMA/CD LAN. The maximum capacity of the LAN for the given average packet size is also calculated.

```
{-----------------------------------------------------}
{> CsmaCD - Calculations for CSMA/CD lan              }
{ Inputs:                                             }
{      LAMBDA      customer arrival rate              }
{      LP          average length of packet, bytes    }
{      D           distribution of packet size        }
{                    1=constant                       }
{                    2=exponential                    }
{      CL          cable length, kilometres           }
{ Outputs:                                            }
{      A           propagation time/packet time       }
{      TS          average packet transmission time   }
{      RHO         LAN utilisation                    }
{      RHOMAX      maximum achievable utilisation     }
{      TQ          average time in system             }
{      VALID       true if calculations valid         }
{   Copyright Mike Tanner 1993                        }
{-----------------------------------------------------}
Procedure CsmaCd(LAMBDA,LP:Qthreal;D:integer;
                 CL:QTHreal;
                 Var A,TS,RHO,RHOMAX,TQ:QTHreal;
                 Var VALID:boolean);
Var Z,F,TAU,CS2:QTHreal;
Const E=2.7182818285;
Const VEL=5E-6;        { velocity in seconds per km }
Const SPD=10;          { transmission speed Mbit/s  }
begin
   {-Packet transmission time and LAN utilisation--}
   TS:=LP*8/(SPD*1E6);
   RHO:=LAMBDA*TS;
   {-Propagation time, maximum utilisation--------}
   TAU:=CL*VEL;
   A:=TAU/TS;
   RHOMAX:=1/(1+6.44*A);
   VALID:=false;
   If RHO>=RHOMAX then Exit;
   VALID:=true;
```

```
    {-Distribution-dependent values----------------}
    Case D of
       1:begin {constant } CS2:=0; F:=Exp(-RHO); end;
       2:begin {exponent } CS2:=1; F:=1/(1+RHO); end;
    end;
    {-Bux/Lam formula-------------------------------}
    TQ:=(1+CS2)+(4*E+2)*A+(5+4*E*(2*E-1))*Sqr(A);
    TQ:=TQ*RHO/(2*(1-RHO*(1+(2*E+1)*A)));
    TQ:=TQ+1+2*E*A+A/2;
    Z:=2*(F*Exp(-(RHO*A+1))-1+Exp(-2*RHO*A));
    TQ:=TQ-((1-Exp(-2*RHO*A))*(2/RHO+2*A/E-6*A))/Z;
    TQ:=TQ*TS;
end;
```

TokRng—transit time for a token-ring LAN

This routine does the corresponding calculations to CsmaCd, but for a token-ring LAN.

```
{------------------------------------------------------}
{> TokRng - Calculations for token-ring lan           }
{ Inputs:                                              }
{      LAMBDA     customer arrival rate                }
{      LP         average length of packet, bytes      }
{      D          distribution of packet size          }
{                    1=constant                        }
{                    2=exponential                     }
{      CL         cable length, kilometres             }
{      RS         transmission speed in Mbit/s         }
{      N          number of active stations            }
{ Outputs:                                             }
{      A          propagation time/packet time         }
{      TS         average packet transmission time     }
{      RHO        LAN utilisation                      }
{      TQ         average time in system               }
{      VALID      true if calculations valid           }
{   Copyright Mike Tanner 1993                         }
{------------------------------------------------------}
Procedure TokRng(LAMBDA,LP:Qthreal;D:integer;
                 CL,RS:QTHreal;N:integer;
                 Var A,TS,RHO,TQ:QTHreal;
                 Var VALID:boolean);
Var Z,TAU,CS2,L,RN:QTHreal;
Const E=2.7182818285;
Const VEL=5E-6;          { velocity in seconds per km }
Const B=1;               { station latency in bits    }
begin
    {-Packet transmission time and LAN utilisation--}
    TS:=LP*8/(RS*1E6);
    RHO:=LAMBDA*TS;
    {-Propagation time-------------------------------}
    RN:=N;
    TAU:=CL*VEL;
    A:=TAU/TS;
```

```
      VALID:=(RHO<=0.99);
      If (not VALID) then Exit;
      L:=TAU+RN*B/(RS*1E6);
      {-Time in system-------------------------------}
      Case D of
         1:CS2:=0;    { constant    }
         2:CS2:=1;    { exponential }
      end;
      TQ:=(L*(1-RHO/RN)
           +RHO*TS*(1+CS2))/(2*(1-RHO))+TS+L/2;
end;
```

ThruPut—connection throughput with flow control

This routine calculates the throughput and packet delay for connection established over several network links. The parameter WMODE selects the flow-control method used. The parameter AMODE selects the arrival pattern. A value of AMODE = 0 means heavy traffic, and corresponds to file transfer or other batch operation. AMODE = 1 is for limited traffic, and is appropriate for interactive traffic.

```
{--------------------------------------------------}
{> ThruPut  -- Calculate connection throughput with }
{>              flow control mechanisms.             }
{ Inputs:                                            }
{     W          window size or pacing count         }
{     M          number of stages or links           }
{     WMODE      windowing, or flow control, mode     }
{                1=sliding window                     }
{                2=ack at end of window               }
{                3=ack at start of window             }
{     AMODE      traffic arrival mode                 }
{                0=heavy traffic                      }
{                1=limited traffic                    }
{     CAP        capacity of each stage (packets/sec) }
{ Outputs:                                           }
{     TQ         average packet transit time, secs    }
{     TPUT       throughput, packets/sec              }
{ Copyright Mike Tanner 1993                         }
{--------------------------------------------------}
Procedure ThruPut(W,M:integer;CAP:QTHreal;
                  WMODE,AMODE:integer;
                  Var TQ,TPUT:QTHreal);
Var RW,RM,RK,RJ,TW,SK,NDASH:QTHreal;
Var J:integer;
begin
   RW:=W; RM:=M;
   {-Calculate factors according to flow control---}
   Case WMODE of
      1:begin  { sliding window  }
           NDASH:=RW;
           RK:=NDASH;
        end;
```

```
2:begin  { ack at end of window }
     NDASH:=(1+RW)/2;
     TW:=0;
     For J:=1 to W do
     begin
         RJ:=J;
         TW:=TW+1/RJ;
     end;
     RK:=RW/TW;
   end;
 3:begin  { SNA pacing }
     NDASH:=(3*RW-1)/2;
     SK:=0;
     For J:=W to (2*W-1) do
     begin
         RJ:=J;
         SK:=SK+1/RJ;
     end;
     RK:=RW/SK;
   end;
 else QTHError('ThruPut invalid WMODE');
end;
{-Calculate transit and throughput--------------}
Case AMODE of
  0:begin  { heavy traffic  }
       TQ:=(1/CAP)*(RM-1+NDASH);
       TPUT:=CAP/(1+(RM-1)/RK);
     end;
  1:begin  { limited traffic }
       TQ:=(1/CAP)*RM*(RM+NDASH)/(RM+1);
       TPUT:=CAP/(1+M/RK);
     end;
  else QTHError('ThruPut invalid AMODE');
 end;
end;
```

Part Ten
Appendices

Appendix 1
Probability distributions

Introduction

This appendix contains basic facts about the probability distributions commonly encountered in queueing theory. Discussion of the role of each distribution will be found in the body of the book. Distributions are included in this appendix in alphabetical order. Programs for calculating the pdf, cdf, and percentiles of the distributions are given later in this appendix. For additional details and other distributions see Hastings and Peacock (1975).

Erlang distribution

The Erlang distribution is discussed in Chapter 11, and the pdf is illustrated in Fig. 11.2.

Defining parameters are μ = mean, k = Erlang parameter

PDF is $f(x) = \dfrac{k}{\mu}(kx)^{k-1}\dfrac{\mathrm{e}^{-kx}}{(k-1)!}$ where $x = \dfrac{t}{\mu}$

CDF is $F(x) = 1 - \mathrm{e}^{-kx}\displaystyle\sum_{i=0}^{k-1}\dfrac{(kx)^i}{i!}$ where $x = \dfrac{t}{\mu}$

Variance is $\sigma^2 = \mu^2/k$

Coeff of variation squared is $C^2 = 1/k \leq 1$

Moments about zero are $s_r = r$th moment about zero $= \left(\dfrac{\mu}{k}\right)^r \displaystyle\prod_{i=0}^{r-1}(k+i)$

Exponential distribution

The exponential distribution is discussed at length in Chapter 6, and the pdf is illustrated in Fig. 6.1.

Defining parameter is μ = mean

PDF is $f(x) = \dfrac{e^{-x/\mu}}{\mu}$

CDF is $F(x) = 1 - e^{-x/\mu}$

Variance is $\sigma^2 = \mu^2$

Coeff of variation squared is $C^2 = 1$

Moments about zero are $s_r = r$th moment about zero $= r!\mu^r$

Gamma distribution

The gamma distribution is discussed in Chapter 12, and the pdf is illustrated in Fig. 12.1. The exponential, Erlang, and hyperexponential distributions are special cases of the gamma distribution.

Defining parameters are $\alpha =$ scale factor, $\beta =$ shape factor

If the mean and variance are given, the parameters α and β can be found from

$\alpha = \sigma^2/\mu$ and $\beta = \mu^2/\sigma^2$

PDF is $f(x) = \left(\dfrac{x}{\alpha}\right)^{\beta-1} \dfrac{e^{-x/\alpha}}{\alpha\Gamma(\beta)}$

CDF is $F(x) = \displaystyle\int_{-\infty}^{x} f(\theta)\, d\theta$

Mean is $\mu = \alpha\beta$

Variance is $\sigma^2 = \alpha^2\beta$

Coeff of variation squared is $C^2 = 1/\beta$

Moments about zero are $s_r = r$th moment about zero $= \alpha^r \displaystyle\prod_{i=0}^{r-1}(\beta + i)$

and in particular

$s_1 = \mu = \alpha\beta$

$s_2 = \alpha^2\beta(1 + \beta)$

$s_3 = \alpha^3\beta(1 + \beta)(2 + \beta)$

Geometric distribution

The geometric distribution is a discrete distribution, and can be regarded as the discrete form of the exponential distribution, since the geometric and exponential distributions are the only ones that have the 'memoryless' property. An example of the pdf and cdf of a geometric distribution are shown in Figs 7.5 and 7.6.

Defining parameters are $\mu =$ mean

We use the parameter $p = 1/\mu$

PDF is $f(n) = p(1-p)^{n-1}$

CDF is $F(n) = 1 - (1-p)^n$

Variance is $\sigma^2 = (1-p)/p^2$

Hyperexponential distribution

The hyperexponential distribution is discussed in Chapter 13, and the pdf is illustrated in Fig. 13.3. The coefficient of variation is always greater than or equal to 1.

Defining parameters are μ = mean, C^2 = coeff of variation squared

Calculate the parameters α_1, α_2, θ_1, θ_2 as follows:

$$\alpha_1 = \frac{1}{2}\left(1 - \sqrt{\frac{C^2-1}{C^2+1}}\right) \text{ and } \alpha_2 = 1 - \alpha_1$$

$$\theta_1 = \frac{\mu}{2\alpha_1} \text{ and } \theta_2 = \frac{\mu}{2\alpha_2}$$

PDF is $f(x) = \dfrac{\alpha_1}{\theta_1}e^{-x/\theta_1} + \dfrac{\alpha_2}{\theta_2}e^{-x/\theta_2}$

CDF is $F(x) = 1 - \alpha_1 e^{-x/\theta_1} - \alpha_2 e^{-x/\theta_2}$

Mean is $\mu = \alpha_1\theta_1 + \alpha_2\theta_2$

Variance is $\sigma^2 = \mu^2 C^2 = 2\alpha_1\theta_1^2 + 2\alpha_2\theta_2^2 - (\alpha_1\theta_1 + \alpha_2\theta_2)^2$

Coeff of variation squared is $C^2 \geq 1$

Moments about zero are $s_r = r$th moment about zero $= r!(\alpha_1\theta_1^r + \alpha_2\theta_2^r)$

Poisson distribution

The Poisson distribution is a discrete distribution. The shape of the pdf for a Poisson distribution is illustrated in Fig. 6.5.

Defining parameter is μ = mean

PDF is $f(n) = \dfrac{\mu^n}{n!}e^{-\mu}$

CDF is $F(n) = \sum_{r=0}^{n} f(n)$

Variance is $\sigma^2 = \mu$

Uniform (rectangular) distribution

The uniform, or rectangular, distribution has values evenly distributed between its minimum and maximum values, as illustrated in Fig. A1.1.

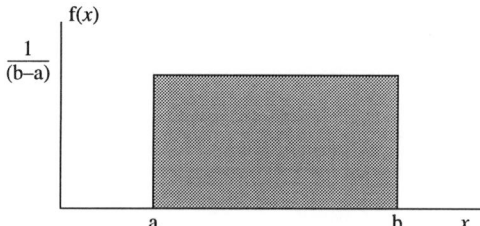

Figure A1.1. The uniform distribution.

Defining parameters are $a = $ minimum, $b = $ maximum

PDF is $f(x) = 1/(b - a)$

CDF is $F(x) = \dfrac{x - a}{b - a}$

Mean is $\mu = (a + b)/2$

Variance is $\sigma^2 = (b - a)^2/12$

Coeff of variation squared is $C^2 = \dfrac{(b - a)^2}{3(b + a)^2}$

Moments about zero are $s_r = r$th moment about zero $= \dfrac{b^{r+1} - a^{r+1}}{(r + 1)(b - a)}$

Programs for distributions

To use the subroutines in this appendix, the following statements are required.

```
{$N+,E+ }
{$I \qthprog\QTHtypes.pas  }
{$I \qthprog\QTHError.pas  }
{$I \qthprog\erlangd.pas   }
{$I \qthprog\expondst.pas  }
{$I \qthprog\gammaln.pas   }
{$I \qthprog\gammadst.pas  }
{$I \qthprog\hypexdst.pas  }
{$I \qthprog\normal.pas    }
{$I \qthprog\poisson.pas   }
```

ErlangPdf/ErlangCdf—pdf/cdf of the Erlang distribution

```
{-------------------------------------------------------}
{> ErlangPdf - Pdf of the Erlang distribution          }
{> ErlangCdf - Cdf of the Erlang distribution          }
{  Inputs: T     value for which pdf/cdf required       }
{          MEAN  mean of distribution                   }
{          K     Erlang parameter                       }
{  Copyright Mike Tanner 1993                           }
{-------------------------------------------------------}
Function ErlangPdf(T,MEAN:QTHreal;K:integer):QTHreal;
Var RK,X,RI,PDF:QTHreal;I:integer;
begin
   RK:=K;  X:=RK*T/MEAN;  PDF:=RK/MEAN;
   For I:=1 to (K-1) do begin
      RI:=I; PDF:=PDF*X/RI;
   end;
   ErlangPdf:=PDF/Exp(X);
end;
{-------------------------------------------------------}
Function ErlangCdf(T,MEAN:QTHreal;K:integer):QTHreal;
var RK,X,RI,S:QTHreal;I:integer;
begin
   RK:=K; X:=RK*T/MEAN; S:=1.0;   T:=1.0;
   For I:=1 to (K-1) do begin
      RI:=I; T:=T*X/RI; S:=S+T;
   end;
   ErlangCdf:=1-S/Exp(X);
end;
```

Erlang Moments—moments about zero for the Erlang distribution

```
{-------------------------------------------------------}
{> ErlangMoments - Calculate moments of Erlang-k        }
{ Inputs:    MEAN   mean                                }
{            K      no. of stages                       }
{ Outputs:   B      moments about the origin            }
{ Copyright Mike Tanner 1993                            }
{-------------------------------------------------------}
Procedure ErlangMoments(MEAN:QTHreal;K:integer;
                        Var B:Moments);
Var RI,RK:QTHreal;I:integer;
begin
   RK:=K;  B.MOM[0]:=1; B.MOM[1]:=MEAN;
   For I:=2 to MaxMoments do begin
      RI:=I;
      B.MOM[I]:=B.MOM[I-1]*MEAN*(RI+RK-1)/RK;
   end;
   B.MVM:=MaxMoments;
end;
```

ExponPdf/ExponCdf—pdf/cdf of exponential distribution

```
{------------------------------------------------------}
{> ExponPdf - Pdf of exponential distribution          }
{ Inputs: X      Value for which pdf/cdf required       }
{         MEAN   mean                                   }
{ Copyright Mike Tanner 1993                            }
{------------------------------------------------------}
Function ExponPdf(X,MEAN:QTHreal):QTHreal;
Var ALPHA:QTHreal;
begin
    ALPHA:=1/MEAN;
    ExponPdf:=ALPHA*Exp(-ALPHA*X);
end;
{------------------------------------------------------}
Function ExponCdf(X,MEAN:QTHreal):QTHreal;
Var ALPHA:QTHreal;
begin
    ALPHA:=1/MEAN;
    ExponCdf:=1-Exp(-ALPHA*X);
end;
```

Gamma Pdf—pdf of the gamma distribution

```
{------------------------------------------------------}
{> GammaPdf - Pdf of Gamma distribution.               }
{ Inputs: X        Value for which pdf required         }
{         MEAN     mean                                 }
{         COEFF2   sqd coeff of variation service       }
{ Copyright Mike Tanner 1993                            }
{------------------------------------------------------}
Function GammaPdf(X,MEAN,COEFF2:QTHreal):QTHreal;
Var ALPHA,BETA,LNPDF:QTHreal;
begin
    {-Calculate parameters of gamma distribution----}
    ALPHA:=MEAN*COEFF2;
    BETA:=1/COEFF2;
    {-Modify value of XX if it is very low---------}
    If X<0.00001 then QTHError('GammaPdf X too small');
    {-Calculate the pdf------------------------------}
    LNPDF:=(BETA-1.0)*Ln(X/ALPHA) - (X/ALPHA)
           - Ln(ALPHA) - GammaLn(BETA);
    GammaPdf:=Exp(LNPDF);
end;
```

GammaMoments—Moments about zero of the gamma distribution

```
{------------------------------------------------------}
{> GammaMomemts -- Moments of Gamma distribution        }
{ Inputs:  MEAN     mean                                }
{          COEFF2   coeff of variation squared          }
{ Outputs: B        moments about the origin            }
{ Copyright Mike Tanner 1993                            }
{------------------------------------------------------}
```

```
Procedure GammaMoments(MEAN,COEFF2:QTHreal;
                           Var B:Moments);
Var RI,ALPHA,BETA:QTHreal;I:integer;
begin
   ALPHA:=MEAN*COEFF2;   BETA:=1/COEFF2;
   B.MOM[0]:=1;
   For I:=1 to MaxMoments do begin
      RI:=I;
      B.MOM[I]:=B.MOM[I-1]*ALPHA*(BETA+RI-1);
   end;
   B.MVM:=MaxMoments;
end;
```

HypexFit—fit a hyperexponential distribution to a given mean and variance

```
{----------------------------------------------------}
{> HypexFit - Calculate parms of hyperexponential  }
{>            distribution for given mean and       }
{            coeff. of variation squared.           }
{   The method used is suggested by Allen in        }
{   "Probability, Statistics, and Queueing Theory"  }
{   Academic Press 1978 pages 246-247.              }
{ Inputs:   MEAN    mean                            }
{           COEFF2  squared coeff of variation      }
{ Outputs: ALPHA   proportion of first exponential  }
{          THETA1  mean of first exponential dist   }
{          THETA2  mean of second exponential dist  }
{   Copyright Mike Tanner 1993                      }
{----------------------------------------------------}
Procedure HypexFit(MEAN,COEFF2:QTHreal;
                Var ALPHA,THETA1,THETA2:QTHreal);
begin
   If COEFF2<1 then QTHError('HypexFit invalid COEFF2');
   ALPHA:=(1-Sqrt((COEFF2-1)/(COEFF2+1)))/2;
   THETA1:=MEAN/(2*ALPHA);
   THETA2:=MEAN/(2*(1-ALPHA));
end;
```

HypexPdf/HypexCdf—pdf/cdf of hyperexponential distribution

```
{----------------------------------------------------}
{> HypexPdf - Calculate the pdf hyperexponential    }
{> HypexCdf - Calculate the cdf hyperexponential    }
{ Inputs: X     value for which pdf/cdf is required }
{         MEAN     mean                             }
{         COEFF2  squared coeff of variation        }
{   Copyright Mike Tanner 1993                      }
{----------------------------------------------------}
Function HypexPdf(X,MEAN,COEFF2:QThreal):QThreal;
Var ALPHA,THETA1,THETA2:QThreal;
```

```
begin
   HypexFit(MEAN,COEFF2,ALPHA,THETA1,THETA2);
   HypexPdf:=ALPHA*Exp(-X/THETA1)/THETA1
             +(1-ALPHA)*Exp(-X/THETA2)/THETA2;
{-------------------------------------------------------}
Function HypexCdf(X,MEAN,COEFF2:QTHreal):QTHreal;
Var ALPHA,THETA1,THETA2:QTHreal;
begin
   HypexFit(MEAN,COEFF2,ALPHA,THETA1,THETA2);
   HypexCdf:=1-ALPHA*Exp(-X/THETA1)
             -(1-ALPHA)*Exp(-X/THETA2);
end;
```

HypexMoments—calculate momements about zero of hyperexponential distribution

```
{-------------------------------------------------------}
{> HypexMomemts - Moments of hyperexponential     }
{ Inputs:  MEAN    mean                           }
{          COEFF2 coeff of variation squared      }
{ Outputs: B       moments about the origin       }
{  Copyright Mike Tanner 1993                     }
{-------------------------------------------------------}
Procedure HypexMoments(MEAN,COEFF2:QTHreal;
                       Var B:Moments);
Var RI,M1,M2,FAC,ALPHA,THETA1,THETA2:QTHreal;I:integer;
begin
   HypexFit(MEAN,COEFF2,ALPHA,THETA1,THETA2);
   B.MOM[0]:=1; M1:=1; M2:=1; FAC:=1;
   For I:=1 to MaxMoments do begin
      RI:=I; FAC:=FAC*RI;
      M1:=M1*THETA1;  M2:=M2*THETA2;
      B.MOM[I]:=FAC*(ALPHA*M1+(1-ALPHA)*M2);
   end;
   B.MVM:=MaxMoments;
end;
```

NormaPdf/NormalCdf—pdf/cdf of the normal distribution

```
{-------------------------------------------------------}
{> NormalPdf - Pdf of the Normal distribution     }
{> NormalCdf - Cdf of the Normal distribution     }
{ Inputs: X      Value for which pdf required     }
{         MEAN    Mean                            }
{         STDEV   Standard deviation              }
{  Copyright Mike Tanner 1993                     }
{-------------------------------------------------------}
Function NormalPdf(X,MEAN,STDEV:QTHreal):QTHreal;
Var Z:QTHreal;
begin
   Z:=(X-MEAN)/STDEV;
   NormalPdf:=(0.39894228/STDEV)*Exp(-0.5*Sqr(Z));
end;
```

```
{--------------------------------------------------------}
Function NormalCdf(X,MEAN,STDEV:QTHreal):QTHreal;
Const A1= 0.319381530; A2=-0.356563782; A3= 1.781477937;
      A4=-1.821255978; A5= 1.330274429;
      C1= 0.2316419;   C2= 0.39894228;
Var P,W,Y,Z:QTHreal;
begin
   Z:=(X-MEAN)/STDEV; Y:=Abs(Z); W:=1/(1+C1*Y);
   Y:=C2*Exp(-0.5*Sqr(Y));
   P:=W*(A1+W*(A2+W*(A3+W*(A4+A5*W))));
   P:=P*Y;
   If Z>0 then P:=1-P;
   NormalCdf:=P;
end;
```

PoissonPdf/PoissonCdf—pdf/cdf of the Poisson distribution

```
{--------------------------------------------------------}
{> PoissonPdf - Pdf of the Poisson distribution   }
{> PoissonCdf - Cdf of the Poisson distribution   }
{ Inputs: K     Value for which pdf/cdf required  }
{         MEAN  mean                              }
{ Copyright Mike Tanner 1993                      }
{--------------------------------------------------------}
Function PoissonPdf(K:integer;MEAN:QTHreal):QTHreal;
Var PDF,RI:QTHreal;I:integer;
begin
   If K<0 then QTHError('PoisnPdf - invalid input');
   PDF:=1.0/Exp(MEAN); I:=0; RI:=0.0;
   While (I<=(K-1)) do begin
      I:=I+1; RI:=RI+1.0;
      PDF:=PDF*MEAN/RI;
   end;
   PoissonPDF:=PDF;
end;
{--------------------------------------------------------}
Function PoissonCdf(K:integer;MEAN:QTHreal):QTHreal;
Var PDF,CDF,RI:QTHreal;I:integer;
begin
   If K<0 then QTHError('PoisnCdf - invalid input');
   PDF:=1.0/Exp(MEAN); CDF:=PDF; I:=0; RI:=0.0;
   While (I<=(K-1)) do begin
      I:=I+1; RI:=RI+1.0;
      PDF:=PDF*MEAN/RI; CDF:=CDF+PDF;
   end;
   PoissonCdf:=CDF;
end;
```

PoissonPct—calculate a percentile of the Poisson distribution

```
{--------------------------------------------------------}
{> PoissonPct -Percentiles of Poisson distribution }
{ Inputs: PCT     Percentile required              }
{         MEAN    mean                              }
```

```
{ Returns: PCTth percentile                               }
{  Copyright Mike Tanner 1993                             }
{----------------------------------------------------------}
Function PoissonPct(PCT,MEAN:QTHreal):integer;
Var PDF,CDF,RK,P,SQRTMEAN:QTHreal;K:integer;
begin
   P:=PCT/100;
   If MEAN>20
   then begin    { use Normal approximation }
            SQRTMEAN:=Sqrt(MEAN); K:=0; RK:=0.0;
            While (CDF<P) do begin
               K:=K+1; RK:=K;
               CDF:=NormalCdf(RK,MEAN,SQRTMEAN);
            end;
        end
   else begin    { explicit evaluation    }
            PDF:=1.0/Exp(MEAN); CDF:=PDF;
            K:=0; RK:=0.0;
            While (CDF<P) do begin
               K:=K+1; RK:=K;
               PDF:=PDF*MEAN/RK; CDF:=CDF+PDF;
            end;
        end;
   PoissonPct:=K;
end;
```

Appendix 2
Special functions

Introduction

This appendix gives a brief description of some special mathematical functions that are used in some queueing-theory formulae, together with subroutines for calculating the functions either directly or by means of approximations. Programs to calculate the functions are given later in the appendix.

Binomial coefficients

Binomial coefficients are the coefficients of successive terms in the algebraic expansion of $(a + b)^n$. For example

$$(a + b)^2 = a^2 + 2ab + b^2 \text{ so coeffs are } 1, 2, 1$$

$$(a + b)^3 = a^3 + 3a^2b + 3ab^2 + b^3 \text{ so coeffs are } 1, 3, 3, 1$$

In general, the coefficient of the rth term in the expansion of $(a + b)^n$ is

$$\binom{n}{r-1} \text{ where } \binom{n}{k} = \frac{n!}{k!(n-k)!}$$

and we define

$$\binom{n}{0} = 1$$

and can write

$$(a + b)^n = \sum_{k=0}^{k=n} \binom{n}{k} a^{n-k} b^k$$

The gamma function

The gamma function is of widespread importance in mathematics. Many other special functions are defined in terms of the gamma function, and many statistical distributions also involve it. For integral values, the gamma function is the factorial function, i.e.

$$\Gamma(n+1) = n! = 1, 2, 3 \ldots (n-2)(n-1)n$$

and in general it is defined by the integral

$$\Gamma(z) = \int_0^\infty t^{z-1} e^{-t} \, dt$$

Just considering integer arguments it is obvious that, except for quite small x, $\Gamma(x)$ will be extremely large. Since when we use the gamma function we are usually using ratios of gamma functions, or at least ratios of some other very large value with a gamma function, it is sensible to calculate the logarithm of $\Gamma(x)$ rather than the actual function value.

Modified Bessel function of the first kind

Bessel functions are important mathematical functions that crop up in the solution of differential equations, and will probably be familiar to readers with an engineering background. There are various forms of Bessel function. We are interested here only in one particular Bessel function that appears in the distribution of the busy period for $M/M/1$. A Bessel function of integral order is defined by

$$J_\nu(z) = \left(\frac{z}{2}\right)^\nu \sum_{k=0}^\infty \frac{(-\frac{1}{4}z^2)^k}{k! \Gamma(\nu + k + 1)}$$

and the modified Bessel function of integar order I_1 is defined by

$$I_1(x) = -i J_1(ix)$$

Poisson ratio function

The Poisson ratio function occurs in several places in queueing theory. It is defined by

$$R(m, u) = 1 - \frac{e^{-u}\left(\frac{u^m}{m!}\right)}{e^{-u} \sum_{k=0}^{k=m} \frac{u^k}{k!}}$$

In the second term the factor e^{-u} in both the numerator and denominator is obviously redundant, but is included to illustrate the relationship with the Poisson distribution, namely

$$R(m, u) = 1 - \frac{f(m, u)}{F(m, u)}$$

where $f(m,u)$ and $F(m,u)$ are the pdf and cdf respectively of the Poisson distribution with mean u for the value m.

Programs for special functions

In order to use the subroutines from this chapter, the following statements are required in a program.

```
{$N+,E+ }
{$I \qthprog\QTHtypes.pas  }
{$I \qthprog\QTHError.pas  }
{$I \qthprog\bicoeff.pas  }
{$I \qthprog\gammaln.pas  }
{$I \qthprog\besseli1.pas  }
{$I \qthprog\pratio.pas }
```

BiCoeff—calculate binomial coefficient

While this is not hard to program, a little care is required because the factorials involved can be very large for quite modest values of n and k.

```
{------------------------------------------------------}
{> BiCoeff - Calculate a binomial coefficient.       }
{ Inputs:                                             }
{   N        number of objects to select from         }
{   K        number of objects to be selected          }
{            service time                              }
{ Returns:                                            }
{   The binomial coefficient                          }
{   Copyright Mike Tanner 1993                        }
{------------------------------------------------------}
Function BiCoeff(N,K:integer):QTHreal;
Var NUM,DEN,BICO:QTHreal;
begin
   If (K=N) or (K=0)
   then BICO:=1
   else begin
           NUM:=N;
           DEN:=K;
           BICO:=1;
           Repeat
              BICO:=BICO*NUM/DEN;
              NUM:=NUM-1;
              If DEN>1 then DEN:=DEN-1;
           until NUM<=(N-K);
         end;
   BiCoeff:=BICO;
end;
```

GammaLn—calculate logarithm of gamma function

```
{------------------------------------------------------}
{> GammaLn -- Log of the gamma function               }
{ Method:                                             }
{   An approximation derived by Lanczos is used.      }
```

```
{   The algorithm used here is based on the        }
{   routine in "Numerical Recipes in              }
{   Turbo-Pascal" by  Press, Flannery, Teukolsky, }
{   and Vetterling, Cambridge 1990 Pg 176-177.    }
{   Copyright Mike Tanner 1993                     }
{-------------------------------------------------}
Function GammaLn(XX:double):double;
Const C1=76.18009173;    C2=-86.50532033;
      C3=24.01409822;    C4=-1.231739516;
      C5=0.120858003E-2; C6=-0.536382E-5;
      P=2.50662827465;
Var X,Z,S:double;{ temporary variables         }
Var F:double;    { correction factor if XX<1    }
begin
    If XX<0 then QTHError('GammaLn X<0');
    {-Ensure XX>1 to get full accuracy-------------}
    If XX<1.0
    then begin
            F:=1.0/XX;
            XX:=XX+1.0;
         end
    else F:=1.0;
    {-Now apply Lanczos approximation--------------}
    X:=XX-1.0;
    Z:=X+5.5;
    S:=1.0+C1/(X+1.0)+C2/(X+2.0)+C3/(X+3.0)
        +C4/(X+4.0)+C5/(X+5.0)+C6/(X+6.0);
    GammaLn:=F*(X+0.5)*Ln(Z)-Z+Ln(P*S);
end;
```

BesselI1—calculate modified Bessel function of order 1

The following algorithm calculates the value of I_1 using a polynomial approximation.

```
{-------------------------------------------------}
{> BesselI1 - Modified Bessel function I1(x)      }
{ Inputs:                                          }
{   X      value for which I1(x) required          }
{ Returns:                                         }
{   Modified Bessel function I1(x)                 }
{ Method:                                          }
{   A polynomial approximation is used. The        }
{   algorithm used here is a recoded and modified  }
{   version of the routine in "Numerical Recipes   }
{   in Turbo-Pascal" by Press, Flannery,           }
{   Teukolsky, and Vetterling.                     }
{   Copyright Mike Tanner 1993                     }
{-------------------------------------------------}
Function BesselI1(X:qthreal):qthreal;
Const A1=0.87890594;    A2=0.51498869;
      A3=0.15084934;    A4=0.2658733e-1;
      A5=0.301532e-2;   A6=0.32411e-3;
      B1=0.2282967e-1;  B2=-0.2895312e-1;
```

```
          B3=0.1787654e-1;   B4=0.420059e-2;
          C1=0.39894228;     C2=-0.3988024e-1;
          C3=-0.362018e-2;   C4=0.163801e-2;
          C5=-0.1031555e-1;
Var AX:qthreal;
Var Y,R:qthreal;
begin
   If Abs(X)<3.75
   then begin
             Y:=Sqr(X/3.75);
             R:=X*(0.5+Y*(A1+Y*(A2+Y*(A3
                             +Y*(A4+Y*(A5+Y*A6))))));
         end
   else begin
             AX:=Abs(X);
             Y:=3.75/AX;
             R:=B1+Y*(B2+Y*(B3-Y*B4));
             R:=C1+Y*(C2+Y*(C3
                             +Y*(C4+Y*(C5+Y*R))));
             R:=(Exp(AX)/Sqrt(AX))*R;
             If X<0.0 then R:=-R;
         end;
   BesselI1:=R;
end;
```

PRatio—calculate Poisson ratio function

The Poisson ratio function could be calculated by using the algorithms for the pdf and cdf of the Poisson distribution which are given in Appendix 1. However, this involves duplicate calculations and the unnecessary evaluation of exponentials. These inefficiencies can be avoided by direct calculation with the following algorithm.

```
{------------------------------------------------------}
{> PRatio - Poisson ratio function                     }
{ Inputs:                                              }
{    M         Value for which ratio required          }
{    MEAN      mean                                     }
{ Returns:                                             }
{    Value of the Poisson ratio function               }
{    Copyright Mike Tanner 1993                         }
{------------------------------------------------------}
Function PRatio(M:integer;MEAN:QTHreal):QTHreal;
Var K,T,S,R:QTHreal;I integer;
Label L1;
begin
   If M<0 then QTHError('PRatio - invalid input');
   {-Calculate ratio directly---------------------}
   K:=0.0;
   S:=1.0;
   T:=1.0;
   for I:=1 to M do
```

```
    begin
        if T<0.000001 then goto L1;
        K:=K+1.0;
        T:=T*MEAN/K;
        S:=S+T;
    end;
L1:PRatio:=1.0-T/S;
end;
```

Appendix 3
The programs

Compiler and system used

The programs in this book were developed and tested using Turbo-Pascal V5.5 running under DOS 5 and Windows 3.1. A PS/2 model P70 was used, which has a 386 chip. No math co-processor was installed. Turbo-Pascal built-in functions that access DOS or Windows internals have been avoided. All routines are supplied in source-code form, with code modules being combined by means of the {$I . . .} directive which imbeds one source module in another. In programming terms the use of Turbo-Pascal units would be neater, but this would be inconvenient for users of earlier versions of Turbo-Pascal, and for users of other Pascal compilers. The author's aim has been to avoid facilities that some users may not have, or that would make conversion to other compilers and systems less convenient.

Programming style and conventions

Names of variables are in upper case. Pascal reserved words are in lower case, with initial letters sometimes in upper case. For example

```
If ALPHA<BETA
then ALPHA:=Sqrt(BETA)
else begin
        BETA:=BETA+1;
        Goto LAB1;
     end;
```

The indentation rules used here place corresponding 'begin's and 'end's in the same column, which makes the program structure more readily apparent. Programming purists will no doubt be horrified to find a few **goto**s in the algorithms, but in the author's considerable experience of programming, limited and judicious use of **goto**s makes programs clearer and more reliable.

Real/double precision variables

The use of single or double precision for real variables is controlled by means of some Pascal type-definitions in the file QTHTYPES.PAS. Two types are defined: QTHreal for variables whose precision is not critical, and QTHdouble for variables that must be of double precision if possible. Most of the variables that actually require double precision are in procedures for numerical approximation of complicated functions. For uniformity, and bearing in mind that very few of the routines in this book are numerically intensive, both QTHreal and QTHdouble specify the use of double-precision variables.

Turbo-Pascal provides software emulation of double-precision arithmetic when a math co-processor is not available. By using the compiler directives {$N+,E+}, the resulting program will use a co-processor if it is present, and use software emulation if not. Very few of the routines in this book are numerically intensive, so the extra overhead of using software emulation of a math co-processor is not a reason for avoiding double-precision variables.

If the reader wishes to use a compiler/system combination that does not provide double-precision arithmetic, then he or she should edit the file QTHTYPES.PAS so that all both QTHreal and QTHdouble types specify single precision.

Common routines and definitions

There is a very small amount of code that is needed in every program that uses any of the supplied routines in this book. The first such item is some Pascal type-definitions in the file QTHTYPES.PAS, which is shown below.

```
{-----------------------------------------------------}
{ Double precision is used throughout the programs }
{ and subroutines in the Queueing Theory Handbook. }
{ This can be changed by editing the type           }
{ definitions below;                                 }
{-----------------------------------------------------}
Type QTHreal=double; { =real; for single precision }
Type QTHdouble=double;
{-----------------------------------------------------}
{ MaxMoments is the maximum number of moments used }
{ Moments is basically an array of size MaxMoments }
{ to hold the moments of a distribution. The MVM   }
{ field is the Maximum Valid Moments                }
{-----------------------------------------------------}
Const MaxMoments=10;
Type Moments=record
        MOM:array[0..MaxMoments] of double;
        MVM:integer;
     end;
```

Another piece of common code is an error routine to be called when, for example, some invalid parameters are passed to a routine. The supplied routine, which is listed below, is trivial, and the user will probably wish to expand it into something more elegant. The routine is in the file QTHERROR.PAS.

```
{--------------------------------------------------}
{ QTHError     - error handling routine.           }
{--------------------------------------------------}
Procedure QTHError(MSG:string);
Var CH:char;
begin
   Writeln(MSG);
   Writeln('Press any key to halt program');
   CH:=ReadKey;
   Halt;
end;
```

In the file QTHFILE.PAS there are some routines used by the programs that tabulate results. These programs are designed to use either the screen as output or a disk file. The first of these routines is QTHOpenFile, which is shown below, together with the declarations of a couple of global variables used to handle the screen or output file. If QTHOpenFile is called with a zero-length string, the screen will be used for output. Otherwise the string is assumed to be the name of the file to be used, complete with directory path and extension.

```
Var OPF:text;      { output file                    }
Var FLSW:boolean; {T=use file, F=use screen         }
{--------------------------------------------------}
{ QTHOpenFile -- open file for results             }
{--------------------------------------------------}
Procedure QTHOpenFile(DNE:string);
begin
   Assign(OPF,DNE);
   If Length(DNE)>0
   then begin
           ClrScr;
           Rewrite(OPF);
           FLSW:=true;
        end
   else FLSW:=false;
end;
```

The next routine for handling output is QTHClrScr, which is listed below. This routine clears the screen if the screen is being used for output. When a disk file is being used, the routine simply writes a blank line to the file, in order to separate different tables of results in the file.

```
{-----------------------------------------------------}
{ QTHClrScr  -- clear screen if in use               }
{-----------------------------------------------------}
Procedure QTHClrScr;
begin
    If FLSW
    then Writeln(OPF,'   ')
    else ClrScr;
end;
```

The next routine for handling output is QTHWaitKey, which is listed below. This routine is intended to be used when a table of results has been written out. If the screen is being used, the routine waits until the user presses a key before continuing. This is to give the user a chance to view the results displayed on the screen. If a disk file is being used, the routine does nothing.

```
{-----------------------------------------------------}
{ QTHWaitKey  - wait for user to press a key if  }
{               screen is being used for results }
{-----------------------------------------------------}
Procedure QTHWaitKey;
Var CH:char;
begin
    If FLSW then Exit;
    Writeln('Press any key to continue');
    CH:=ReadKey;
end;
```

The final routine in QTHFILE.PAS is QTHCloseFile, which is listed below. QTHCloseFile closes the disk file if one is being used; otherwise the routine does nothing.

```
{-----------------------------------------------------}
{ QTHCloseFile -- close results file if open   }
{-----------------------------------------------------}
Procedure QTHCloseFile;
begin
    If FLSW then Close(OPF);
    FLSW:=false;
end;
```

Ordering program diskettes

For readers outside the US, programs from this book and a copy of Q-Calc can be obtained by sending:

i)   a formatted diskette
ii)   a cheque for £5.95 sterling
iii) a copy of the form below to

McGraw-Hill Book Company (UK) Ltd
Shoppenhangers Road
Maidenhead, Berks SL6 2QL
England.

Although every effort has been made to ensure the reliability and accuracy of the disk, the Publisher cannot guarantee that the disk, when used for its intended or any other use, is free from error or defect.

Distribution of the machine-readable programs (either as created by you or from the associated diskette) to any other person is not authorized. You are authorized to make use of the programs in this book for your personal and professional activities, but not to incorporate the programs into other programs which are then sold or otherwise distributed. Purchasing one copy of this book in effect gives you a licence to use one copy of the programs and other associated materials.

Contact the author via the publisher or on Compuserve id 100341,425 for permission to use and distribute the programs in other ways, e.g. departmental, site, and company distribution. Permission will not be unreasonably or arbitrarily withheld.

Copyright does not protect ideas, but only the expression of those ideas in a particular form. In the case of a computer program, the ideas consist of the program's methodology and algorithm. The expression of those ideas is the source code of the program and the object code derived from the source code.

Whatever the legal rights and wrongs, it is not ethical to deprive a program's author of compensation for, and acknowledgement of, the creative effort involved in developing a program.

      Name . . . .

      Address . . . .

      . . . .

      . . . .

      . . . .

      Telephone . . . .

Appendix 4
The Q-Calc package

Introduction

Not all readers will wish to build programs from the subroutines in this book. As an alternative to programming, Q-Calc provides a means of doing the calculations for most of the queueing models discussed. Q-Calc is as easy to use as a hand-calculator or simple spreadsheet.

Q-Calc main menu

When Q-Calc is started, it first displays a title panel and then presents the user with a straightforward menu as illustrated in Fig. A4.1. This is a 'bounce-bar' menu. The currently selected item is indicated by the whole of that line being highlighted. The user moves the highlight line by means of the cursor-up and cursor-down keys. Pressing the enter key then invokes the selected queueing model, an example of which is described in the next section. When the user exits from the selected queueing model, the main menu is again displayed. Pressing ESC when in the main menu terminates Q-Calc.

```
┌─────────────────────Select queueing model─────────────────────┐
│  M/M/1 Single server,random arrivals,exponential service       │
│  M/G/1 Single server, random arrivals, general service         │
│  G/G/1 Single server, general arrivals, exponential service    │
│  M/M/infinity - Unlimited number of servers                    │
│  M/M/m Multiserver with queueing (Erlang-C)                    │
│  M/M/m/m Multiserver loss (Erlang-B)                           │
│  M/M/m/K Multiserver with limited waiting                      │
│  G/G/m Multiserver, general arrivals, general service          │
│  Single server priority queue ████████████████████████████     │
│  M/M/m/K/K Multiple repairmen, finite number of customers      │
│  Engset - Multiserver loss with finite sources                 │
│  Erlang-B extended - Multiserver loss with retries             │
└──F1=Help───────────────────────────────────────────────────────┘
```

Figure A4.1. Q-Calc main menu panel.

Panel for a selected queueing model

Once a queueing model has been invoked, two panels are displayed. An example, for the G/M/1 model, is shown in Fig. A4.2. The top panel is the list of input parameters for the model. The bottom panel is the output from the model, and lists all the potentially useful results that can be calculated. The user can type in new input parameter values, or select new values from a drop-down list. Whenever an input parameter is changed, the results are updated automatically.

The user selects which input parameter to change using the same bounce-bar technique as used for the main menu. For a numeric parameter, the user simply types in the new number and presses 'enter'. Keystrokes are checked for validity as they are entered, so that it is impossible to enter an invalid numeric format. Values such as the type of distribution, in the example 'Hypergeometric', are selected from a list. The list is displayed as a drop-down list if the space bar is pressed, and the user selects from the list with a bounce-bar, and presses 'enter'. Menus and lists may be longer than the number of available lines in a panel, in which case the menu or list is scrolled.

```
————————————————G/M/1 input parameters——————————————
  Inter-arrival distribution     Hyperexponential
  Mean arrival rate    ████████   0.600  ██████████████
  Inter-arrival C2                1.500
  Mean service time               1.000
——F1=Help—————————————————————————————————————————————
```

```
————————————————————G/M/1 results—————————————————
  Server utilisation                          0.6000
  Theta                                       0.650556
  Time in system -- average                   2.8617
      " "        -- standard deviation        2.8617
      " "        -- 90th pctile               6.5893
      " "        -- 95th pctile               8.5729
  Waiting time   -- average                   1.8617
      " "        -- standard deviation        2.6813
      " "        -- 90th pctile               5.3590
      " "        -- 95th pctile               7.3425
  Customers in system -- average              1.7170
           " "        -- std. deviation       2.2720
  Customers waiting   -- average              1.1170
           " "        -- std. deviation       2.0071
  Customers on arrival - average              1.8617
           " "         - std. deviation       2.3082
```

Figure A4.2. Q-Calc panels for the G/M/1 model.

Obtaining and running Q-Calc

Q-Calc can be obtained using the order form in Appendix 3. Q-Calc is written in Turbo-Pascal V5.5, and runs on an IBM-compatible PC under DOS. This environment was chosen as being the most widely available, rather than using OS/2 or Windows in native mode. Q-Calc has been successfully installed under each of DOS, OS/2, and Windows (in DOS mode, of course, with OS/2 and Windows).

Appendix 5
Bibliography and references

Allen, A.O. *Probability, Statistics, and Queueing Theory with Computer Science Applications*. Academic Press, New York, 1978.

Bear, D. *Principles of Telecommunications Traffic Engineering*, 3rd edition, Peter Peregrinus, London, 1988.

Bohl, M. *Introduction to IBM Direct Access Storage Devices*. Science Research Associates, Chicago, 1981.

Chaudrry, M.L. and Templeton, J.G.C. *A First Course in Bulk Queues*. Wiley, Chichester, 1983.

Hastings, N.A.J. and Peacock, J.B. *Statistical Distributions*. Butterworth, Guildford, 1975.

Houtekamer, G.E. and Artis, H.P. *MVS I/O Subsystems: Configuration Management and Performance Analysis*. McGraw-Hill, London, 1991.

Keiser, G.E. *Local Area Networks*. McGraw-Hill, London, 1989.

King, P.J.B. *Computer and Communication Systems Performance Modelling*. Prentice-Hall, Englewood Cliffs, NJ, 1990.

Kleinrock, L. *Queueing Systems*. Volume I: Theory. Wiley, New York, 1975.

Kleinrock, L. *Queueing Systems*. Volume II: Computer Applications. Wiley, New York, 1976.

Lavenberg, S.S. (ed.). *Computer Performance Modelling Handbook*. Academic Press, London, 1983.

Lazowska, E.D., Zahoran, J., Scott Graham, G. and Sevcik, K.C. *Quantitative System Performance (Computer System Analysis Using Queueing Network Models)*. Prentice-Hall, Englewood Cliffs, NJ, 1984.

Medhi, J. *Recent Developments in Bulk Queueing Models*. Wiley Eastern, New Delhi, 1984.

Newell, G.F. *Applications of Queueing Theory*. (2nd edn). Chapman & Hall, London, 1982.

Press, W.H., Flannery, B.P., Teukolsky, S.A. and Vetterling, W.T. *Numerical Recipes in PASCAL (The Art of Scientific Computing)*. Cambridge University Press, 1989.

Rickert, J.B. Jr. Evaluating MAC-layer bridges—beyond filtering and forwarding. *Data Communications International*. May 1990.

Schwartz, M. *Computer-Communication Network Design and Analysis*. Prentice-Hall, Englewood Cliffs, NJ, 1977.

Schwartz, M. *Telecommunication Networks: Protocols, Modeling and Analysis*. Addison-Wesley, New York, 1987.

Seelen, L.P., Tijms, H.C. and VanHoorn, M.H. *Tables for Multi-Server Queues*. Elsevier, Amsterdam, 1985.

Spiegel, M.R., 'Mathematical Handbook', Schaum's Outline Series. McGraw-Hill, New York.

Index to subroutines and programs

Index

Further Titles in the IBM McGraw-Hill Series

OS/2 Presentation Manager Programming Hints and Tips	Bryan Goodyer
PC User's Guide Simple Steps to Powerful Personal Computing	Peter Turner
The IBM RISC System/6000	Clive Harris
The IBM RISC System/6000 User Guide	Mike Leaver Hardev Sanghera
MVS Systems Programming	Dave Elder-Vass
CICS Concepts and Uses A Management Guide	Jim Geraghty
Dynamic Factory Automation	Alastair Ross